Springer Handbook
of Electronic and Photonic Materials

Safa Kasap, Peter Capper (Eds.)

 Springer

Springer Handbook of Electronic and Photonic Materials
Organization of the Handbook

Each chapter has a concise summary that provides a general overview of the subject in the chapter in a clear language. The chapters begin at fundamentals and build up towards advanced concepts and applications. Emphasis is on physical concepts rather than extensive mathematical derivations. Each chapter is full of clear color illustrations that convey the concepts and make the subject matter enjoyable to read and understand. Examples in the chapters have practical applications. Chapters also have numerous extremely useful tables that summarize equations, experimental techniques, and most importantly, properties of various materials. The chapters have been divided into five parts. Each part has chapters that form a coherent treatment of a given area. For example,

Part A contains chapters starting from basic concepts and build up to up-to-date knowledge in a logical easy to follow sequence. Part A would be equivalent to a graduate level treatise that starts from basic structural properties to go onto electrical, dielectric, optical, and magnetic properties. Each chapter starts by assuming someone who has completed a degree in physics, chemistry, engineering, or materials science.

Part B provides a clear overview of bulk and single-crystal growth, growth techniques (epitaxial crystal growth: LPE, MOVPE, MBE), and the structural, chemical, electrical and thermal characterization of materials. Silicon and II–VI compounds and semiconductors are especially emphasized.

Part C covers specific materials such as crystalline Si, microcrystalline Si, GaAs, high-temperature semiconductors, amorphous semiconductors, ferroelectric materials, and thin and thick films.

Part A Fundamental Properties
2 Electrical Conduction in Metals and Semiconductors
3 Optical Properties of Electronic Materials: Fundamentals and Characterization
4 Magnetic Properties of Electronic Materials
5 Defects in Monocrystalline Silicon
6 Diffusion in Semiconductors
7 Photoconductivity in Materials Research
8 Electronic Properties of Semiconductor Interfaces
9 Charge Transport in Disordered Materials
10 Dielectric Response
11 Ionic Conduction and Applications

Part B Growth and Characterization
12 Bulk Crystal Growth-Methods and Materials
13 Single-Crystal Silicon: Growth and Properties
14 Epitaxial Crystal Growth: Methods and Materials
15 Narrow-Bandgap II–VI Semiconductors: Growth
16 Wide-Bandgap II–VI Semiconductors: Growth and Properties
17 Structural Characterization
18 Surface Chemical Analysis
19 Thermal Properties and Thermal Analysis: Fundamentals, Experimental Techniques and Applications
20 Electrical Characterization of Semiconductor Materials and Devices

Part C Materials for Electronics
21 Single-Crystal Silicon: Electrical and Optical Properties
22 Silicon-Germanium: Properties, Growth and Applications
23 Gallium Arsenide
24 High-Temperature Electronic Materials: Silicon Carbide and Diamond
25 Amorphous Semiconductors: Structure, Optical, and Electrical Properties
26 Amorphous and Microcrystalline Silicon
27 Ferroelectric Materials
28 Dielectric Materials for Microelectronics
29 Thin Films
30 Thick Films

Part D examines materials that have applications in optoelectronics and photonics. It covers some of the state-of-the-art developments in optoelectronic materials, and covers III–V Ternaries, III–Nitrides, II–VI compounds, quantum wells, photonic crystals, glasses for photonics, nonlinear photonic glasses, nonlinear organic, and luminescent materials.

Part D Materials for Optoelectronics and Photonics
31 III–V Ternary and Quaternary Compounds
32 Group III Nitrides
33 Electron Transport within the III–V Nitride Semiconductors, GaN, AlN, and InN: A Monte Carlo Analysis
34 II–IV Semiconductors for Optoelectronics: CdS, CdSe, CdTe
35 Doping Aspects of Zn-Based Wide-Band-Gap Semiconductors
36 II–VI Narrow-Bandgap Semiconductors for Optoelectronics
37 Optoelectronic Devices and Materials
38 Liquid Crystals
39 Organic Photoconductors
40 Luminescent Materials
41 Nano-Engineered Tunable Photonic Crystals in the Near-IR and Visible Electromagnetic Spectrum
42 Quantum Wells, Superlattices, and Band-Gap Engineering
43 Glasses for Photonic Integration
44 Optical Nonlinearity in Photonic Glasses
45 Nonlinear Optoelectronic Materials

Part E provides a survey on novel materials and applications such as information recording devices (CD, video, DVD) as well as phase-change optical recording. The chapters also include applications such as solar cells, sensors, photoconductors, and carbon nanotubes. Both ends of the spectrum from research to applications are represented in chapters on molecular electronics and packaging materials.

Part E Novel Materials and Selected Applications
46 Solar Cells and Photovoltaics
47 Silicon on Mechanically Flexible Substrates for Large-Area Electronics
48 Photoconductors for X-Ray Image Detectors
49 Phase-Change Optical Recording
50 Carbon Nanotubes and Bucky Materials
51 Magnetic Information-Storage Materials
52 High-Temperature Superconductors
53 Molecular Electronics
54 Organic Materials for Chemical Sensing
55 Packaging Materials

Glossary of Defining Terms There is a glossary of *Defining Terms* at the end of the handbook that covers important terms that are used throughout the handbook. The terms have been defined to be clear and understandable by an average reader not directly working in the field.

使用说明

1.《电子与光子材料手册》原版为一册,分为A、B、C、D、E五部分。考虑到各部分内容相对独立完整,为使用方便,影印版按部分分为5册。

2.各册在页脚重新编排页码,该页码对应中文目录。保留了原书页眉及页码,其页码对应原书目录及主题索引。

3.各册均有完整5册书的内容简介。

4.作者及其联系方式、缩略语表各册均完整呈现。

5.名词术语表、主题索引安排在第5册。

6.文前页基本采用中英文对照形式,方便读者快速浏览。

材料科学与工程图书工作室
联系电话 0451-86412421
         0451-86414559
邮　　箱 yh_bj@yahoo.com.cn
         xuyaying81823@gmail.com
         zhxh6414559@yahoo.com.cn

Springer 手册精选系列

电子与光子材料手册

电子与光子材料基础性质

【第1册】

Springer
Handbook of
Electronic
and Photonic
Materials

〔加拿大〕Safa Kasap
〔英　国〕Peter Capper 主编

（影印版）

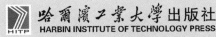
哈尔滨工业大学出版社
HARBIN INSTITUTE OF TECHNOLOGY PRESS

黑版贸审字08-2012-031号

Reprint from English language edition:
Springer Handbook of Electronic and Photonic Materials
by Safa Kasap and Peter Capper
Copyright © 2007 Springer US
Springer US is a part of Springer Science+Business Media
All Rights Reserved

This reprint has been authorized by Springer Science & Business Media for distribution in China Mainland only and not for export there from.

图书在版编目（CIP）数据

电子与光子材料手册. 第1册, 电子与光子材料基础性质=Handbook of Electronic and Photonic Materials I Fundamental Properties：英文／（加）卡萨普（Kasap S.），（英）卡珀（Capper P.）主编. —影印本. —哈尔滨：哈尔滨工业大学出版社，2013.1

（Springer手册精选系列）
ISBN 978-7-5603-3760-9

Ⅰ. ①电… Ⅱ. ①卡…②卡… Ⅲ. ①电子材料–性质–手册–英文②光学材料–性质–手册–英文 Ⅳ. ①TN04-62②TB34-62

中国版本图书馆CIP数据核字（2012）第189775号

责任编辑　杨　桦　许雅莹
出版发行　哈尔滨工业大学出版社
社　　址　哈尔滨市南岗区复华四道街10号　邮编 150006
传　　真　0451-86414749
网　　址　http://hitpress.hit.edu.cn
印　　刷　哈尔滨市石桥印务有限公司
开　　本　787mm×960mm　1/16　印张 16.5
版　　次　2013年1月第1版　2013年1月第1次印刷
书　　号　ISBN 978-7-5603-3760-9
定　　价　48.00元

（如因印刷质量问题影响阅读，我社负责调换）

序 言

　　本书的编辑、作者、出版人都将庆祝这本卓著书籍的出版,这对于电子与光子材料领域的工作者也将是无法衡量的好消息。从以往编辑的系列手册看,我认为本书的出版是值得的,坚持出版这样一本书也是必要的。本书之所以显得特别重要,是因为它在这个领域,内容覆盖范围广泛,涉及的方法也是当今的最新研究进展。在这样一个迅速发展的领域,这是一个相当大的挑战,它已经赢得了人们的敬意。

　　早期的手册和百科全书也都注重阐述半导体材料的发展趋势,而且必须覆盖半导体材料广泛的研究范围和所涉及的现象。这是可以理解的,原因在于半导体材料在电子领域中的主导地位。但没有多少人有足够的勇气预测未来的发展趋势。1992年,Mahajan和Kimerling在其《简明半导体材料百科全书和相关技术》一书的引言中做了尝试,并且预测未来的挑战将是纳米电子领域、低位错密度的III-V族衬底技术、半绝缘III-V族衬底技术、III-V族图形外延技术、替换电介质和硅接触技术、离子注入和扩散技术的发展。这些预测或多或少地成为了现实,但是这也同样说明做出这样的预测是多么的困难。

　　十年前没有多少人会想到III族氮化物在这本书中将成为重要的部分。与制备相关的问题是,作为高熔点材料,在受欢迎的能在光谱蓝端作光发射器的材料中它们的熔点并不高。这是一个很有意思的话题,至少与解决早期光谱红端的固体激光器工作寿命短的问题一样有趣。总地说来,光电子学和光子学在前十年中已经呈现出一些令人瞩目的研究进展,这些在本书中得到了体现,范围从可见光发光器件材料到红外线材料。书中Part D的内容范围很宽,包括III-V族和II-VI族光电子材料和能带隙工程,以及光子玻璃、液态晶体、有机光电导体和光子晶体的新领域。整个部分反映了材料的光产生、工艺、光传输和光探测,包括所有用光取代电子的必要内容。

　　在电子材料这一章(Part C)探讨了硅的进展。毋庸置疑地是,硅是占据了电子功能和电子电路整个范围主导地位的材料,包括新电介质和其他关于缩减电路和器件的几何尺寸以实现更高密度的封装方面的内容,以及其他书很少涉及的领域、薄膜、高温电子材料、非晶和微晶材料。增加硅使用寿命的新技术成果(包括硅/锗合金)在书中也有介绍,并且又一次提出了同样问题,即,预测硅过时时间是否过于超前!铁电体——一类与硅非常有效结合的材料同样也出现在书中。

Part E章节中（新型材料和选择性的应用）使用了一些极好的新方法开辟了新领地。我们大都知道且频繁使用信息记录器件，但是很少知道，涉及器件使用的材料或原理，比如说CD、视频、DVD等。本书介绍了磁信息存储材料，同样介绍了相变光记录材料，使我们充分与当前发展步伐保持同步更新。该章也同样介绍了太阳能电池、传感器、光导体和碳纳米管的应用，这样大量的工作也体现出编写内容汲取到了世界范围的广度。本章各节中的分子电子和封装材料从研究到应用都得到了呈现。

  本书的突出优点在于它的内容覆盖了从基础科学（Part A）到材料的制备、特性（Part B）再到材料的应用（Part C～E）。实际上，书中介绍了涉及的所有材料的广泛应用，这就是本书为什么将会实用的原因之一。就像我之前提及的那样，我们之中没有多少人能够成功地预测未来的发展方向和趋势，在未来十年占领这个领域的主导地位。但是，本书教给我们关于材料的基本性能，可用它们去满足将来的需要。我热切地把这本书推荐给你们。

Prof. Arthur Willoughby
Materials Research Group,
University of Southampton,
UK

前　言

不同学科各种各样的手册,例如电子工程、电子学、生物医学工程、材料科学等手册被广大学生、教师、专业人员很好地使用着,大部分的图书馆也都藏有这些手册。这类手册一般包含许多章(至少50章)内容,在已确定的学科内覆盖广泛的课题;学科选材和论述水平吸引着本科生、研究生、研究员,乃至专业工程人员;最新课题提供广泛的信息,这对该领域所有初学者和研究人员是非常有帮助的;每隔几年,就会有增加新内容的新版本更新之前的版本。

电子和光子材料领域没有类似手册的出版,我们出版这本《电子与光子材料手册》的想法是源自于对手册的需求。它广泛覆盖当今材料领域内的课题,在工程学、材料科学、物理学和化学中都有需要。电子和光子材料真正是一门跨学科的学问,它包含了一些传统的学科,如材料科学、电子工程、化学工程、机械工程、物理学和化学。不难发现,机械工程人员对电子封装实施研究,而电子工程人员对半导体特性进行测量。只有很少的几所大学创建了电子材料或光子材料系。一般来说,电子材料作为一个"学科"是以研究组或跨学科的活动出现在"学院"中。有人可能会对此有异议,因为它事实上是一个跨学科领域,非常需要既包括基础学科又要有最新课题介绍的手册,这就是出版本手册的原因。

本手册是一部关于电子和光子材料的综合论述专著,每一章都是由该领域的专家编写的。本手册针对于大学四年级学生或研究生、研究人员和工作在电子、光电子、光子材料领域的专业人员。书中提供了必要的背景知识和内容广泛的更新知识。每一章都有对内容的一个介绍,并且有许多清晰的说明和大量参考文献。清晰的解释和说明使手册对所有层次的研究者有很大的帮助。所有的章节内容都尽可能独立。既有基础又有前沿的章节内容将吸引不同背景的读者。本手册特别重要的一个特点就是跨学科。例如,将会有这样一些读者,其背景(第一学历)是学化学工程的,工作在半导体工艺线上,而想要学习半导体物理的基础知识;第一学历是物理学的另外一些读者需要尽快更新材料科学的新概念,例如,液相外延等。只要可能,本手册尽量避免采用复杂的数学公式,论述将以半定量的形式给出。手册给出了名词术语表(Glossary of Defining Terms),可为读者提供术语定义的快速查找——这对跨学科工具书来说是必须的。

编者非常感激所有作者们卓越的贡献和相互合作，以及在不同阶段对撰写这本手册的奉献。真诚地感谢Springer Boston的Greg Franklin在文献整理以及手册出版的漫长的工作中给予的支持和帮助。Dr.Werner Skolaut在Springer Heidelberg非常熟练地处理了无数个出版问题，涉及审稿、绘图、书稿的编写和校样的修改，我们真诚地感谢他和他所做出的工作——使得手册能够吸引读者。他是我们见过的最有奉献精神和有效率的编者。

　　感谢Arthur Willoughby教授的诸多建设性意见使得本手册更加完善。他在材料科学杂志（Journal of Materials Science）积累了非常丰富的编辑经验：电子材料这一章在书中起着重要作用，不仅仅是选取章节，而且还要适应读者需要。

　　最后，编者感谢所有的成员（Marian, Samuel and Tomas; and Nicollette）在全部工作中的支持和付出的特别耐心。

Dr. Peter Capper
Materials Team Leader,
SELEX Sensors and Airborne Systems,
Southampton, UK

Prof. Safa Kasap
Professor and Canada Research Chair,
Electrical Engineering Department,
University of Saskatchewan,
Canada

Foreword

The Editors, Authors, and Publisher are to be congratulated on this distinguished volume, which will be an invaluable source of information to all workers in the area of electronic and photonic materials. Having made contributions to earlier handbooks, I am well aware of the considerable, and sustained work that is necessary to produce a volume of this kind. This particular handbook, however, is distinguished by its breadth of coverage in the field, and the way in which it discusses the very latest developments. In such a rapidly moving field, this is a considerable challenge, and it has been met admirably.

Previous handbooks and encyclopaedia have tended to concentrate on semiconducting materials, for the understandable reason of their dominance in the electronics field, and the wide range of semiconducting materials and phenomena that must be covered. Few have been courageous enough to predict future trends, but in 1992 Mahajan and Kimerling attempted this in the Introduction to their Concise Encyclopaedia of Semiconducting Materials and Related Technologies (Pergamon), and foresaw future challenges in the areas of nanoelectronics, low dislocation-density III-V substrates, semi-insulating III-V substrates, patterned epitaxy of III-Vs, alternative dielectrics and contacts for silicon technology, and developments in ion-implantation and diffusion. To a greater or lesser extent, all of these have been proved to be true, but it illustrates how difficult it is to make such a prediction.

Not many people would have thought, a decade ago, that the III-nitrides would occupy an important position in this book. As high melting point materials, with the associated growth problems, they were not high on the list of favourites for light emitters at the blue end of the spectrum! The story is a fascinating one – at least as interesting as the solution to the problem of the short working life of early solid-state lasers at the red end of the spectrum. Optoelectronics and photonics, in general, have seen one of the most spectacular advances over the last decade, and this is fully reflected in the book, ranging from visible light emitters, to infra-red materials. The book covers a wide range of work in Part D, including III-V and II-VI optoelectronic materials and band-gap engineering, as well as photonic glasses, liquid crystals, organic photoconductors, and the new area of photonic crystals. The whole Part reflects materials for light generation, processing, transmission and detection – all the essential elements for using light instead of electrons.

In the Materials for Electronics part (Part C) the book charts the progress in silicon – overwhelmingly the dominant material for a whole range of electronic functions and circuitry – including new dielectrics and other issues associated with shrinking geometry of circuits and devices to produce ever higher packing densities. It also includes areas rarely covered in other books – thick films, high-temperature electronic materials, amorphous and microcrystalline materials. The existing developments that extend the life of silicon technology, including silicon/germanium alloys, appear too, and raise the question again as to whether the predicted timetable for the demise of silicon has again been declared too early!! Ferroelectrics – a class of materials used so effectively in conjunction with silicon – certainly deserve to be here.

The chapters in Part E (Novel Materials and Selected Applications), break new ground in a number of admirable ways. Most of us are aware of, and frequently use, information recording devices such as CDs, videos, DVDs etc., but few are aware of the materials, or principles, involved. This book describes magnetic information storage materials, as well as phase-change optical recording, keeping us fully up-to-date with recent developments. The chapters also include applications such as solar cells, sensors, photoconductors, and carbon nanotubes, on which such a huge volume of work is presently being pursued worldwide. Both ends of the spectrum from research to applications are represented in chapters on molecular electronics and packaging materials.

A particular strength of this book is that it ranges from the fundamental science (Part A) through growth and characterisation of the materials (Part B) to

Prof. Arthur Willoughby
Materials Research Group,
University of Southampton,
UK

applications (Parts C–E). Virtually all the materials covered here have a wide range of applications, which is one of the reasons why this book is going to be so useful. As I indicated before, few of us will be successful in predicting the future direction and trends, occupying the high-ground in this field in the coming decade, but this book teaches us the basic principles of materials, and leaves it to us to adapt these to the needs of tomorrow. I commend it to you most warmly.

Preface

Other handbooks in various disciplines such as electrical engineering, electronics, biomedical engineering, materials science, etc. are currently available and well used by numerous students, instructors and professionals. Most libraries have these handbook sets and each contains numerous (at least 50) chapters that cover a wide spectrum of topics within each well-defined discipline. The subject and the level of coverage appeal to both undergraduate and postgraduate students and researchers as well as to practicing professionals. The advanced topics follow introductory topics and provide ample information that is useful to all, beginners and researchers, in the field. Every few years, a new edition is brought out to update the coverage and include new topics.

There has been no similar handbook in electronic and photonic materials, and the present Springer Handbook of Electronic and Photonic Materials (SHEPM) idea grew out of a need for a handbook that covers a wide spectrum of topics in materials that today's engineers, material scientists, physicists, and chemists need. Electronic and photonic materials is a truly interdisciplinary subject that encompasses a number of traditional disciplines such as materials science, electrical engineering, chemical engineering, mechanical engineering, physics and chemistry. It is not unusual to find a mechanical engineering faculty carrying out research on electronic packaging and electrical engineers carrying out characterization measurements on semiconductors. There are only a few established university departments in electronic or photonic materials. In general, electronic materials as a "discipline" appears as a research group or as an interdisciplinary activity within a "college". One could argue that, because of the very fact that it is such an interdisciplinary field, there is a greater need to have a handbook that covers not only fundamental topics but also advanced topics; hence the present handbook.

This handbook is a comprehensive treatise on electronic and photonic materials with each chapter written by experts in the field. The handbook is aimed at senior undergraduate and graduate students, researchers and professionals working in the area of electronic, optoelectronic and photonic materials. The chapters provide the necessary background and up-to-date knowledge in a wide range of topics. Each chapter has an introduction to the topic, many clear illustrations and numerous references. Clear explanations and illustrations make the handbook useful to all levels of researchers. All chapters are as self-contained as possible. There are both fundamental and advanced chapters to appeal to readers with different backgrounds. This is particularly important for this handbook since the subject matter is highly interdisciplinary. For example, there will be readers with a background (first degree) in chemical engineering and working on semiconductor processing who need to learn the fundamentals of semiconductors physics. Someone with a first degree in physics would need to quickly update himself on materials science concepts such as liquid phase epitaxy and so on. Difficult mathematics has been avoided and, whenever possible, the explanations have been given semiquantitatively. There is a "*Glossary of Defining Terms*" at the end of the handbook, which can serve to quickly find the definition of a term – a very necessary feature in an interdisciplinary handbook.

Dr. Peter Capper
Materials Team Leader,
SELEX Sensors and Airborne Systems,
Southampton, UK

Prof. Safa Kasap
Professor and Canada Research Chair,
Electrical Engineering Department,
University of Saskatchewan, Canada

The editors are very grateful to all the authors for their excellent contributions and for their cooperation in delivering their manuscripts and in the various stages of production of this handbook. Sincere thanks go to Greg Franklin at Springer Boston for all his support and help throughout the long period of commissioning, acquiring the contributions and the production of the handbook. Dr. Werner Skolaut at Springer Heidelberg has very skillfully handled the myriad production issues involved in copy-editing, figure redrawing and proof preparation and correction and our sincere thanks go to him also for all his hard work in making the handbook attractive to read. He is the most dedicated and efficient editor we have come across.

It is a pleasure to thank Professor Arthur Willoughby for his many helpful suggestions that made this a better handbook. His wealth of experience as editor of the Journal of Materials Science: Materials in Electronics played an important role not only in selecting chapters but also in finding the right authors.

Finally, the editors wish to thank all the members of our families (Marian, Samuel and Thomas; and Nicollette) for their support and particularly their endurance during the entire project.

Peter Capper and Safa Kasap
Editors

List of Authors

Martin Abkowitz
1198 Gatestone Circle
Webster, NY 14580, USA
e-mail: *mabkowitz@mailaps.org*,
abkowitz@chem.chem.rochester.edu

Sadao Adachi
Gunma University
Department of Electronic Engineering,
Faculty of Engineering
Kiryu-shi 376-8515
Gunma, Japan
e-mail: *adachi@el.gunma-u.ac.jp*

Alfred Adams
University of Surrey
Advanced Technology Institute
Guildford, Surrey, GU2 7XH,
Surrey, UK
e-mail: *alf.adams@surrey.ac.uk*

Guy J. Adriaenssens
University of Leuven
Laboratorium voor Halfgeleiderfysica
Celestijnenlaan 200D
B-3001 Leuven, Belgium
e-mail: *guy.adri@fys.kuleuven.ac.be*

Wilfried von Ammon
Siltronic AG
Research and Development
Johannes Hess Strasse 24
84489 Burghausen, Germany
e-mail: *wilfried.ammon@siltronic.com*

Peter Ashburn
University of Southampton
School of Electronics and Computer Science
Southampton, SO17 1BJ, UK
e-mail: *pa@ecs.soton.ac.uk*

Mark Auslender
Ben-Gurion University of the Negev Beer Sheva
Department of Electrical
and Computer Engineering
P.O.Box 653
Beer Sheva 84105, Israel
e-mail: *marka@ee.bgu.ac.il*

Darren M. Bagnall
University of Southampton
School of Electronics and Computer Science
Southampton, SO17 1BJ, UK
e-mail: *dmb@ecs.soton.ac.uk*

Ian M. Baker
SELEX Sensors and Airborne Systems Infrared Ltd.
Southampton, Hampshire SO15 0EG, UK
e-mail: *ian.m.baker@selex-sas.com*

Sergei Baranovskii
Philipps University Marburg
Department of Physics
Renthof 5
35032 Marburg, Germany
e-mail: *baranovs@staff.uni-marburg.de*

Mark Baxendale
Queen Mary, University of London
Department of Physics
Mile End Road
London, E1 4NS, UK
e-mail: *m.baxendale@qmul.ac.uk*

Mohammed L. Benkhedir
University of Leuven
Laboratorium voor Halfgeleiderfysica
Celestijnenlaan 200D
B-3001 Leuven, Belgium
e-mail: *MohammedLoufti.Benkhedir
@fys.kuleuven.ac.be*

Monica Brinza
University of Leuven
Laboratorium voor Halfgeleiderfysica
Celestijnenlaan 200D
B-3001 Leuven, Belgium
e-mail: *monica.brinza@fys.kuleuven.ac.be*

Paul D. Brown
University of Nottingham
School of Mechanical, Materials and
Manufacturing Engineering
University Park
Nottingham, NG7 2RD, UK
e-mail: *paul.brown@nottingham.ac.uk*

Mike Brozel
University of Glasgow
Department of Physics and Astronomy
Kelvin Building
Glasgow, G12 8QQ, UK
e-mail: *mikebrozel@beeb.net*

Lukasz Brzozowski
University of Toronto
Sunnybrook and Women's Research Institute,
Imaging Research/
Department of Medical Biophysics
Research Building, 2075 Bayview Avenue
Toronto, ON, M4N 3M5, Canada
e-mail: *lukbroz@sten.sunnybrook.utoronto.ca*

Peter Capper
SELEX Sensors and Airborne Systems Infrared Ltd.
Materials Team Leader
Millbrook Industrial Estate, PO Box 217
Southampton, Hampshire SO15 0EG, UK
e-mail: *pete.capper@selex-sas.com*

Larry Comstock
San Jose State University
6574 Crystal Springs Drive
San Jose, CA 95120, USA
e-mail: *Comstock@email.sjsu.edu*

Ray DeCorby
University of Alberta
Department of Electrical
and Computer Engineering
7th Floor, 9107-116 Street N.W.
Edmonton, Alberta T6G 2V4, Canada
e-mail: *rdecorby@trlabs.ca*

M. Jamal Deen
McMaster University
Department of Electrical
and Computer Engineering (CRL 226)
1280 Main Street West
Hamilton, ON L8S 4K1, Canada
e-mail: *jamal@mcmaster.ca*

Leonard Dissado
The University of Leicester
Department of Engineering
University Road
Leicester, LE1 7RH, UK
e-mail: *lad4@le.ac.uk*

David Dunmur
University of Southampton
School of Chemistry
Southampton, SO17 1BJ, UK
e-mail: *d.a.dunmur@soton.ac.uk*

Lester F. Eastman
Cornell University
Department of Electrical
and Computer Engineering
425 Phillips Hall
Ithaca, NY 14853, USA
e-mail: *lfe2@cornell.edu*

Andy Edgar
Victoria University
School of Chemical and Physical Sciences SCPS
Kelburn Parade/PO Box 600
Wellington, New Zealand
e-mail: *Andy.Edgar@vuw.ac.nz*

Brian E. Foutz
Cadence Design Systems
1701 North Street, Bldg 257-3
Endicott, NY 13760, USA
e-mail: *foutz@cadence.com*

Mark Fox
University of Sheffield
Department of Physics and Astronomy
Hicks Building, Hounsefield Road
Sheffield, S3 7RH, UK
e-mail: *mark.fox@shef.ac.uk*

Darrel Frear
RF and Power Packaging Technology Development,
Freescale Semiconductor
2100 East Elliot Road
Tempe, AZ 85284, USA
e-mail: *darrel.frear@freescale.com*

Milan Friesel
Chalmers University of Technology
Department of Physics
Fysikgränd 3
41296 Göteborg, Sweden
e-mail: *friesel@chalmers.se*

Jacek Gieraltowski
Université de Bretagne Occidentale
6 Avenue Le Gorgeu, BP: 809
29285 Brest Cedex, France
e-mail: *Jacek.Gieraltowski@univ-brest.fr*

Yinyan Gong
Columbia University
Department of Applied Physics
and Applied Mathematics
500 W. 120th St.
New York, NY 10027, USA
e-mail: *yg2002@columbia.edu*

Robert D. Gould[†]
Keele University
Thin Films Laboratory, Department of Physics,
School of Chemistry and Physics
Keele, Staffordshire ST5 5BG, UK

Shlomo Hava
Ben-Gurion University of the Negev Beer Sheva
Department of Electrical
and Computer Engineering
P.O. Box 653
Beer Sheva 84105, Israel
e-mail: *hava@ee.bgu.ac.il*

Colin Humphreys
University of Cambridge
Department of Materials Science and Metallurgy
Pembroke Street
Cambridge, CB2 3!Z, UK
e-mail: *colin.Humphreys@msm.cam.ac.uk*

Stuart Irvine
University of Wales, Bangor
Department of Chemistry
Gwynedd, LL57 2UW, UK
e-mail: *sjc.irvine@bangor.ac.uk*

Minoru Isshiki
Tohoku University
Institute of Multidisciplinary Research
for Advanced Materials
1-1, Katahira, 2 chome, Aobaku
Sendai, 980-8577, Japan
e-mail: *isshiki@tagen.tohoku.ac.jp*

Robert Johanson
University of Saskatchewan
Department of Electrical Engineering
57 Campus Drive
Saskatoon, SK S7N 5A9, Canada
e-mail: *johanson@engr.usask.ca*

Tim Joyce
University of Liverpool
Functional Materials Research Centre,
Department of Engineering
Brownlow Hill
Liverpool, L69 3BX, UK
e-mail: *t.joyce@liv.ac.uk*

M. Zahangir Kabir
Concordia University
Department of Electrical and Computer Engineering
Montreal, Quebec S7N5A9, Canada
e-mail: *kabir@encs.concordia.ca*

Safa Kasap
University of Saskatchewan
Department of Electrical Engineering
57 Campus Drive
Saskatoon, SK S7N 5A9, Canada
e-mail: *safa.kasap@usask.ca*

Alexander Kolobov
National Institute of Advanced Industrial Science and Technology
Center for Applied Near-Field Optics Research
1-1-1 Higashi, Tsukuba
Ibaraki, 305-8562, Japan
e-mail: *a.kolobov@aist.go.jp*

Cyril Koughia
University of Saskatchewan
Department of Electrical Engineering
57 Campus Drive
Saskatoon, SK S7N 5A9, Canada
e-mail: *kik486@mail.usask.ca*

Igor L. Kuskovsky
Queens College, City University of New York (CUNY)
Department of Physics
65-30 Kissena Blvd.
Flushing, NY 11367, USA
e-mail: *igor_kuskovsky@qc.edu*

Geoffrey Luckhurst
University of Southampton
School of Chemistry
Southampton, SO17 1BJ, UK
e-mail: *g.r.luckhurst@soton.ac.uk*

Akihisa Matsuda
Tokyo University of Science
Research Institute for Science and Technology
2641 Yamazaki, Noda-shi
Chiba, 278-8510, Japan
e-mail: *amatsuda@rs.noda.tus.ac.jp,
a.matsuda@aist.go.jp*

Naomi Matsuura
Sunnybrook Health Sciences Centre
Department of Medical Biophysics,
Imaging Research
2075 Bayview Avenue
Toronto, ON M4N 3M5, Canada
e-mail: *matsuura@sri.utoronto.ca*

Kazuo Morigaki
University of Tokyo
C-305, Wakabadai 2-12, Inagi
Tokyo, 206-0824, Japan
e-mail: *k.morigaki@yacht.ocn.ne.jp*

Hadis Morkoç
Virginia Commonwealth University
Department of Electrical and Computer Engineering
601 W. Main St., Box 843072
Richmond, VA 23284-3068, USA
e-mail: *hmorkoc@vcu.edu*

Winfried Mönch
Universität Duisburg-Essen
Lotharstraße 1
47048 Duisburg, Germany
e-mail: *w.moench@uni-duisburg.de*

Arokia Nathan
University of Waterloo
Department of Electrical and Computer Engineering
200 University Avenue W.
Waterloo, Ontario N2L 3G1, Canada
e-mail: *anathan@uwaterloo.ca*

Gertrude F. Neumark
Columbia University
Department of Applied Physics
and Applied Mathematics
500W 120th St., MC 4701
New York, NY 10027, USA
e-mail: *gfn1@columbia.edu*

Stephen K. O'Leary
University of Regina
Faculty of Engineering
3737 Wascana Parkway
Regina, SK S4S 0A2, Canada
e-mail: *stephen.oleary@uregina.ca*

Chisato Ogihara
Yamaguchi University
Department of Applied Science
2-16-1 Tokiwadai
Ube, 755-8611, Japan
e-mail: *ogihara@yamaguchi-u.ac.jp*

Fabien Pascal
Université Montpellier 2/CEM2-cc084
Centre d'Electronique
et de Microoptoélectronique de Montpellier
Place E. Bataillon
34095 Montpellier, France
e-mail: *pascal@cem2.univ-montp2.fr*

Michael Petty
University of Durham
Department School of Engineering
South Road
Durham, DH1 3LE, UK
e-mail: *m.c.petty@durham.ac.uk*

Asim Kumar Ray
Queen Mary, University of London
Department of Materials
Mile End Road
London, E1 4NS, UK
e-mail: *a.k.ray@qmul.ac.uk*

John Rowlands
University of Toronto
Department of Medical Biophysics
Sunnybrook and Women's College
Health Sciences Centre
S656-2075 Bayview Avenue
Toronto, ON M4N 3M5, Canada
e-mail: *john.rowlands@sri.utoronto.ca*

Oleg Rubel
Philipps University Marburg
Department of Physics
and Material Sciences Center
Renthof 5
35032 Marburg, Germany
e-mail: *oleg.rubel@physik.uni-marburg.de*

Harry Ruda
University of Toronto
Materials Science and Engineering,
Electrical and Computer Engineering
170 College Street
Toronto, M5S 3E4, Canada
e-mail: *ruda@ecf.utoronto.ca*

Edward Sargent
University of Toronto
Department of Electrical
and Computer Engineering
ECE, 10 King's College Road
Toronto, M5S 3G4, Canada
e-mail: *ted.sargent@utoronto.ca*

Peyman Servati
Ignis Innovation Inc.
55 Culpepper Dr.
Waterloo, Ontario N2L 5K8, Canada
e-mail: *pservati@uwaterloo.ca*

Derek Shaw
Hull University
Hull, HU6 7RX, UK
e-mail: *DerekShaw1@compuserve.com*

Fumio Shimura
Shizuoka Institute of Science and Technology
Department of Materials and Life Science
2200-2 Toyosawa
Fukuroi, Shizuoka 437-8555, Japan
e-mail: *shimura@ms.sist.ac.jp*

Michael Shur
Renssellaer Polytechnic Institute
Department of Electrical, Computer,
and Systems Engineering
CII 9017, RPI, 110 8th Street
Troy, NY 12180, USA
e-mail: *shurm@rpi.edu*

Jai Singh
Charles Darwin University
School of Engineering and Logistics,
Faculty of Technology, B-41
Ellengowan Drive
Darwin, NT 0909, Australia
e-mail: *jai.singh@cdu.edu.au*

Tim Smeeton
Sharp Laboratories of Europe
Edmund Halley Road, Oxford Science Park
Oxford, OX4 4GB, UK
e-mail: *tim.smeeton@sharp.co.uk*

Boris Straumal
Russian Academy of Sciences
Institute of Sold State Physics
Institutskii prospect 15
Chernogolovka, 142432, Russia
e-mail: *straumal@issp.ac.ru*

Stephen Sweeney
University of Surrey
Advanced Technology Institute
Guildford, Surrey GU2 7XH, UK
e-mail: *s.sweeney@surrey.ac.uk*

David Sykes
Loughborough Surface Analysis Ltd.
PO Box 5016, Unit FC, Holywell Park, Ashby Road
Loughborough, LE11 3WS, UK
e-mail: *d.e.sykes@lsaltd.co.uk*

Keiji Tanaka
Hokkaido University
Department of Applied Physics,
Graduate School of Engineering
Kita-ku, N13 W8
Sapporo, 060-8628, Japan
e-mail: *keiji@eng.hokudai.ac.jp*

Charbel Tannous
Université de Bretagne Occidentale
LMB, CNRS FRE 2697
6 Avenue Le Gorgeu, BP: 809
29285 Brest Cedex, France
e-mail: *tannous@univ-brest.fr*

Ali Teke
Balikesir University
Department of Physics, Faculty of Art and Science
Balikesir, 10100, Turkey
e-mail: *ateke@balikesir.edu.tr*

Junji Tominaga
National Institute of Advanced Industrial
Science and Technology, AIST
Center for Applied Near-Field Optics Research,
CAN-FOR
Tsukuba Central 4 1-1-1 Higashi
Tsukuba, 3.5-8562, Japan
e-mail: *j-tomonaga@aist.go.jp*

Dan Tonchev
University of Saskatchewan
Department of Electrical Engineering
57 Campus Drive
Saskatoon, SK S7N 5A9, Canada
e-mail: *dan.tonchev@usask.ca*

Harry L. Tuller
Massachusetts Institute of Technology
Department of Materials Science and Engineering,
Crystal Physics and Electroceramics Laboratory
77 Massachusetts Avenue
Cambridge, MA 02139, USA
e-mail: *tuller@mit.edu*

Qamar-ul Wahab
Linköping University
Department of Physics,
Chemistry, and Biology (IFM)
SE-581 83 Linköping, Sweden
e-mail: *quw@ifm.liu.se*

Robert M. Wallace
University of Texas at Dallas
Department of Electrical Engineering
M.S. EC 33, P.O.Box 830688
Richardson, TX 75083, USA
e-mail: *rmwallace@utdallas.edu*

Jifeng Wang
Tohoku University
Institute of Multidisciplinary Research
for Advanced Materials
1-1, Katahira, 2 Chome, Aobaku
Sendai, 980-8577, Japan
e-mail: *wang@tagen.tohoku.ac.jp*

David S. Weiss
NexPress Solutions, Inc.
2600 Manitou Road
Rochester, NY 14653-4180, USA
e-mail: *David_Weiss@Nexpress.com*

Rainer Wesche
Swiss Federal Institute of Technology
Centre de Recherches en Physique des Plasmas
CRPP (c/o Paul Scherrer Institute), WMHA/C31,
Villigen PS
Lausanne, CH-5232, Switzerland
e-mail: *rainer.wesche@psi.ch*

Roger Whatmore
Tyndall National Institute
Lee Maltings, Cork , Ireland
e-mail: *roger.whatmore@tyndall.ie*

Neil White
University of Southampton
School of Electronics and Computer Science
Mountbatten Building
Highfield, Southampton SO17 1BJ, UK
e-mail: *nmw@ecs.soton.ac.uk*

Magnus Willander
University of Gothenburg
Department of Physics
SE-412 96 Göteborg, Sweden
e-mail: *mwi@fy.chalmers.se*

Jan Willekens
University of Leuven
Laboratorium voor Halfgeleiderfysica
Celestijnenlaan 200D
B-3001 Leuven, Belgium
e-mail: *jan.willekens@kc.kuleuven.ac.be*

Acknowledgements

1 Perspectives on Electronic and Optoelectronic Materials
by Tim Smeeton, Colin Humphreys
Many thanks to Dr Ron Broom for his helpful comments on this manuscript.

A.5 Defects in Monocrystalline Silicon
by Wilfried von Ammon
The author is greatly indebted to W. Haeckl, E. Dornberger, D. Gräf and R. Schmolke for many helpful discussions and for providing some of the figures.

A.7 Photoconductivity in Materials Research
by Monica Brinza, Jan Willekens, Mohammed L. Benkhedir, Guy J. Adriaenssens
The authors are grateful to the *Fonds voor Wetenschappelijk Onderzoek – Vlaanderen* for its financial support of their research.

A.9 Charge Transport in Disordered Materials
by S. D. Baranovskii, O. Rubel
The authors are indebted to numerous colleagues for stimulating and enlightening discussions. Among those are Boris Shklovski (University of Minnesota), Alexei Efros and Michael Raikh (Utah University), Hellmut Fritzsche (Chicago University), Peter Thomas, Walther Fuhs and Heinz Bässler (Philipps-University Marburg), Igor Zvyagin (Moscow State University) and many other colleagues. Financial support of the Deutsche Forschungsgemeinschaft is gratefully acknowledged.

A.11 Ionic Conduction and Applications
by Harry L. Tuller
Support from the National Science Foundation (Grant Nos. DMR-0243993 and ECS-0428696) and ARO-MURI under grant DAAD-0101-0566 for topics related to this work are highly appreciated. In assembling this work, I drew on earlier journal and proceedings articles published by myself or in conjunction with colleagues. In particular, I wish to acknowledge my collaborator in Ref. 2, Prof. P. Knauth of the Université de Provence, Marseille, France.

目 录

缩略语

引 言

1 电子和光电子材料背景 ... 3
    1.1 前期 ... 4
    1.2 硅时代 ... 4
    1.3 化合物半导体 ... 8
    1.4 从法拉第到今天 ... 14
    参考文献 ... 14

Part A 基本特性

2 金属和半导体中的导电体 ... 19
    2.1 基本原理:漂移速度、迁移率和电导率 ... 20
    2.2 马蒂定则 ... 22
    2.3 金属的电阻率 ... 23
    2.4 固溶体和诺德海姆定则 ... 26
    2.5 半导体中载流子散射 ... 28
    2.6 玻耳兹曼传导方程 ... 29
    2.7 多晶薄膜电阻率 ... 30
    2.8 非均匀介质、等效介质近似 ... 32
    2.9 霍尔效应 ... 35
    2.10 高电场传输 ... 37
    2.11 雪崩 ... 38
    2.12 二维电子气 ... 39
    2.13 一维电导率 ... 41
    2.14 量子霍尔效应 ... 42
    参考文献 ... 44

3 电子材料的光特性：基本原理与特性描述 47
3.1 光常量 47
3.2 折射率 50
3.3 光吸收 53
3.4 薄膜光学 70
3.5 光材料 74
参考文献 76

4 电子材料的磁特性 79
4.1 传统磁学 81
4.2 非传统磁学 93
参考文献 99

5 单晶硅的缺陷 101
5.1 本征点缺陷聚集的工艺影响 102
5.2 本征点缺陷的热物理特性 103
5.3 本征点缺陷聚合物 104
5.4 OSF环的形成 115
参考文献 117

6 半导体中的扩散 121
6.1 基本概念 122
6.2 扩散机理 122
6.3 扩散机制 123
6.4 内建电场 126
6.5 扩散系数的测量 126
6.6 半导体中的氢 127
6.7 IV族半导体组的扩散 128
6.8 III-V族化合物的扩散 130
6.9 II-VI族化合物的扩散 131
6.10 结论 133
6.11 课外读物和参考文献 133
参考文献 133

7 材料中的光导率研究 137
7.1 稳态光导率方法 138
7.2 瞬态光导率实验 142
参考文献 146

8 半导体界面电子特性 147
8.1 实验数据库 149
8.2 IFIGS 和电负性理论 153
8.3 实验和理论对比 155
8.4 结论 159
参考文献 159

9 无序材料电荷传输 161
9.1 无序材料电荷传输的一般理论 163
9.2 扩展态下的无序材料电荷传输 167
9.3 局部态下的无序材料跳跃电荷传输 169
9.4 结论 184
参考文献 185

10 介电响应 187
10.1 介电响应的定义 188
10.2 与频率相关的线性响应 190
10.3 松弛响应中的信息内容 196
10.4 电荷传输 208
10.5 一些结论 211
参考文献 211

11 离子传导与应用 213
11.1 离子固态传导 214
11.2 快速离子传导 216
11.3 混合离子-电子传导 221
11.4 应用 223
11.5 未来趋势 226
参考文献 226

Contents

List of Abbreviations

Introduction

1 **Perspectives on Electronic and Optoelectronic Materials** 3
    1.1   The Early Years 4
    1.2   The Silicon Age 4
    1.3   The Compound Semiconductors 8
    1.4   From Faraday to Today 14
    References 14

Part A Fundamental Properties

2 **Electrical Conduction in Metals and Semiconductors** 19
    2.1   Fundamentals: Drift Velocity, Mobility and Conductivity 20
    2.2   Matthiessen's Rule 22
    2.3   Resistivity of Metals 23
    2.4   Solid Solutions and Nordheim's Rule 26
    2.5   Carrier Scattering in Semiconductors 28
    2.6   The Boltzmann Transport Equation 29
    2.7   Resistivity of Thin Polycrystalline Films 30
    2.8   Inhomogeneous Media. Effective Media Approximation 32
    2.9   The Hall Effect 35
    2.10   High Electric Field Transport 37
    2.11   Avalanche 38
    2.12   Two-Dimensional Electron Gas 39
    2.13   One Dimensional Conductance 41
    2.14   The Quantum Hall Effect 42
    References 44

3 **Optical Properties of Electronic Materials: Fundamentals and Characterization** 47
    3.1   Optical Constants 47
    3.2   Refractive Index 50
    3.3   Optical Absorption 53
    3.4   Thin Film Optics 70
    3.5   Optical Materials 74
    References 76

4	**Magnetic Properties of Electronic Materials**	79
	4.1 Traditional Magnetism	81
	4.2 Unconventional Magnetism	93
	References	99
5	**Defects in Monocrystalline Silicon**	101
	5.1 Technological Impact of Intrinsic Point Defects Aggregates	102
	5.2 Thermophysical Properties of Intrinsic Point Defects	103
	5.3 Aggregates of Intrinsic Point Defects	104
	5.4 Formation of OSF Ring	115
	References	117
6	**Diffusion in Semiconductors**	121
	6.1 Basic Concepts	122
	6.2 Diffusion Mechanisms	122
	6.3 Diffusion Regimes	123
	6.4 Internal Electric Fields	126
	6.5 Measurement of Diffusion Coefficients	126
	6.6 Hydrogen in Semiconductors	127
	6.7 Diffusion in Group IV Semiconductors	128
	6.8 Diffusion in III–V Compounds	130
	6.9 Diffusion in II–VI Compounds	131
	6.10 Conclusions	133
	6.11 General Reading and References	133
	References	133
7	**Photoconductivity in Materials Research**	137
	7.1 Steady State Photoconductivity Methods	138
	7.2 Transient Photoconductivity Experiments	142
	References	146
8	**Electronic Properties of Semiconductor Interfaces**	147
	8.1 Experimental Database	149
	8.2 IFIGS-and-Electronegativity Theory	153
	8.3 Comparison of Experiment and Theory	155
	8.4 Final Remarks	159
	References	159
9	**Charge Transport in Disordered Materials**	161
	9.1 General Remarks on Charge Transport in Disordered Materials	163
	9.2 Charge Transport in Disordered Materials via Extended States	167
	9.3 Hopping Charge Transport in Disordered Materials via Localized States	169
	9.4 Concluding Remarks	184
	References	185

10 Dielectric Response 187
- 10.1 Definition of Dielectric Response 188
- 10.2 Frequency-Dependent Linear Responses 190
- 10.3 Information Contained in the Relaxation Response 196
- 10.4 Charge Transport 208
- 10.5 A Few Final Comments 211
- References 211

11 Ionic Conduction and Applications 213
- 11.1 Conduction in Ionic Solids 214
- 11.2 Fast Ion Conduction 216
- 11.3 Mixed Ionic–Electronic Conduction 221
- 11.4 Applications 223
- 11.5 Future Trends 226
- References 226

List of Abbreviations

2DEG	two-dimensional electron gas

A

AC	alternating current
ACCUFET	accumulation-mode MOSFET
ACRT	accelerated crucible rotation technique
AEM	analytical electron microscopes
AES	Auger electron spectroscopy
AFM	atomic force microscopy
ALD	atomic-layer deposition
ALE	atomic-layer epitaxy
AMA	active matrix array
AMFPI	active matrix flat-panel imaging
AMOLED	amorphous organic light-emitting diode
APD	avalanche photodiode

B

b.c.c.	body-centered cubic
BEEM	ballistic-electron-emission microscopy
BEP	beam effective pressure
BH	buried-heterostructure
BH	Brooks–Herring
BJT	bipolar junction transistor
BTEX	m-xylene
BZ	Brillouin zone

C

CAIBE	chemically assisted ion beam etching
CB	conduction band
CBE	chemical beam epitaxy
CBED	convergent beam electron diffraction
CC	constant current
CCD	charge-coupled device
CCZ	continuous-charging Czochralski
CFLPE	container-free liquid phase epitaxy
CKR	cross Kelvin resistor
CL	cathodoluminescence
CMOS	complementary metal-oxide-semiconductor
CNR	carrier-to-noise ratio
COP	crystal-originated particle
CP	charge pumping
CPM	constant-photocurrent method
CR	computed radiography
CR-DLTS	computed radiography deep level transient spectroscopy
CRA	cast recrystallize anneal
CTE	coefficient of thermal expansion
CTO	chromium(III) trioxalate
CuPc	copper phthalocyanine
CuTTBPc	tetra-tert-butyl phthalocyanine
CV	chemical vapor
CVD	chemical vapor deposition
CVT	chemical vapor transport
CZ	Czochralski
CZT	cadmium zinc telluride

D

DA	Drude approximation
DAG	direct alloy growth
DBP	dual-beam photoconductivity
DC	direct current
DCPBH	double-channel planar buried heterostructure
DET	diethyl telluride
DFB	distributed feedback
DH	double heterostructure
DIL	dual-in-line
DIPTe	diisopropyltellurium
DLC	diamond-like carbon
DLHJ	double-layer heterojunction
DLTS	deep level transient spectroscopy
DMCd	dimethyl cadmium
DMF	dimethylformamide
DMOSFET	double-diffused MOSFET
DMS	dilute magnetic semiconductors
DMSO	dimethylsulfoxide
DMZn	dimethylzinc
DOS	density of states
DQE	detective quantum efficiency
DSIMS	dynamic secondary ion mass spectrometry
DTBSe	ditertiarybutylselenide
DUT	device under test
DVD	digital versatile disk
DWDM	dense wavelength-division multiplexing
DXD	double-crystal X-ray diffraction

E

EBIC	electron beam induced conductivity
ED	electrodeposition
EDFA	erbium-doped fiber amplifier
EELS	electron energy loss spectroscopy
EFG	film-fed growth
EHP	electron–hole pairs
ELO	epitaxial lateral overgrowth
ELOG	epitaxial layer overgrowth
EM	electromagnetic
EMA	effective media approximation

ENDOR	electron–nuclear double resonance		IFIGS	interface-induced gap states
EPD	etch pit density		IFTOF	interrupted field time-of-flight
EPR	electron paramagnetic resonance		IGBT	insulated gate bipolar transistor
ESR	electron spin resonance spectroscopy		IMP	interdiffused multilayer process
EXAFS	extended X-ray absorption fine structure		IPEYS	internal photoemission yield spectroscopy
			IR	infrared
			ITO	indium-tin-oxide

F

FCA	free-carrier absorption
f.c.c.	face-centered cubic
FET	field effect transistor
FIB	focused ion beam
FM	Frank–van der Merwe
FPA	focal plane arrays
FPD	flow pattern defect
FTIR	Fourier transform infrared
FWHM	full-width at half-maximum
FZ	floating zone

J

JBS	junction barrier Schottky
JFET	junction field-effect transistors
JO	Judd–Ofelt

K

KCR	Kelvin contact resistance
KKR	Kramers–Kronig relation
KLN	$K_3Li_2Nb_5O_{12}$
KTPO	$KTiOPO_4$

G

GDA	generalized Drude approximation
GDMS	glow discharge mass spectrometry
GDOES	glow discharge optical emission spectroscopy
GF	gradient freeze
GMR	giant magnetoresistance
GOI	gate oxide integrity
GRIN	graded refractive index
GSMBE	gas-source molecular beam epitaxy
GTO	gate turn-off

L

LB	Langmuir–Blodgett
LD	laser diodes
LD	lucky drift
LDD	lightly doped drain
LEC	liquid-encapsulated Czochralski
LED	light-emitting diodes
LEIS	low-energy ion scattering
LEL	lower explosive limit
LF	low-frequency
LLS	laser light scattering
LMA	law of mass action
LO	longitudinal optical
LPE	liquid phase epitaxy
LSTD	laser light scattering tomography defect
LVM	localized vibrational mode

H

HAADF	high-angle annular dark field
HB	horizontal Bridgman
HBT	hetero-junction bipolar transistor
HDC	horizontal directional solidification crystallization
HEMT	high electron mobility transistor
HF	high-frequency
HOD	highly oriented diamond
HOLZ	high-order Laue zone
HPc	phthalocyanine
HPHT	high-pressure high-temperature
HRXRD	high-resolution X-ray diffraction
HTCVD	high-temperature CVD
HVDC	high-voltage DC
HWE	hot-wall epitaxy

M

MBE	molecular beam epitaxy
MCCZ	magnetic field applied continuous Czochralski
MCT	mercury cadmium telluride
MCZ	magnetic field applied Czochralski
MD	molecular dynamics
MEED	medium-energy electron diffraction
MEM	micro-electromechanical systems
MESFET	metal-semiconductor field-effect transistor
MFC	mass flow controllers
MIGS	metal-induced gap states
ML	monolayer
MLHJ	multilayer heterojunction
MOCVD	metal-organic chemical vapor deposition
MODFET	modulation-doped field effect transistor

I

IC	integrated circuit
ICTS	isothermal capacitance transient spectroscopy
IDE	interdigitated electrodes

MOMBE	metalorganic molecular beam epitaxy		PL	photoluminescence
MOS	metal/oxide/semiconductor		PM	particulate matter
MOSFET	metal/oxide/semiconductor field effect transistor		PMMA	poly(methyl-methacrylate)
			POT	poly(n-octyl)thiophene
MOVPE	metalorganic vapor phase epitaxy		ppb	parts per billion
MPc	metallophthalocyanine		ppm	parts per million
MPC	modulated photoconductivity		PPS	polyphenylsulfide
MPCVD	microwave plasma chemical deposition		PPY	polypyrrole
MQW	multiple quantum well		PQT-12	poly[5,5'-bis(3-alkyl-2-thienyl)-2,2'-bithiophene]
MR	magnetoresistivity			
MS	metal–semiconductor		PRT	platinum resistance thermometers
MSRD	mean-square relative displacement		PSt	polystyrene
MTF	modulation transfer function		PTC	positive temperature coefficient
MWIR	medium-wavelength infrared		PTIS	photothermal ionisation spectroscopy
			PTS	1,1-dioxo-2-(4-methylphenyl)-6-phenyl-4-(dicyanomethylidene)thiopyran

N

NDR	negative differential resistance
NEA	negative electron affinity
NeXT	nonthermal energy exploration telescope
NMOS	n-type-channel metal–oxide–semiconductor
NMP	N-methylpyrrolidone
NMR	nuclear magnetic resonance
NNH	nearest-neighbor hopping
NSA	naphthalene-1,5-disulfonic acid
NTC	negative temperature coefficient
NTD	neutron transmutation doping

O

OLED	organic light-emitting diode
OSF	oxidation-induced stacking fault
OSL	optically stimulated luminescence
OZM	overlap zone melting

P

PAE	power added efficiency
PAni	polyaniline
pBN	pyrolytic boron nitride
Pc	phthalocyanine
PC	photoconductive
PCA	principal component analysis
PCB	printed circuit board
PDMA	poly(methylmethacrylate)/poly(decyl methacrylate)
PDP	plasma display panels
PDS	photothermal deflection spectroscopy
PE	polysilicon emitter
PE BJT	polysilicon emitter bipolar junction transistor
PECVD	plasma-enhanced chemical vapor deposition
PEN	polyethylene naphthalate
PES	photoemission spectroscopy
PET	positron emission tomography
pHEMT	pseudomorphic HEMT

PTV	polythienylene vinylene
PV	photovoltaic
PVD	physical vapor transport
PVDF	polyvinylidene fluoride
PVK	polyvinylcarbazole
PVT	physical vapor transport
PZT	lead zirconate titanate

Q

QA	quench anneal
QCL	quantum cascade laser
QCSE	quantum-confined Stark effect
QD	quantum dot
QHE	quantum Hall effect
QW	quantum well

R

RAIRS	reflection adsorption infrared spectroscopy
RBS	Rutherford backscattering
RCLED	resonant-cavity light-emitting diode
RDF	radial distribution function
RDS	reflection difference spectroscopy
RE	rare earth
RENS	resolution near-field structure
RF	radio frequency
RG	recombination–generation
RH	relative humidity
RHEED	reflection high-energy electron diffraction
RIE	reactive-ion etching
RIU	refractive index units
RTA	rapid thermal annealing
RTD	resistance temperature devices
RTS	random telegraph signal

S

SA	self-assembly
SAM	self-assembled monolayers

SAW	surface acoustic wave		TMA	trimethyl-aluminum
SAXS	small-angle X-ray scattering		TMG	trimethyl-gallium
SCH	separate confinement heterojunction		TMI	trimethyl-indium
SCVT	seeded chemical vapor transport		TMSb	trimethylantimony
SE	spontaneous emission		TO	transverse optical
SEM	scanning electron microscope		TOF	time of flight
SIMS	secondary ion mass spectrometry		ToFSIMS	time of flight SIMS
SIPBH	semi-insulating planar buried heterostructure		TPC	transient photoconductivity
			TPV	thermophotovoltaic
SIT	static induction transistors		TSC	thermally stimulated current
SK	Stranski–Krastanov		TSL	thermally stimulated luminescence
SNR	signal-to-noise ratio			

U

ULSI	ultra-large-scale integration
UMOSFET	U-shaped-trench MOSFET
UPS	uninterrupted power systems
UV	ultraviolet

SO	small outline
SOA	semiconductor optical amplifier
SOC	system-on-a-chip
SOFC	solid oxide fuel cells
SOI	silicon-on-insulator
SP	screen printing
SPECT	single-photon emission computed tomography
SPR	surface plasmon resonance
SPVT	seeded physical vapor transport
SQW	single quantum wells
SSIMS	static secondary ion mass spectrometry
SSPC	steady-state photoconductivity
SSR	solid-state recrystallisation
SSRM	scanning spreading resistance microscopy
STHM	sublimation traveling heater method
SVP	saturated vapor pressure
SWIR	short-wavelength infrared

V

VAP	valence-alternation pairs
VB	valence band
VCSEL	vertical-cavity surface-emitting laser
VCZ	vapor-pressure-controlled Czochralski
VD	vapor deposition
VFE	vector flow epitaxy
VFET	vacuum field-effect transistor
VGF	vertical gradient freeze
VIS	visible
VOC	volatile organic compounds
VPE	vapor phase epitaxy
VRH	variable-range hopping
VUVG	vertical unseeded vapor growth
VW	Volmer–Weber

T

TAB	tab automated bonding
TBA	tertiarybutylarsine
TBP	tertiarybutylphosphine
TCE	thermal coefficient of expansion
TCNQ	tetracyanoquinodimethane
TCR	temperature coefficient of resistance
TCRI	temperature coefficient of refractive index
TDCM	time-domain charge measurement
TE	transverse electric
TED	transient enhanced diffusion
TED	transmission electron diffraction
TEGa	triethylgallium
TEM	transmission electron microscope
TEN	triethylamine
TFT	thin-film transistors
THM	traveling heater method
TL	thermoluminescence
TLHJ	triple-layer graded heterojunction
TLM	transmission line measurement
TM	transverse magnetic

W

WDX	wavelength dispersive X-ray
WXI	wide-band X-ray imager

X

XAFS	X-ray absorption fine-structure
XANES	X-ray absorption near-edge structure
XEBIT	X-ray-sensitive electron-beam image tube
XPS	X-ray photon spectroscopy
XRD	X-ray diffraction
XRSP	X-ray storage phosphor

Y

YSZ	yttrium-stabilized zirconia

Introduction

1 Perspectives on Electronic and Optoelectronic Materials
Tim Smeeton, Oxford, UK
Colin Humphreys, Cambridge, UK

1. Perspectives on Electronic and Optoelectronic Materials

This opening chapter will concentrate on the changes in the world of semiconducting materials and devices over the latter half of the twentieth century. Within this field we have chosen to concentrate on a few developments and cannot claim to cover all of the major areas. What we plan to do is give a sense of perspective of how the science and technology of these materials has come to its current state and to present a brief overview of why certain materials are chosen for particular device applications.

We start by identifying some of the earliest developments in our understanding of electronic materials; follow the development of silicon technology from the first demonstration of the transistor through to today's integrated circuit; track some of the key electronic and optoelectronic uses of the conventional III–V semiconductors; and end with a review of the last decade's explosion of

1.1	The Early Years	4
1.2	The Silicon Age	4
	1.2.1 The Transistor and Early Semiconductor Materials Development	4
	1.2.2 The Integrated Circuit	6
1.3	The Compound Semiconductors............	8
	1.3.1 High Speed Electronics..............	9
	1.3.2 Light Emitting Devices...............	10
	1.3.3 The III–Nitrides	12
1.4	From Faraday to Today........................	14
References		14

interest in the III–nitride materials. The band gaps of the semiconductors encountered in this chapter are shown in Fig. 1.1 – a figure which will be frequently referred to in explaining the choice of materials for specific applications.

We sometimes forget how remarkable electronic and optoelectronic materials are. Take the light emitting diode (LED) as an example: an electric current is passed through nothing more than a tiny stack of layers of slightly different compounds and brilliant coloured light is emitted. Of course the stack in question is carefully designed, and there are theories to explain the behaviour, but that should not detract from the initial moment of wonderment that it works at all! Devices from LEDs through to microprocessors containing tens of millions of transistors influence life at the start of this twenty first century to an incalculable extent. These complex devices are only available today because an array of electronic materials have been developed using increasingly sophisticated research methods over the past six decades.

Fig. 1.1 Parameter perspective: band gaps and lattice parameters of selected semiconductors discussed in the text. The important wavelengths for optical storage (CD, DVD and Blu–Ray) and the 1.55 μm used for efficient data transmission through optical fibres are labelled

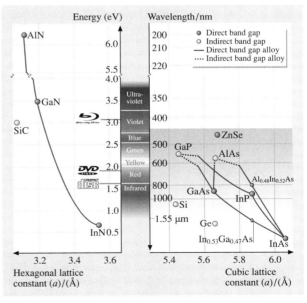

Most semiconductor-based appliances which affect us every day are made using silicon, but many key devices depend on a number of different compound semiconductors. For example GaP-based LEDs in digital displays; GaAs-based HEMTs which operate in our satellite television receivers; AlGaAs and AlInGaP lasers in our CD and DVD players; and the InP-based lasers delivering the internet and telecommunications along optic fibres.

None of these devices could be manufactured without a basic understanding (either fundamental or empirical) of the materials science of the components. At the same time the overwhelming reason for scientific study of the materials is to elicit some improvement in the performance of the devices based on them. The rest of this book concentrates on the *materials* more than their *devices* but to give some idea of how a world largely untouched by electronic materials in the 1940s has become so changed by them we will consider the developments in the two fields in parallel in this chapter. Often they are inseparable anyway.

1.1 The Early Years

The exploitation of electronic materials in solid-state devices principally occurred in the second half of the twentieth century but the first serendipitous observations of semiconducting behaviour took place somewhat earlier than this. In 1833, *Faraday* found that silver sulphide exhibited a negative temperature coefficient [1.1]. This property of a decrease in electrical resistivity with increasing temperature was to be deployed in thermistor components a century later. In the 1870s scientists discovered and experimented with the photoconductivity (decreased resistivity of a material under incident light) of selenium [1.2, 3]. Amorphous selenium was to be used for this very property in the first Xerox copying machines of the 1950s.

While these discoveries had limited immediate impact on scientific understanding, more critical progress was made such as Hall's 1879 discovery of what was to become known as the Hall Effect. The discovery of quantum mechanics was of fundamental importance for our understanding of semiconductors. Based on the advances in quantum theory in the early 1900s a successful theory to explain semiconductor behaviour was formulated in 1931 [1.4, 5]. However, the semiconductors of the 1930s were too impure to allow the theory to be compared with experiment. For example it was believed at the time that silicon, which was to become the archetypal semiconducting material, was a substance belonging to a group of materials which were "good metallic conductors in the pure state and ... therefore to be classed as metals" [1.5]!

However, a solid theoretical understanding of semiconductors was in place by the 1940s. Hence when the device development focus of the second world war-time research was replaced by peace-time research into the fundamental understanding of real semiconductors, the foundations had been laid for working devices based on elemental semiconductors to be realised.

1.2 The Silicon Age

1.2.1 The Transistor and Early Semiconductor Materials Development

As its name suggests, electronics is about the control of electrons to produce useful properties; electronic materials are the media in which this manipulation takes place. Exactly fifty years after J.J. Thompson had discovered the electron in 1897, mankind's ability to control them underwent a revolution due to the discovery of the transistor effect. It could be said that the world began to change in the final couple of weeks of 1947 when John Bardeen and Walter Brattain used germanium to build and demonstrate the first "semi-conductor triode" (a device later to be named the point contact transistor to reflect its transresistive properties). This success at Bell Laboratories was obtained within just a few years of the post-war establishment of a research group led by William Shockley focussing on the understanding of semiconducting materials. It was to earn Brattain, Bardeen and Shockley the 1956 Nobel Prize for Physics.

The first point-contact transistor was based around three contacts onto an n-doped germanium block: when a small current passed between the "base" and "emitter", an amplified current would flow between the "collector" and "emitter" [1.6]. The emitter and collector contacts needed to be located very close to one another (50–250 μm) and this was achieved by evaporating gold onto the corner of a plastic triangle, cutting the film with a razor blade and touching this onto the germanium – the two isolated strips of gold serving as the two contacts [1.7]. At about one centimetre in height, based on relatively impure polycrystalline germanium and adopting a different principle of operation, the device bears barely any resemblance to today's integrated circuit electronics components. Nonetheless it was the first implementation of a solid-state device capable of modulating (necessary for signal amplification in communications) and switching (needed for logic operations in computing) an electric current. In a world whose electronics were delivered by the thermionic vacuum tube, the transistor was immediately identified as a component which could be "employed as an amplifier, oscillator, and for other purposes for which vacuum tubes are ordinarily used" [1.6].

In spite of this, after the public announcement of the invention at the end of June 1948 the response of both the popular and technical press was somewhat muted. It was after all still "little more than a laboratory curiosity" [1.8] and ultimately point-contact transistors were never suited to mass production. The individual devices differed significantly in characteristics, the noise levels in amplification were high and they were rapidly to be superseded by improved transistor types.

A huge range of transistor designs have been introduced from the late 1940s through to today. These successive generations either drew upon, or served as a catalyst for, a range of innovations in semiconductor materials processing and understanding. There are many fascinating differences in device design but from a materials science point of view the three most striking differences between the first point contact transistor and the majority of electronics in use today are the choice of semiconductor, the purity of this material and its crystalline quality. Many of the key electronic materials technologies of today derive from the developments in these fields in the very early years of the post-war semiconductor industry.

Both germanium and silicon had been produced with increasing purity throughout the 1940s [1.9]. Principally because of germanium's lower melting temperature (937 °C compared with 1415 °C) and lower chemical reactivity its preparation had always proved easier and was therefore favoured for the early device manufacture such as the first transistor. However, the properties of silicon make it a much more attractive choice for solid state devices. While germanium is expensive and rare, silicon is, after oxygen, the second most abundant element. Silicon has a higher breakdown field and a greater power handling ability; its semiconductor band gap (1.1 eV at 300 K; Fig. 1.1) is substantially higher then germanium's (0.7 eV) so silicon devices are able to operate over a greater range of temperatures without intrinsic conductivity interfering with performance.

The two materials competed with one another in device applications until the introduction of novel doping techniques in the mid-1950s. Previously p- and n-doping had been achieved by the addition of dopant impurities to the semiconductor melt during solidification. A far more flexible technique involved the diffusion of dopants from the vapour phase into the solid semiconductor surface [1.10]. It became possible to dope with a degree of two-dimensional precision when it was discovered that silicon's oxide served as an effective mask to dopant atoms and that a photoresist could be used to control the etching away of the oxide [1.11, 12]. Successful diffusion masks could not be found for germanium and it was soon abandoned for mainstream device manufacture. Dopant diffusion of this sort has since been superseded by the implantation of high-energy ions which affords greater control and versatility.

Shockley was always aware that the material of the late 1940s was nothing like pure enough to make reliable high performance commercial devices. Quantum mechanics suggested that to make a high quality transistor out of the materials it was necessary to reduce the impurity level to about one part in 10^{10}. This was a far higher degree of purity than existed in any known material. However, William Pfann, who worked at Bell Laboratories, came up with the solution. He invented a technique called zone-refining to solve this problem, and showed that repeated zone refining of germanium and silicon reduced the impurities to the level required. The work of *Pfann* is not widely known but was a critical piece of materials science which enabled the practical development of the transistor [1.13, 14].

At a similar time great progress was being made in reducing the crystalline defect density of semiconducting materials. Following initial hostility by some of the major researchers in the field it was rapidly accepted that transistor devices should adopt single crystalline material [1.15]. Extended single crystals of germanium several centimetres long and up to two centimetres in diameter [1.15, 16] and later similar silicon crystals [1.17] were produced using the *Czochralski* technique of pulling a seed crystal from a high purity melt [1.18]. The majority of material in use today is derived from this route. To produce silicon with the very lowest impurity concentration, an alternative method called float zoning was developed where a polycrystalline rod was converted to a single crystal by the passage of a surface tension confined molten zone along its length [1.19–21]. No crucible is required in the process so there are fewer sources of impurity contamination. Float zoning is used to manufacture some of the purest material in current use [1.22]. The early Czochralski material contained dislocation densities of $10^5 - 10^6$ cm^{-2} but by the start of the 1960s dislocation free material was obtained [1.23–26]. Initially most wafers were on the silicon (111) plane, which was easiest to grow, cut and polish [1.27]. For field-effect devices, which are discussed below, use of the (100) plane was found to offer preferable properties so this was introduced in the same decade. The impurity concentration in dislocation-free silicon has been continually reduced up to the present day and wafer diameters have increased almost linearly (though accelerating somewhat in recent years) from about 10 mm in the early 1960s to the "dinner plate" 300 mm today [1.22]. These improvements represent one of the major achievements in semiconductor materials growth and processing.

A series of generations of transistors followed in rapid succession after Brattain and Bardeen's first triumph. Here we only mention a few of the major designs whose production have traits in common with technology today. Early in 1948 *Shockley* developed a detailed formulation of the theory of p–n junctions that concluded with the conception of the junction transistor [1.28, 29]. This involved a thin n-doped base layer sandwiched between p-doped emitter and collector layers (or vice versa). This p–n–p (n–p–n) structure is the simplest form of the bipolar transistor (so-called because of its use of both positive and negative charge carriers), a technology which remains important in analogue and high-speed digital integrated circuits today. In April 1950, by successively adding arsenic and gallium (n- and p-type dopants respectively) impurities to the melt, n–p–n junction structures with the required p-layer thickness ($\approx 25 \, \mu$m) were formed from single crystal germanium. When contacts were applied to the three regions the devices behaved much as expected from *Shockley's* theory [1.28, 30]. Growth of junction transistors in silicon occurred shortly afterwards and they entered production by Texas Instruments in 1954 [1.15].

By the later years of the 1950s, the diffusion doping technique was used to improve the transistor's speed response by reducing the thickness of the base layer in the diffused base transistor [1.31]. This began the trend of manufacturing a device in situ on a substrate material so in a sense it was the foundation for all subsequent microelectronic structures. Soon afterwards, epitaxial growth techniques were introduced [what would today be described as vapour phase epitaxy (VPE)] which have since become central to both silicon and compound semiconductor technology. Gas phase precursors were reacted to produce very high quality and lightly doped crystalline silicon on heavily doped substrate wafers to form epitaxial diffused transistors. Since the collector contact was made through the thickness of the wafer, the use of highly doped (low resistance) wafers reduced the series resistance and therefore increased the frequency response [1.32].

For some years the highest performance devices were manufactured using the so-called "mesa" process where the emitter and diffused base were raised above the collector using selective etching of the silicon [1.28]. The planar process (which is still at the heart of device production today) was subsequently developed, in which the p–n junctions were all formed inside the substrate using oxide masking and diffusion from the surface. This resulted in a flat surface to which contacts could be made using a patterned evaporated film [1.33]. This processing technique was combined with some exciting thoughts at the end of the 1950s and led to the application of transistor devices and other components in a way which was to transform the world: the integrated circuit.

1.2.2 The Integrated Circuit

With the benefit of hindsight, the integrated circuit concept is quite simple. The problem faced by the electronics industry in the 1950s was the increasing difficulty of physically fitting into a small device all of the discrete electronic components (transistors, diodes, resistors and capacitors), and then connecting them together. It was clear that this problem would eventually limit the complexity, reliability and speed of circuits which could be created. Transistors and diodes were manufactured from semiconductors but resistors and capacitors were best formed from alternative materials. Even though they would not deliver the levels of performance achievable from the traditional materials, functioning capacitors and resistors *could* be manufactured from semiconductors so, in principle, all of the components of a circuit could be prepared on a single block of semiconducting material. This reasoning had been proposed by Englishman G.W.A. Dummer at a conference in 1952 [1.34] but small-scale attempts to realise circuits had failed, largely because they were based on connecting together layers in grown-junction transistors [1.35]. In 1958, however, Jack Kilby successfully built a simple oscillator and "flip-flop" logic circuits from components formed in situ on a germanium block and interconnected to produce circuits. He received the Nobel Prize in 2000 for "his part in the invention of the integrated circuit".

Kilby's circuits were the first built on a single semiconductor block, but by far the majority of the circuit's size was taken up by the wires connecting together the components. Robert Noyce developed a truly integrated circuit (IC) in the form that it was later to be manufactured. While Kilby had used the mesa technique with external wiring, Noyce applied the planer technique to form transistors on silicon and photolithographically defined gold or aluminium interconnects. This was more suited to batch processing in production and was necessary for circuits with large numbers of components.

Most integrated circuits manufactured today are based around a transistor technology distinct from the bipolar device used in the first chips but one still dating from the 1960s. In 1960 the first metal oxide semiconductor field effect transistor (MOSFET) was demonstrated [1.36]. In this device a "gate" was deposited onto a thin insulating oxide layer on the silicon. The application of a voltage to the gate resulted in an inversion layer in the silicon below the oxide thereby modifying the conducting channel between "source" and "drain" contacts. This structure was a p-MOS device (current transfer between the collector and emitter was by hole conduction) grown on (111) silicon using an aluminium gate. Earlier attempts at such a device had failed because of trapped impurities and charges in the gate oxide – this new structure had reduced the density of these to below tolerable levels but the device still could not compete with the bipolar transistors of its time [1.27]. By 1967, however, (100) silicon (which offered lower densities of states at the Si/SiO_2 interface) was used together with a polycrystalline silicon gate to construct a more effective and more easily processed device with advantages over the bipolar transistor. In the early 1970s the n-MOS device, which was even less tolerant to the positive gate oxide charges, was realised thanks to much improved cleanliness in the production environment. With conduction occurring by the transfer of electrons rather than holes these were capable of faster operation than similar p-MOS structures (the mobility of electrons in silicon is about three times that of holes). By the 1980s, these two devices were combined in the complementary MOS (CMOS) device which afforded much lower power consumption and simplified circuit design [1.37]. This remains the principal structure used in microelectronics today. Of course now it is much smaller and significantly faster thanks to a range of further advances. These include improved control of the doping and oxidation of silicon and developments in optical lithographic techniques [1.37]. The minimum dimension of components which can be lithographically patterned on an integrated circuit is ultimately limited by the wavelength of radiation used in the process and this has continually been decreased over the past few decades. In the late 1980s wavelengths of 365 nm were employed; by the late 1990s 248 nm were common and today 193 nm is being used. Research into extreme ultraviolet lithography at 13.5 nm may see this being adopted within the next decade enabling feature sizes perhaps as low as 25 nm.

Though ICs implementing CMOS devices are the foundations for computing, silicon-based bipolar transistors maintain a strong market position today in radio frequency applications. In particular germanium is making something of a comeback as a constituent of the latest generation of SiGe bipolar devices [1.38]. The combination of the two forms of technology on a single chip (BiCMOS) offers the potential for computing and communications to be integrated together in the wireless devices of the coming decade.

The development in complexity and performance of silicon devices, largely due to materials science progress, is unparalleled in the history of technology. Never before could improvements be measured in terms of a logarithmic scale for such a sustained period. This is often seen as the embodiment of "Moore's Law". Noting a doubling of the number of components fitted onto integrated circuits each year between 1959 and 1965, *Moore* predicted that this rate of progress would continue until at least ten years later [1.39]. From the early 1970s, a modified prediction of doubling the number of components every couple of years has been sustained to the current day. Since the goals for innovation have often been defined assuming the continuation of the trend, it should perhaps be viewed more as a self-fulfilling prophecy. A huge variety of statistics relating to the silicon microelectronics industry follow a logarithmically scaled improvement from the late 1960s to the current day: the number of transistors shipped per year (increasing); average transistor price (decreasing); and number of transistors on a single chip (increasing) are examples [1.40]. The final member of this list is

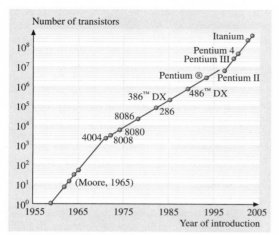

Fig. 1.2 The realisation of Moore's law by commercial Intel®processors: The logarithmic increase in the number of transistors in each processor chip

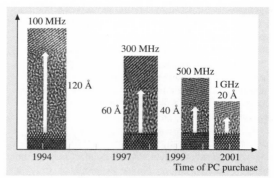

Fig. 1.3 The relentless march to zero thickness. HRTEM images of SiO_2 gate oxide thickness used in personal computers (After [1.41] with permission from the Materials Research Society). (*Courtesy of Prof D.A. Muller*)

plotted in Fig. 1.2. A graphic example of the continuous scaling down of IC component dimensions is the reduction in the thickness of the MOSFET gate insulator. Figure 1.3 shows high-resolution transmission electron microscopy images of typical commercial devices since the mid-1990s: improved device performance has predominately been achieved by reducing all dimensions of the IC components. With state-of-the-art gates now approaching the sub-nanometre thicknesses at which SiO_2 breaks down as an insulator, it is clear that achieving device performance by scaling alone cannot be sustained for much longer. Overcoming this limit, perhaps through the substitution of SiO_2 for a material with a higher dielectric constant, is one of the major challenges in silicon IC research today.

1.3 The Compound Semiconductors

It has been said that silicon is to electronics what steel is to mechanical engineering [1.42]. Steel is very effectively used for most of the world's construction but there are some tasks which it is incapable of performing and others for which alternative structural materials are better suited. In the same way there are some crucial applications – such as optoelectronics and very high speed electronics – that silicon cannot usually deliver but which a wide range of compound semiconductors are better equipped to perform.

Silicon's band gap is indirect (an electron-hole recombination across the band gap must be accompanied by an interaction with a phonon in the lattice) which severely limits the potential efficiency of light emission from the material. Many of the important compound semiconductors, such as the alloys $Al_xGa_{1-x}As$, $In_xGa_{1-x}N$, $Al_yGa_{1-y}N$, and $Al_xIn_yGa_{1-x-y}P$ exhibit direct band gaps (no phonon interaction is required) so can efficiently emit brilliant light in light emitting diodes (LEDs) and laser diodes. Furthermore, in these alloy systems, where the band gap can be adjusted by changing the composition, there is a means of selecting the energy released when an electron and hole recombine across the gap and therefore controlling the wavelength of the photons emitted. From the $Al_xIn_yGa_{1-x-y}As$ and $Al_yIn_xGa_{1-x-y}N$ alloy systems there is, in principle, a continuous range of direct band gaps from deep in the infra red (InAs; $\lambda = 3.5\,\mu m$) to far into the ultraviolet (AlN; $\lambda = 200\,nm$). The semiconductor band gaps of these materials and the corresponding photon wavelengths are put into context with the visible spectrum in Fig. 1.1.

Compounds are also very useful in high speed electronics applications. One of the determining factors in the speed of a transistor is the velocity of the charge carriers in the semiconductor. In GaAs the electron drift velocity is much higher than in silicon so its transistors are able to operate at significantly higher frequencies. The electron velocity in InAs is higher still. Furthermore, in the same way that silicon was preferred to germanium, devices manufactured using semiconductors such as GaN, which have much wider band gaps than silicon (3.4 eV compared with 1.1 eV), are capable of operating in much higher temperature environments.

Aside from these advantageous properties of compound semiconductors, the use of different alloy compositions, or totally different semiconductors, in a single device introduces entirely new possibilities. In silicon, most device action is achieved by little more than careful control of dopant impurity concentrations. In structures containing thin layers of semiconductors with different band gaps (heterostructures) there is the potential to control more fundamental parameters such as the band gap width, mobilities and effective masses of the carriers [1.42]. In these structures, important new features become available which can be used by the device designer to tailor specific desired properties. Hebert Kroemer and Zhores Alferov shared the Nobel prize in 2000 "for developing semiconductor heterostructures used in high-speed- and opto-electronics".

We will mainly consider the compounds formed between elements in Group III of the periodic table and those in Group V (the III–V semiconductors); principally those based around GaAs and InP which were developed over much of the last forty years, and GaN and its related alloys which have been most heavily studied only during the last decade. Other families are given less attention here though they also have important applications (for example the II–VI materials in optoelectronic applications). It can be hazardous to try and consider the "compound semiconductors" as a single subject. Though lessons can be learnt from the materials science of one of the compounds and transferred to another, each material is unique and must be considered on its own (that is, of course, the purpose of the specialised chapters which follow in this handbook!).

It is worth repeating that the power of the compound semiconductors lies in their use as the constituent layers in heterostructures. The principal contribution from chemistry and materials science to enable successful devices has been in the manufacture of high-quality bulk single crystal substrates and the creation of techniques to reliably and accurately produce real layered structures on these substrates from the plans drawn up by a device theorist. In contrast to silicon, the compound semiconductors include volatile components so encapsulation has been required for the synthesis of low-defect InP and GaAs substrates such as in the liquid encapsulated Czochralski technique [1.43, 44]. The size and crystalline quality of these substrates lag some way behind those available in silicon. Crucial to the commercialisation of electronic and optoelectronic heterostructures were the improvements over the last few decades in the control of epitaxial growth available to the crystal grower. The first successful heterostructures were manufactured using deposition onto a substrate from the liquid phase (liquid phase epitaxy; LPE) – "a beautifully simple technology but with severe limitations" [1.42]. However, the real heterostructure revolution had to wait for the 1970s and the introduction of molecular beam epitaxy (MBE) and metalorganic chemical vapour deposition (MOCVD) – also known as metalorganic vapour phase epitaxy (MOVPE) provided that the deposition is epitaxial.

MBE growth occurs in an ultra-high vacuum with the atoms emitted from effusion cells forming "beams" which impinge upon, and form compounds at the substrate surface. It derives from pioneering work at the start of the 1970s [1.45]. MOCVD relies on chemical reactions occurring on the substrate involving metalorganic vapour phase precursors and also stems from initial work at this time [1.46]. In contrast to LPE, these two techniques permit the combination of a wide range of different semiconductors in a single structure and offer a high degree of control over the local composition, in some cases on an atomic layer scale. The successful heterostructure devices of the late 1970s and 1980s would not have been achievable without these two tools and they still dominate III–V device production and research today.

1.3.1 High Speed Electronics

The advantages of the III–V materials over silicon for use in transistors capable of operating at high frequencies were identified early in the semiconductor revolution [1.47]. Shockley's first patent for p-n junction transistors had included the proposal to use a wide-gap emitter layer to improve performance and in the 1950s *Kroemer* presented a theoretical design for a heterostructure transistor [1.48]. Some years later the structure of a GaAs metal semiconductor FET (MESFET) was proposed and realised soon afterwards [1.49, 50]. In these devices a Schottky barrier surface potential was used to modulate the conductivity of the GaAs channel. One of the earliest applications of the III–V's was as low noise amplifiers in microwave receivers which offered substantial improvements relative to the silicon bipolar transistors of the time. The devices were later used to demonstrate sub-nanosecond switching in monolithic digital ICs [1.51]. Today they form the core of the highest speed digital circuits and are used in high speed electronics in microwave radar systems and wireless communications which incorporate monolithic integrated circuits.

For at least 30 years there have been repeated attempts to replicate the MOSFET, the dominant transistor form in silicon ICs, on GaAs material. These attempts have been frustrated by the difficulty of reproducibly forming a high quality stoichiometric oxide on GaAs. In direct analogy with the initial failure of constructing working n-MOSFETs on silicon, the GaAs devices have consistently been inoperable because of poor quality gate oxides with a high density of surface states at the GaAs-insulator interface [1.44]. One of the research efforts focussed on realising this device was, however, to be diverted and resulted in the discovery of probably the most important III–V electronic device: the high electron mobility transistor (HEMT).

The background to this invention lies in the beautiful concept of modulation doping of semiconductors which was first demonstrated in 1978 [1.52]. One of the tenets of undergraduate semiconductor courses is the demonstration that as the dopant density in a semiconductor increases, the mobility of the carriers is reduced because the carriers are scattered more by the ionised dopants. It was found that in a multilayer of repeating n-AlGaAs layers and undoped GaAs layers, the electrons supplied by donor atoms in the AlGaAs moved into the adjacent potential wells of the lower-band gap GaAs layers. In the GaAs these suffered from substantially less ionised impurity scattering and therefore demonstrated enhanced mobility.

While working in a group attempting to create GaAs MOSFETs (and seemingly despairing at the task! [1.53]), *Mimura* heard of these results and conceived of a field effect transistor where the conducting channel exploited the high mobility associated with a modulation doped structure. In essence, a doped AlGaAs layer was formed above the undoped GaAs channel of the transistor. Donated carriers gathered in the GaAs immediately below the interface where they did not suffer from as much ionised impurity scattering and so their mobility would approach that of an ultra-pure bulk semiconductor. The current was conducted from the source to the drain by these high mobility carriers and so the devices were able to operate in higher frequency applications [1.53]. Realisation of the structure required a very abrupt interface between the GaAs and AlGaAs and was considered beyond the capability of MOCVD of the time [1.53]. However, following the advances made in MBE procedures during the 1970s the structure was achieved by that technique within a few months of the original conception [1.53, 54]. The first operational HEMT chips were produced on 24th December 1980: by pleasing coincidence this was the anniversary of Brattain and Bardeen's demonstration of their point contact resistor to the management of Bell labs in 1947! Structures based on the same principle as Mimura's device were realised in France very shortly afterwards [1.55].

The commercialisation of the HEMT became significant in the late 1980s thanks to broadcasting satellite receivers. The improved performance of the devices compared with the existing technology allowed the satellite parabolic dish size to be reduced by at least a factor of two. Structures similar to these have since played a crucial role in the massive expansion in mobile telephones.

The evolution in HEMT structures since the early 1980s is a fine example of how fundamental compound semiconductor properties have been exploited as the technology has become available to realise new device designs. The electron mobility in InAs is much higher than in GaAs and rises as the indium content in $In_xGa_{1-x}As$ is increased [1.56]. The introduction of an InGaAs (as opposed to GaAs) channel to the HEMT structure resulted both in increased electron mobility and a higher density of carriers gathering from the doped AlGaAs layer (because of the larger difference in energy between the conduction band minima of InGaAs and AlGaAs than between GaAs and AlGaAs). This so called pseudomorphic HEMT (pHEMT) demonstrates state of the art power performance at microwave and millimetre wave frequencies [1.43]. The indium content and thickness of the channel is limited by the lattice mismatch with the GaAs (Fig. 1.1). If either is increased too much then misfit dislocations are formed within the channel. The restriction is reduced by growing lattice matched structures on InP, rather than GaAs, substrates. $Al_{0.48}In_{0.52}As$ and $In_{0.53}Ga_{0.47}As$ are both lattice matched to the InP (Fig. 1.1) and their conduction band minimum energies are well separated so that in the InGaAs below the interface between the two compounds a high density of electrons with a very high mobility is formed. Compared with the pHEMTs these InP based HEMTs exhibit significant improvements, have been shown to exhibit gain at over 200 GHz and are established as the leading transistor for millimetre-wave low noise applications such as radar [1.43].

1.3.2 Light Emitting Devices

LEDs and laser diodes exploit the direct band gap semiconductors to efficiently convert an electric current into photons of light. Work on light emission from semiconductor diodes was carried out in the early decades of the twentieth century [1.57] but the start of the modern era of semiconductor optoelectronics traces from the demonstration of LED behaviour and lasing from p-n junctions in GaAs [1.58, 59] and $GaAs_{1-x}P_x$ [1.60]. The efficiency of these LEDs was low and the lasers had large threshold currents and only operated at low temperatures. A year later, in 1963, *Kroemer* and *Alferov* independently proposed the concept of the double heterostructure (DH) laser [1.61, 62]. In the DH device, a narrow band gap material was to be sandwiched between layers with a wider gap so that there would be some degree of confinement of carriers in the "active layer". By the end of the decade DH devices had been constructed which exhibited continuous lasing at room temperature [1.63, 64]. Alferov's laser was grown by LPE on a GaAs substrate with a 0.5 µm GaAs active layer confined between 3 µm of $Al_{0.25}Ga_{0.75}As$ on either side. The launch of the Compact Disc in 1982 saw this type of device, or at least its offspring, becoming taken for granted in the households of the world.

One of the major challenges in materials selection for heterostructure manufacture has always been avoiding the formation of misfit dislocations to relieve the strain associated with lattice parameter mismatch between the layers. $Al_xGa_{1-x}As$ exhibits a direct band gap for $x < 0.45$ and the early success and sustained dominance of the AlGaAs/GaAs system derives significantly from the very close coincidence of the AlAs and GaAs lattice parameters (5.661 Å and 5.653 Å – see Fig. 1.1). This allows relatively thick layers of AlGaAs with reasonably high aluminium content to be grown lattice matched onto GaAs substrates with no misfit dislocation formation. The use of the quaternary alloy solid solution $In_xGa_{1-x}As_yP_{1-y}$ was also suggested in 1970 [1.63] to offer the independent control of lattice parameters and band gaps. Quaternaries based on three Group III elements have since proved very powerful tools for lattice matching within heterostructures. $(Al_xGa_{1-x})_{0.5}In_{0.5}P$ was found to be almost perfectly lattice matched to GaAs and additionally have a very similar thermal expansion coefficient (which is important to avoid strain evolution when cooling after growth of heteroepitaxial layers at high temperatures). By varying x in this compound, direct band gaps corresponding to light between red and green could be created [1.65]. Lasers based on this alloy grown by MOCVD are a common choice for the red wavelengths (650 nm) used in DVD reading.

Obtaining lattice matching is not so crucial for layers thinner than the critical thickness for dislocation production and can be less of an issue these days because of probably the most important development in the history of optoelectronic devices: the introduction of the quantum well. In some ways a quantum well structure is an evolution of the double heterostructure but with a very much thinner active layer. It is the chosen design for most solid state light emitting devices today. With the accurate control available from MBE or MOCVD, and following from some early work on superlattices [1.66], very thin layers of carefully controlled composition could be deposited within heterostructure superlattice stacks. It became possible to grow GaAs layers much less than 10 nm thick within AlGaAs–GaAs heterostructures. The carriers in the GaAs were found to exhibit quantum mechanical confinement within the one dimensional potential well [1.67, 68]. Lasing from $GaAs/Al_{0.2}Ga_{0.8}As$ quantum wells was reported the following year, in 1975, [1.69] but it was a few years before the performance matched that achievable from DH lasers of the time [1.70] and the quantum well laser was further advanced to significantly outperform the competition by researchers in the 1980s [1.71].

The introduction of heterostructures with layer thicknesses on the nanometre scale represents the final stage in scaling down of these devices. Similarly Brattain and Bardeen's centimetre-sized transistor has evolved into today's microprocessors with sub-micron FETs whose gate oxide thicknesses are measured in Angstroms. Throughout this evolution, materials characterisation techniques have contributed heavily to the progress in our understanding of electronic materials and deserve a brief detour here. As the dimensions have been reduced over the decades, the cross-sectional images of device structures published in the literature have changed from a period where optical microscopy techniques were sufficient [1.31] to a time when scanning electron microscopy (SEM) images were used [1.70] and to today's high resolution transmission electron microscopy (TEM) analysis of ultra-thin layers (e.g. Fig. 1.3). For each new material family, understanding of defects and measurement of their densities (e.g. by TEM and X-ray topography) have contributed to improvements in quality. Huge improvements in X-ray optics have seen high-resolution X-ray diffraction techniques develop to become a cornerstone of heterostructure research and production quality control [1.72]. Scanning-probe techniques such as scanning tunnelling and atomic force microscopy have become crucial to the understanding of MBE and MOCVD growth. Chemically sensitive techniques such as secondary ion mass spectroscopy and Rutherford

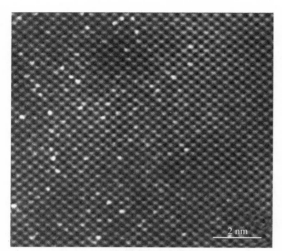

Fig. 1.4 Annular dark field-scanning transmission electron microscope (ADF-STEM) image of Sb-doped Si. The undoped region (*right*) shows atomic columns of uniform intensity. The brightest columns in the doped region (*left*) contain at least one Sb atom. The image is smoothed and background subtracted (After [1.76], with permission Elsevier Amsterdam). (*Courtesy of Prof D.A. Muller*)

backscattering have improved to provide information on doping concentrations and compositions in layered structures with excellent depth resolution. Meanwhile recently developed techniques such as energy-filtered TEM [1.73] afford chemical information at extremely high spatial resolutions. The characterisation of doping properties is also coming of age with more quantitative measurement of dopant contrast in the SEM [1.74], analysis of biased junctions in situ in the TEM [1.75] and the recent exciting demonstration of imaging of single impurity atoms in a silicon sample using scanning TEM, Fig. 1.4. The materials characterisation process remains a very important component of electronic materials research.

Two commercial-product oriented aims dominate semiconductor laser research: the production of more effective emitters of infra-red wavelengths for transmission of data along optic fibres; and the realisation of shorter wavelength devices for reading optical storage media. In the first of these fields devices based on InP have proven to be extremely effective because of its fortuitous lattice parameter match with other III–V alloys which have band gaps corresponding to the low-absorption "windows" in optic fibres. While remaining lattice matched to InP, the $In_xGa_{1-x}As_yP_{1-y}$ quaternary can exhibit band gaps corresponding to infra-red wavelengths of $1.3\,\mu m$ and $1.55\,\mu m$ at which conventional optic fibres absorb the least of the radiation (the absolute minimum is for $1.55\,\mu m$). Room temperature continuous lasing of $1.1\,\mu m$ radiation was demonstrated from the material in 1976 [1.77] and InP based lasers and photodiodes have played a key role in the optical communications industry since the 1980s [1.43].

We have already mentioned the AlGaAs infra-red ($\lambda = 780\,nm$) emitters used to read compact discs and the AlGaInP red ($\lambda = 650\,nm$) devices in DVD readers (see Fig. 1.1). As shorter wavelength lasers have become available the optical disc's surface pits (through which bits of data are stored) could be made smaller and the storage density increased. Though wide band gap II–VI compounds, principally ZnSe, have been researched for many decades for their potential in green and blue wavelengths, laser operation in this part of the visible spectrum proved difficult to realise [1.78]. In the early 1990s, following improvements in the p-doping of ZnSe, a blue-green laser was demonstrated [1.79] but such devices remain prone to rapid deterioration during operation and tend to have lifetimes measured in, at most, minutes. However, also in the early years of the 1990s, a revolution began in wide band gap semiconductors which is ongoing today: the exploitation of GaN and its related alloys $In_xGa_{1-x}N$ and $Al_yGa_{1-y}N$. These materials represent the future for optoelectronics over a wide range of previously inaccessible wavelengths and the next generation of optical storage, the "Blu-ray" disc, will be read using an InGaN blue-violet laser ($\lambda = 405\,nm$).

1.3.3 The III-Nitrides

The relevance of the $In_xGa_{1-x}N$ alloy for light emitting devices is clear from Fig. 1.1. The InN and GaN direct band gaps correspond to wavelengths straddling the visible spectrum and the alloy potentially offers access to all points in-between. The early commercially successful blue light emitters were marketed by Nichia Chemical Industries following the research work of *Nakamura* who demonstrated the first InGaN DH LEDs [1.80] and blue InGaN quantum well LEDs and laser diodes soon after [1.81]. Since this time the global research interest in the GaN material family has expanded rapidly and the competing technology (SiC and ZnSe for blue LEDs and lasers respectively) has largely been replaced.

The development of the III-nitride materials has much in common with the early research of other III–V systems. For example MOCVD and MBE technology could be adapted for the nitride systems (the former has to date been more suited for creating optical devices) and one of the obstacles limiting early device development was achieving sufficiently high p-type doping. However, in some ways they are rather different from the other compound semiconductors. It is important to realise that while all III–Vs mentioned previously share the same cubic crystallographic structure, the nitrides most readily form in a hexagonal allotrope. Most significant in terms of device development over the past decade has been the difficulty in obtaining bulk GaN substrates. Due to the very high pressures necessary to synthesise the compound only very small pieces of bulk GaN have been produced and though they have been used to form functioning lasers [1.82] they remain unsuited as yet to commercial device production. There has consequently been a reliance on heavily lattice mismatched heteroepitaxial growth.

Many materials have been used as substrates for GaN growth. These include SiC, which has one of the lowest lattice mismatches with the nitride material and would be more widely used if it was less expensive, and silicon, which has considerable potential as a substrate if problems associated with cracking during cooling from the growth temperature can be overcome (there is a large difference between the thermal expansion coefficients of GaN and silicon). The dominant choice, however, remains sapphire (α-Al$_2$O$_3$) which itself is by no means ideal: it is electrically insulating (so electrical contacts cannot be made to the device through the substrate material) and, most significantly, has a lattice mismatch of $\approx 16\%$ with the GaN [1.83]. This mismatch is relieved by the formation of misfit dislocations which give rise to dislocations threading through the GaN into the active layers (e.g. InGaN quantum wells) of the devices. The key discovery for reducing the defect densities to tolerable levels was the use of nucleation layers at the interface with the sapphire [1.83] but densities of $\approx 10^9$ cm^{-2} remain typical. More recently epitaxial lateral overgrowth (ELOG) techniques have allowed the dislocation densities to be reduced to $\approx 10^6$ cm^{-2} in local regions [1.84]. However, perhaps the most interesting aspect of GaN-based optoelectronic devices is that they emit light so efficiently in spite of dislocation densities orders of magnitude greater than those tolerated in conventional semiconductors.

Even though InGaN based light emitters have been commercially available for several years, the precise mechanism of luminescence from the alloy is still not fully understood. Having so far discussed the evolution of semiconductors with the benefit of hindsight we can now for a moment consider an unresolved issue which, no doubt, will be solved in the coming years. It is widely believed that the tolerance of InGaN optoelectronic devices to high densities of defects is caused by the presence of low-energy sites within the layers at which electrons and holes are localised. They are thus prevented from interacting with the dislocations at which they would recombine in a non-radiative manner. The origin of localisation remains a matter of debate. One popular explanation is that the InGaN alloy has a tendency for phase segregation [1.85] and indium-rich "clusters" form and cause the localisation. However, there is now evidence [1.86] that the results of some of the measurements used to detect the indium rich regions could be misleading so the clustering explanation is being re-assessed. InGaN remains a fascinating and mysterious alloy.

Solid-state lighting will be a huge market for III-nitride materials in the coming decades. LEDs are perfectly suited to coloured light applications: their monochromatic emission is very much more energy efficient than the doubly wasteful process of colour filtering power hungry filament white light bulbs. InGaN LEDs are now the device of choice for green traffic signals worldwide and offer significant environmental benefits in the process. In principle there is also the opportunity to create white light sources for the home which are more efficient than the tungsten filament light bulbs used today and a variety of promising schemes have been developed for converting the coloured output of III-nitride LEDs to white sources. These include the use of three colour (red, green and blue) structures and blue InGaN or ultraviolet AlGaN based LEDs coated with a range of phosphor materials to generate a useful white spectrum [1.65]. In particular ultraviolet LEDs coated with a three-way phosphor (red, green and blue) can produce high quality white light that mimics sunlight in its visible spectrum. The main disadvantages preventing the widespread use of LEDs in white lighting are their high cost and the relatively low output powers from single devices but these obstacles are rapidly being overcome.

Many other applications for the III-nitrides are being investigated including the use of (Al,Ga,In)N solar cells which could offer more efficient conversion of light into electric current than silicon based devices [1.87]; the possibility of lasers and optical switches operating at the crucial 1.55 μm wavelength based on intersubband (between the discrete quantised energy levels of the wells) transitions in AlGaN/GaN quantum well structures; and the use of the compact InGaN LEDs to fluoresce labelled cancerous cells and aid detection of affected areas. The wide band gap is also very attractive for many electronic device applications–particularly in high-temperature, high-power applications. Exploiting its high thermal conductivity and insensitivity to high operating temperatures, GaN-based HEMTs may extend the power of mobile phone base stations and it has even been suggested that GaN devices could be used as an alternative source of ignition in car engines. There is also, of course, the possibility of monolithically integrating electronic and optoelectronic action onto a single chip. GaN-related materials should prove to have a huge impact on the technology of the coming decades.

1.4 From Faraday to Today

So, we have come 170 years from Faraday's nineteenth century observation of semiconductivity to a world dominated by electronic materials and devices. The balance of power between the different semiconductor families is an unstable and unpredictable one. For example if inexpensive, high quality, low defect density GaN substrates can be produced this will revolutionise the applications of GaN-based materials in both optoelectronics and electronics. The only inevitable fact is that the electronics revolution will continue to be crucially dependent on electronic *materials* understanding and improvement. And while reading the more focussed chapters in this book and concentrating on the very important minutiae of a particular field, it can be a good idea to remember the bigger picture and the fact that electronic materials *are* remarkable!

Cambridge, October 2003

Further References

In particular we recommend the transcripts of the Nobel Lectures given by Brattain, Bardeen, Shockley, Kilby, Kroemer and Alferov. Available in printed form as set out below and, for the latter three in video, at www.nobel.se.

References

1.1 M. Faraday: *Experimental Researches in Electricity, Vol I and II* (Dover, New York 1965) pp. 122–125 and pp. 426–427
1.2 W. Smith: J. Soc. Telegraph Eng. **2**, 31 (1873)
1.3 W. G. Adams: Proc. R. Soc. London **25**, 113 (1876)
1.4 A. H. Wilson: Proc. R. Soc. London, Ser. A **133**, 458 (1931)
1.5 A. H. Wilson: Proc. R. Soc. London, Ser. A **134**, 277 (1931)
1.6 J. Bardeen, W. H. Brattain: Phys. Rev., **74**, 230 (1948)
1.7 J. Bardeen: *Nobel Lecture, Physics, 1942-1962* (Elsevier, Amsterdam 1956) www.nobel.se/physics/laureates/1956/shockley-lecture.html
1.8 E. Braun, S. MacDonald: *Revolution in Miniature*, 2 edn. (Cambridge Univ. Press, Cambridge 1982)
1.9 W. H. Brattain: *Nobel Lecture, Physics, 1942-1962* (Elsevier, Amsterdam 1956) www.nobel.se/physics/laureates/1956/brattain-lecture.html
1.10 C. S. Fuller: Phys. Rev. (Ser 2) **86**, 136 (1952)
1.11 I. Derick, C. J. Frosh: ,US Patent 2 802 760 (1955)
1.12 J. Andrus, W. L. Bond: ,US Patent 3 122 817 (1957)
1.13 W. G. Pfann: Trans. Am. Inst. Mech. Eng. **194**, 747 (1952)
1.14 W. G. Pfann: *Zone Melting* (Wiley, New York 1958)
1.15 G. K. Teal: IEEE Trans. Electron. Dev. **23**, 621 (1976)
1.16 G. K. Teal, J. B. Little: Phys. Rev. (Ser 2) **78**, 647 (1950)
1.17 G. K. Teal, E. Buehler: Phys. Rev. **87**, 190 (1952)
1.18 J. Czochralski: Z. Phys. Chem. **92**, 219 (1917)
1.19 P. H. Keck, M. J. E. Golay: Phys. Rev. **89**, 1297 (1953)
1.20 H. C. Theurer: ,US Patent 3 060 123 (1952)
1.21 H. C. Theurer: Trans. Am. Inst. Mech. Eng. **206**, 1316 (1956)
1.22 K. A. Jackson (Ed.): *Silicon Devices* (Wiley, Weinheim 1998)
1.23 W. C. Dash: J. Appl. Phys. **29**, 736 (1958)
1.24 W. C. Dash: J. Appl. Phys. **30**, 459 (1959)
1.25 W. C. Dash: J. Appl. Phys. **31**, 736 (1960)
1.26 G. Ziegler: Z. Naturforsch. **16a**, 219 (1961)
1.27 M. Grayson (Ed.): *Encyclopedia of Semiconductor Technology* (Wiley, New York 1984) p. 734
1.28 I. M. Ross: Bell Labs Tech. J. **2**(4), 3 (1997)
1.29 W. Schockley: Bell Syst. Tech. J. **28**(4), 435 (1949)
1.30 W. Shockley, M. Sparks, G. K. Teal: Phys. Rev. **83**, 151 (1951)
1.31 M. Tanenbaum, D. E. Thomas: Bell Syst. Tech. J. **35**, 1 (1956)
1.32 C. M. Melliar-Smith, D. E. Haggan, W. W. Troutman: Bell Labs Tech. J. **2**(4), 15 (1997)
1.33 J. A. Hoerni: IRE Trans. Electron. Dev. **7**, 178 (1960)
1.34 J. S. Kilby: IEEE Trans. Electron. Dev. **23**, 648 (1976)
1.35 J. S. Kilby: *Nobel Lectures in Physics: 1996-2000* (Imperial College Press, London 2000) www.nobel.se/physics/laureates/2000/kilby-lecture.html

1.36 D. Kahng, M. M. Atalla: *Silicon–Silicon Dioxide Field Induced Surface Devices* (Solid State Research Conference, Pittsburgh, Pennsylvania 1960)
1.37 J. T. Clemens: Bell Labs Tech. J. **2**(4), 76 (1997)
1.38 T. H. Ning: IEEE Trans. Electron. Dev. **48**, 2485 (2001)
1.39 G. E. Moore: Electronics **38**(8) (1965)
1.40 G. E. Moore: *International Solid State Circuits Conference* (2003)
1.41 F. H. Baumann: Mater. Res. Soc. Symp. **611**, C4.1.1–C4.1.12 (2000)
1.42 H. Kroemer: *Nobel Lectures in Physics: 1996–2000* (Imperial College Press, London 2000) www.nobel.se/physics/laureates/2000/kroemer-lecture.html
1.43 O. Wada, H. Hasegawa (Eds.): *InP-Based Materials and Devices* (Wiley, New York 1999)
1.44 C. Y. Chang, F. Kai: *GaAs High-Speed Devices* (Wiley, New York 1994)
1.45 A. Y. Cho: J. Vac. Sci. Technol., **8**, S31 (1971)
1.46 H. M. Manasevit: Appl. Phys. Lett. **12**, 156 (1968)
1.47 H. J. Welker: IEEE Trans. Electron. Dev. **23**, 664 (1976)
1.48 H. Kroemer: RCA Rev. **18**, 332 (1957)
1.49 C. A. Mead: Proc IEEE **54**, 307 (1966)
1.50 W. W. Hooper, W. I. Lehrer: Proc IEEE **55**, 1237 (1967)
1.51 R. van Tuyl, C. Liechti: IEEE Spectrum **14**(3), 41 (1977)
1.52 R. Dingle: Appl. Phys. Lett. **33**, 665 (1978)
1.53 T. Mimura: IEEE Trans. Microwave Theory Tech. **50**, 780 (2002)
1.54 T. Mimura: Jpn. J. Appl. Phys. **19**, L225 (1980)
1.55 D. Delageabeaudeuf: Electron. Lett. **16**, 667 (1980)
1.56 D. Chattopadhyay: J. Phys. C **14**, 891 (1981)
1.57 E. E. Loebner: IEEE Trans. Electron. Dev. **23**, 675 (1976)
1.58 R. N. Hall: Phys. Rev. Lett. **9**, 366 (1962)
1.59 M. I. Nathan: Appl. Phys. Lett. **1**, 62 (1962)
1.60 N. Holonyak: Appl. Phys. Lett. **1**, 82 (1962)
1.61 H. Kroemer: Proc. IEEE **51**, 1782 (1963)
1.62 Z. I. Alferov: *Nobel Lectures in Physics: 1996–2000* (Imperial College Press, London 2000) www.nobel.se/physics/laureates/2000/alferov-lecture.html
1.63 Z. I. Alferov: Fiz. Tekh. Poluprovodn. **4**, 1826 (1970) Translated in: Sov. Phys. – Semicond. **4**, 1573 (1971)
1.64 I. Hayashi: Appl. Phys. Lett. **17**, 109 (1970)
1.65 A. Žukauskas: *Introduction to Solid-State Lighting* (Wiley, New York 2002)
1.66 L. Esaki, R. Tsu: IBM J. Res. Dev. **14**, 61 (1970)
1.67 L. L. Chang: Appl. Phys. Lett. **24**, 593 (1974)
1.68 R. Dingle: Phys. Rev. Lett. **33**, 827 (1974)
1.69 J. P. van der Ziel: Appl. Phys. Lett. **26**, 463 (1975)
1.70 R. Dupuis: Appl. Phys. Lett. **32**, 295 (1978)
1.71 Z. I. Alferov: Semicond. **32**, 1 (1998)
1.72 D. K. Bowen, B. K. Tanner: *High Resolution X-ray Diffractometry and Topography* (Taylor Francis, London 1998)
1.73 L. Reimer, C. Deininger: *Energy-filtering Transmission Electron Microscopy* (Springer, Berlin, Heidelberg 1995)
1.74 C. Schönjahn: Appl. Phys. Lett. **83**, 293 (2003)
1.75 A. C. Twitchett: Phys. Rev. Lett. **88**, 238302 (2002)
1.76 P. M. Voyles: Ultramicrosc. **96**, 251–273 (2003)
1.77 J. J. Hsieh: Appl. Phys. Lett. **28**, 709 (1976)
1.78 H. E. Ruda (Ed.): *Widegap II–VI Compounds for Opto-electronic Applications* (Chapman Hall, London 1992)
1.79 M. A. Haase: Appl. Phys. Lett. **59**, 1272 (1991)
1.80 S. Nakamura: Appl. Phys. Lett. **64**, 1687 (1994)
1.81 S. Nakamura: Jpn. J. Appl. Phys. **35**, L74 (1996)
1.82 P. Prystawko: Phys. Status Solidi (a) **192**, 320 (2002)
1.83 S. Nakamura, G. Fasol: *The Blue Laser Diode* (Springer, Berlin, Heidelberg 1997)
1.84 B. Beaumont: Phys. Status Solidi (b) **227**, 1 (2001)
1.85 I. Ho, G. B. Stringfellow: Appl. Phys. Lett. **69**, 2701 (1996)
1.86 T. M. Smeeton: Appl. Phys. Lett. **83**, 5419 (2003)
1.87 J. Hogan: New Scientist, 24 (7th December 2002)

Part A Fundamental Properties

2 Electrical Conduction in Metals and Semiconductors
Safa Kasap, Saskatoon, Canada
Cyril Koughia, Saskatoon, Canada
Harry Ruda, Toronto, Canada
Robert Johanson, Saskatoon, Canada

3 Optical Properties of Electronic Materials: Fundamentals and Characterization
Safa Kasap, Saskatoon, Canada
Cyril Koughia, Saskatoon, Canada
Jai Singh, Darwin, Australia
Harry Ruda, Toronto, Canada
Stephen K. O'Leary, Regina, Canada

4 Magnetic Properties of Electronic Materials
Charbel Tannous, Brest Cedex, France
Jacek Gieraltowski, 29285 Brest Cedex, France

5 Defects in Monocrystalline Silicon
Wilfried von Ammon, Burghausen, Germany

6 Diffusion in Semiconductors
Derek Shaw, Hull, UK

7 Photoconductivity in Materials Research
Monica Brinza, Leuven, Belgium
Jan Willekens, Leuven, Belgium
Mohammed L. Benkhedir, Leuven, Belgium
Guy J. Adriaenssens, Leuven, Belgium

8 Electronic Properties of Semiconductor Interfaces
Winfried Mönch, Duisburg, Germany

9 Charge Transport in Disordered Materials
Sergei Baranovskii, Marburg, Germany
Oleg Rubel, Marburg, Germany

10 Dielectric Response
Leonard Dissado, Leicester, UK

11 Ionic Conduction and Applications
Harry L. Tuller, Cambridge, USA

2. Electrical Conduction in Metals and Semiconductors

Electrical transport through materials is a large and complex field, and in this chapter we cover only a few aspects that are relevant to practical applications. We start with a review of the semi-classical approach that leads to the concepts of drift velocity, mobility and conductivity, from which Matthiessen's Rule is derived. A more general approach based on the Boltzmann transport equation is also discussed. We review the conductivity of metals and include a useful collection of experimental data. The conductivity of nonuniform materials such as alloys, polycrystalline materials, composites and thin films is discussed in the context of Nordheim's rule for alloys, effective medium theories for inhomogeneous materials, and theories of scattering for thin films. We also discuss some interesting aspects of conduction in the presence of a magnetic field (the Hall effect). We present a simplified analysis of charge transport in semiconductors in a high electric field, including a modern avalanche theory (the theory of "lucky" drift). The properties of low-dimensional systems are briefly reviewed, including the quantum Hall effect.

2.1	Fundamentals: Drift Velocity, Mobility and Conductivity	20
2.2	Matthiessen's Rule	22
2.3	Resistivity of Metals	23
	2.3.1 General Characteristics	23
	2.3.2 Fermi Electrons	25
2.4	Solid Solutions and Nordheim's Rule	26
2.5	Carrier Scattering in Semiconductors	28
2.6	The Boltzmann Transport Equation	29
2.7	Resistivity of Thin Polycrystalline Films	30
2.8	Inhomogeneous Media. Effective Media Approximation	32
2.9	The Hall Effect	35
2.10	High Electric Field Transport	37
2.11	Avalanche	38
2.12	Two-Dimensional Electron Gas	39
2.13	One Dimensional Conductance	41
2.14	The Quantum Hall Effect	42
References		44

A good understanding of charge carrier transport and electrical conduction is essential for selecting or developing electronic materials for device applications. Of particular importance are the drift mobility of charge carriers in semiconductors and the conductivity of conductors and insulators. Carrier transport is a broad field that encompasses both traditional 'bulk' processes and, increasingly, transport in low dimensional or quantized structures. In other chapters of this handbook, Baranovskii describes hopping transport in low mobility solids such as insulators, Morigaki deals with the electrical properties of amorphous semiconductors and Gould discusses in detail conduction in thin films. In this chapter, we outline a semi-quantitative theory of charge transport suitable for a wide range of solids of use to materials researchers and engineers. We introduce theories of "bulk" transport followed by processes pertinent to ultra-fast transport and quantized transport in lower dimensional systems. The latter covers such phenomena as the Quantum Hall Effect, and Quantized Conductance and Ballistic Transport in Quantum Wires that has potential use in new kinds of devices. There are many more rigorous treatments of charge transport; those by *Rossiter* [2.1] and *Dugdale* [2.2] on metals, and and *Nag* [2.3] and *Blatt* [2.4] on semiconductors are highly recommended.

2.1 Fundamentals: Drift Velocity, Mobility and Conductivity

Basic to the theory of the electronic structure of solids are the solutions to the quantum mechanical problem of an electron in a periodic potential known as Bloch waves. These wavefunctions are traveling waves and provide the physical basis for conduction. In the semi-classical approach to conduction in materials, an electron wavepacket made up of a superposition of Bloch waves can in principle travel unheaded in an ideal crystal. No crystal is ideal, however, and the imperfections cause scattering of the wavepacket. Since the interaction of the electron with the potential of the ions is incorporated in the Bloch waves, one can concentrate on the relatively rare scattering events which greatly simplifies the theory. The motion of the electrons between scattering events is essentially free (with certain provisos such as no interband transitions) subject only to external forces, usually applied electric or magnetic fields. A theory can then be developed that relates macroscopic and measurable quantities such as conductivity or mobility to the microscopic scattering processes. Principle in such a theory is the concept of *mean free time* τ which is the average time between scattering events. τ is also known as the *conductivity relaxation time* because it represents the time scale for the momentum gained from an external field to relax. Equivalently, $1/\tau$ is the average probability per unit time that an electron is scattered.

There are two important velocity quantities that must be distinguished. The first is the *mean speed* u or *thermal velocity* v_{th} which as the name implies is the average speed of the electrons. u is quite large being on the order of $\sqrt{3k_B T/m_e^*} \approx 10^5$ m/s for electrons in a nondegenerate semiconductor and $\sqrt{2E_F/m_e^*} \approx 10^6$ m/s for electrons in a metal, where k_B is Boltzmann's constant, T is the temperature, E_F is the Fermi energy of the metal, and m_e^* is the electron effective mass. The distance an electron travels between scattering events is called the free path. It is straightforward to show that the average or *mean free path* for an electron is simply $\ell = u\tau$. The second velocity is the mean or *drift velocity* $\bm{v}_d$ (variables in boldface are vectors) which is simply the vector average over the velocities of all N electrons,

$$\bm{v}_d = \frac{1}{N} \sum_{i=1}^{N} \bm{v}_i . \tag{2.1}$$

With no external forces applied to the solid, the electron motion is random and thus the drift velocity is zero. When subject to external forces like an electric field, the electrons acquire a net drift velocity. Normally, the magnitude of the drift velocity is much smaller than u so that the mean speed of the electron is not affected to any practical extent by the external forces. An exception is charge transport in semiconductors in high electric fields, where $|\bm{v}_d|$ becomes comparable to u.

The drift velocity gives rise to an electric current. If the density of electrons is n then the current density $\bm{J}_e$ is

$$\bm{J}_e = -en\bm{v}_d \tag{2.2}$$

where e is the fundamental unit of electric charge. For the important case of an applied electric field $\bm{E}$, the solutions of the semi-classical equations give a drift velocity that is proportional to $\bm{E}$. The proportionality constant is the *drift mobility* μ_e

$$\bm{v}_d = -\mu_e \bm{E} . \tag{2.3}$$

The drift mobility might be a constant or it might depend on the applied field (usually only if the field is large). Ohm's Law defines the conductivity σ of a material $\bm{J} = \sigma \bm{E}$ resulting in a simple relation to the drift mobility

$$\sigma = en\mu_e . \tag{2.4}$$

Any further progress requires some physical theory of scattering. A useful model results from the simple assumption that the scattering randomizes the electron's velocity (taking into proper account the distribution of electrons and the Pauli Exclusion Principle). The equation of motion for the drift velocity then reduces to a simple form

$$\frac{d\bm{v}_d}{dt} = \frac{\bm{F}(t)}{m_e^*} - \frac{\bm{v}_d}{\tau} , \tag{2.5}$$

where $\bm{F}(t)$ is the sum of all external forces acting on the electrons. The effect of the scattering is to introduce a frictional term into what otherwise would be just Newton's Law. Solutions of (2.5) depend on $\bm{F}(t)$. In the simplest case of a constant applied electric field, the steady-state solution is trivial,

$$\bm{v}_d = \frac{-e\bm{E}\tau}{m_e^*} . \tag{2.6}$$

The conductivity and drift mobility can now be related to the scattering time [2.5],

$$\mu_e = e\tau/m_e^* \quad \text{and} \quad \sigma = ne^2\tau/m_e^* . \tag{2.7}$$

More sophisticated scattering models lead to more accurate but more complicated solutions.

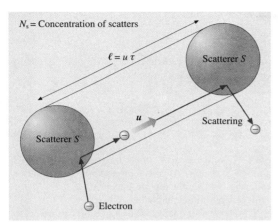

Fig. 2.1 Scattering of an electron from a scattering center. The electron travels a mean distance $\ell = u\tau$ between collisions

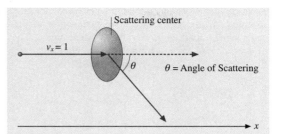

Fig. 2.2 An electron moving in the x-direction becomes scattered though an angle θ with respect to the original direction

The simple expression (2.7) can be used to explain qualitative features of conduction in materials once a physical origin for the scattering is supplied. For any scattering site, the effective area is the *scattering cross-section S* as depicted in Fig. 2.1. The scattering cross section is related to the mean free path since the volume $S\ell$ must contain one scattering center. If there are N_S scattering centers per unit volume then

$$\ell = \frac{1}{SN_S}, \qquad (2.8)$$

or, rewriting in terms of τ,

$$\tau = \frac{1}{SN_S u}. \qquad (2.9)$$

Once the cross-section for each physically relevant scattering mechanism is known then the effect on the scattering time and conductivity is readily calculated.

An overly restrictive assumption in the above analysis is that the electron's velocity is completely randomized every time it is scattered. On the other hand, suppose that ν collisions are required to completely destroy the directional velocity information. That is only after an average of ν collisions do all traces of correlation between the initial and the final velocities disappear. The *effective* mean free path ℓ_{eff} traversed by the electron until its velocity is randomized will now be larger than ℓ; to first order $\ell_{\text{eff}} = \nu\ell$. ℓ_{eff} is termed the *effective* or the *conduction mean free path*. The corresponding *effective scattering cross* section is

$$S_{\text{eff}} = \frac{1}{N_S \ell_{\text{eff}}}. \qquad (2.10)$$

The expressions for mobility and conductivity become

$$\mu = \frac{e v \tau}{m_e^*} = \frac{e v \ell}{m_e^* u} = \frac{e \ell_{\text{eff}}}{m_e^* u} = \frac{e}{m_e^* u N_s S_{\text{eff}}} \qquad (2.11)$$

and

$$\sigma = \frac{e^2 v \tau}{m_e^*} = \frac{e^2 n v \ell}{m_e^* u} = \frac{e^2 n \ell_{\text{eff}}}{m_e^* u} = \frac{e^2 n}{m_e^* u N_s S_{\text{eff}}}. \qquad (2.12)$$

Suppose that in a collision the electron is scattered at an angle θ to its original direction of travel as shown in Fig. 2.2. It is convenient to introduce a quantity $S_\theta(\theta)$, called the *differential scattering cross section*, defined so that $2\pi \sin\theta S_\theta d\theta$ represents the probability of scattering at an angle between θ and $\theta + d\theta$ with respect to the original direction. If the magnitude of the velocity is not changed then the fractional change in component of the velocity along the original direction is $1 - \cos\theta$. The average number of collisions ν required to randomize the velocity is then

$$\nu = \frac{1}{\langle 1 - \cos\theta \rangle}, \qquad (2.13)$$

where the average is given by

$$\langle 1 - \cos\theta \rangle = \frac{\int_0^\pi (1 - \cos\theta) S_\theta(\theta) \sin\theta \, d\theta}{\int_0^\pi S_\theta(\theta) \sin\theta \, d\theta}. \qquad (2.14)$$

The effective cross sectional area S_{eff} is then

$$S_{\text{eff}} = 2\pi \int_{\theta_m}^\pi (1 - \cos\theta) S_\theta(\theta) \sin\theta \, d\theta. \qquad (2.15)$$

As an example, consider conduction electrons scattering from charged impurities in a nondegenerate semiconductor where small angle deviations predominate. The differential cross section for coulombic scattering from a charged impurity center with charge $+Ze$ is

$$S_\theta(\theta) = \frac{16k^2}{u^4} \times \frac{1}{\theta^4}; \quad k = \frac{Ze^2}{4\pi\varepsilon m_e^*}, \quad (2.16)$$

where $k = Ze^2/4\pi\varepsilon_0\varepsilon_r m_e^*$ and ε_r is the relative permittivity of the semiconductor. Let r_m be the maximum effective radius of action for the impurity at which the minimum scattering angle occurs θ_m. Then the integral in (2.15) evaluates to

$$S_{\text{eff}} = \frac{2\pi k^2}{u^4} \ln\left(1 + \frac{r_m^2 u^4}{k^2}\right). \quad (2.17)$$

In a nondegenerate semiconductor, the equipartition theorem links velocity to temperature, $m_e^* u^2/2 = 3kT/2$, so that $u \propto T^{1/2}$. Thus,

$$S_{\text{eff}} = \frac{A}{T^2} \ln\left(1 + BT^2\right), \quad (2.18)$$

where A and B are constants. The drift mobility due to scattering from ionized impurities becomes

$$\mu_I = \frac{e}{m_e^* u S_{\text{eff}} N_I} \propto \frac{1}{T^{1/2}} \frac{T^2}{A \ln(1 + BT^2)} \frac{1}{N_I}$$

$$\approx \frac{CT^{3/2}}{N_I}, \quad (2.19)$$

where N_I is the density of ionized impurities and C is a new constant. At low temperatures where lattice scattering is insignificant, we expect $\mu_I \propto T^{3/2}/N_I$ for nondegenerate semiconductors.

The above semiquatitative description is sufficient to understand the basic principles of conduction. A more rigorous approach involves solving the Boltzmann charge transport equation and is addressed in Sect. 2.6.

2.2 Matthiessen's Rule

In general, the conduction electron whether in a metal or in a semiconductor can be scattered by a number of mechanisms, such as lattice vibrations, impurities, lattice defects such as dislocations, grain boundaries, vacancies, surfaces, or any other deviation from a perfectly periodic lattice. All these scattering processes increase the overall resistivity of the substance by reducing the mean scattering time. The relation between the types of scattering and the total scattering time can be obtained by considering scattering from lattice vibrations and impurities as shown in Figure 3.1. We define two mean free times τ_L and τ_I: τ_L is the mean free time considering only scattering from lattice vibrations (phonons) and τ_I is the mean free time considering only collisions with impurities. In a small unit of time dt, the total probability of scattering (dt/τ) is simply the sum of the probability for phonon scattering (dt/τ_L) and the probability for impurity scattering (dt/τ_I), and thus

$$\frac{1}{\tau} = \frac{1}{\tau_L} + \frac{1}{\tau_I}. \quad (2.20)$$

We have assumed that neither τ_L nor τ_I is affected by the presence of the other scattering mechanism, that is each type of scattering is independent. The above expression can be generalized to include all types of independent scattering yielding

$$\frac{1}{\tau} = \sum_i \frac{1}{\tau_i}, \quad (2.21)$$

where τ_i is the mean scattering time considering the ith scattering process alone. Since the drift mobility is proportional to τ, (2.21) can be written in terms of the drift mobilities determined by the various scattering mechanisms. In other words,

$$\frac{1}{\mu_d} = \sum_i \frac{1}{\mu_i} \quad (2.22)$$

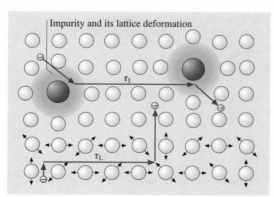

Fig. 2.3 Scattering from lattice vibrations alone with a mean scattering time τ_L, and from impurities alone with a mean scattering time τ_I

where μ_i is the drift mobility limited just by the ith scattering process. Finally, since the resistivity is inversely proportional to the drift mobility, the relation for resistivity is

$$\rho = \sum_i \rho_i \qquad (2.23)$$

where ρ_i is the resistivity of the material if only the ith scattering process were active. Equation (2.23) is known as Matthiessen's rule. For nearly perfect, pure crystals the resistivity is dominated by phonon scattering ρ_T. If impurities or defects are present, however, there are an additional resistivities ρ_I from the scattering off the impurities and ρ_D from defect scattering, and $\rho = \rho_T + \rho_I + \rho_D$.

Matthiessen's rule is indispensable for predicting the resistivities of many types of conductors. In some cases like thin films, the rule is obeyed only approximately, but it is nonetheless still useful for an initial (often quite good) estimate.

2.3 Resistivity of Metals

2.3.1 General Characteristics

The effective resistivity of a metal, by virtue of Matthiessen's rule, is normally written as

$$\rho = \rho_T + \rho_R , \qquad (2.24)$$

where ρ_R is called the *residual resistivity* and is due to the scattering of electrons by impurities, dislocations, interstitial atoms, vacancies, grain boundaries and so on. The residual resistivity shows very little temperature dependence whereas ρ_T is nearly linear in absolute temperature. ρ_T will be the main resistivity term for many good-quality, pure, crystalline metals. In classical terms, we can take the thermal vibrations of a lattice atom with mass M as having a mean kinetic energy KE of $(1/2)Ma^2\omega^2$, where a and ω are the amplitude and frequency of the vibrations. This mean KE must be of the order of kT so that the amplitude $a \propto T^{1/2}$. Thus the electron scattering cross section $S = \pi a^2 \propto T$. Since the mean speed of conduction electron in a metal is the Fermi speed and is temperature independent, $\mu \propto \tau \propto 1/S \propto T^{-1}$, and hence the resistivity $\rho \propto T$. Most nonmagnetic pure metals obey this relationship except at very low temperatures. Figure 2.4 shows the resistivity of Cu as a function of temperature where above ≈ 100 K, $\rho \propto T$.

Frequently, the resistivity vs. temperature behavior of pure metals can be empirically represented by a power law of the form,

$$\rho = \rho_0 \left(\frac{T}{T_0}\right)^n , \qquad (2.25)$$

where ρ_0 is the resistivity at the reference temperature, T_0, and n is a characteristic index that best fits the data. For the nonmagnetic metals, n is close to unity whereas it is close to 2 for the magnetic metals Fe and Ni [2.5]. Figure 2.5 shows ρ vs. T for various metals. Table 2.1 summarizes the values ρ_0 and n for various metals.

As apparent from Fig. 2.4, below ≈ 100 K, the $\rho \propto T$ behavior fails, and $\rho \propto T^5$. The reason is that, as the temperature is lowered, the scattering by phonons becomes less efficient, and it takes many more collisions to fully randomize the initial velocity of the electron. The mean number of collisions ν required the randomize the velocity scales with T^{-2} [2.5], and at low temperatures,

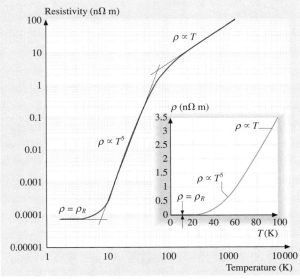

Fig. 2.4 The resistivity of copper from low to high temperatures (near its melting temperature, 1358 K) on a log–log plot. Above about 100 K, $\rho \propto T$, whereas at low temperatures, $\rho \propto T^5$, and at the lowest temperatures ρ approaches the residual resistivity ρ_R. The inset shows the ρ vs. T behavior below 100 K on a linear plot. (ρ_R is too small to see on this scale.) After [2.5]

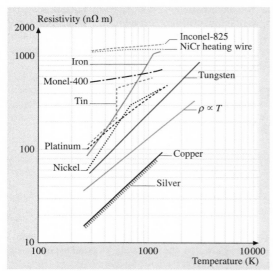

Fig. 2.5 The resistivities of various metals as a function of temperature above 0 °C. Tin melts at 505 K, whereas nickel and iron go through a magnetic-to-nonmagnetic (Curie) transformation at about 627 K and 1043 K, respectively. The theoretical behavior ($\rho \propto T$) is shown for reference. After [2.5]

the concentration n_{phonon} of phonons increases as T^3. Thus,

$$\sigma \propto v\tau \propto \frac{v}{n_{\text{phonon}}} \propto \frac{T^{-2}}{T^3} \propto T^{-5} \qquad (2.26)$$

which explains the low-temperature $\rho - T$ behavior in Fig. 2.4. The low temperature $\rho \propto T^5$ and high temperature $\rho \propto T$ regimes are roughly separated by the *Debye temperature* T_D. For $T > T_D$ we expect $\rho \propto T$, and For $T < T_D$ we expect $\rho \propto T^5$.

In the case of metals with impurities and for alloys, we need to include the ρ_R contribution to the overall resistivity. For $T > T_D$, the overall resistivity is

$$\rho \approx AT + \rho_R, \qquad (2.27)$$

where A is a constant, and the AT term in (2.27) arises from scattering from lattice vibrations. Normally, ρ_R has very little temperature dependence, and hence very roughly ρ vs. T curves shift to higher values as ρ_R is increases due to the addition of impurities, alloying or cold working the sample (mechanical deformation that generates dislocations) as illustrated for Cu–Ni alloys in Fig. 2.6.

Resistivity vs. temperature behavior of nearly all metals is characterized by the *temperature coefficient of resistivity* (TCR) α_0 which is defined as the fractional change in the resistivity per unit temperature increase at the reference temperature T_0, i.e.

$$\alpha_0 = \frac{1}{\rho_0}\left(\frac{d\rho}{dT}\right)_{T=T_0}, \qquad (2.28)$$

where ρ_0 is the resistivity at the reference temperature T_0, usually at 273 K (0 °C) or 293 K (20 °C), $d\rho = \rho - \rho_0$ is the change in the resistivity due to a small increase, $dT = T - T_0$, in temperature. Assuming that α_0 is temperature independent over a small range from T_0 to T, we can integrate (2.28), which leads to the well known equation,

$$\rho = \rho_0[1 + \alpha_0(T - T_0)]. \qquad (2.29)$$

Equation (2.29) is actually only valid when α_0 is *constant* over the temperature range of interest which requires (2.27) to hold. Over a limited temperature range this will usually be the case. Although it is not obvious from (2.28), we should, nonetheless, note that α_0 depends on the reference temperature, T_0 by virtue of ρ_0 depending on T_0.

It is instructive to mention that if $\rho \approx AT$ as we expect for an ideal pure metal, then $\alpha_0 = T_0^{-1}$. If we take the reference temperature T_0 as 273 K (0 °C), then α_0 should ideally be 1/(273 K) or 3.66×10^{-3} K^{-1}. Examination of a_0 for various metals shows that $\rho \propto T$ is not a bad approximation for some of the familiar

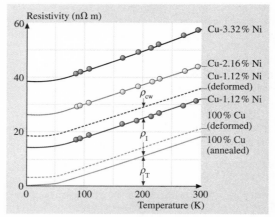

Fig. 2.6 Resistivities of annealed and cold-worked (deformed) copper containing various amounts of Ni (given in atomic percentages) versus temperature

Table 2.1 Resistivities at 293 K (20 °C) ρ_0 and thermal coefficients of resistivity α_0 at 0–100 °C for various metals. The resistivity index n in $\rho = \rho_0(T/T_0)^n$ is also shown. Data was compiled from [2.6, 7]

Metal	$\rho_0(n\Omega m)$	n	$\alpha_0 \times 10^{-3}$ (K^{-1})
Aluminium, Al	26.7	1.20	4.5
Barium, Ba	600	1.57	
Beryllium, Be	33	1.84	9
Bismuth, Bi	1170	0.98	4.6
Cadmium, Cd	73	1.16	4.3
Calcium, Ca	37	1.09	4.57
Cerium, Ce	854	1.35	8.7
Cesium, Cs	200	1.16	4.8
Cromium, Cr	132	1.04	2.14
Cobalt, Co	63	1.80	6.6
Copper, Cu	16.94	1.15	4.3
Gallium, Ga	140		
Gold, Au	22	1.11	4
Hafnium, Ha	322	1.20	4.4
Indium, In	88	1.40	5.2
Iridium, Ir	51	1.17	4.5
Iron, Fe	101	1.73	6.5
Lead, Pb	206	1.13	4.2
Lithium, Li	92.9	1.23	4.35
Magnesium, Mg	4.2	1.09	4.25
Molybdenum, Mo	57	1.26	4.35
Nickel, Ni	69	1.64	6.8
Niobium, Nb	160	0.80	2.6
Osmium, Os	88	1.10	4.1
Palladium, Pd	108	0.96	4.2
Platinum, Pt	105.8	1.02	3.92
Potassium, K	68	1.38	5.7
Rhodium, Rh	47	1.21	4.4
Rubidium, Rb	121	1.41	4.8
Ruthenium, Ru	77	1.15	4.1
Silver, Ag	16.3	1.13	4.1
Sodium, Na	47	1.31	5.5
Strontium, Sr	140	0.99	3.2
Tantalum, Ta	135	1.01	3.5
Tin, Sn	126	1.4	4.6
Titanium, Ti	540	1.27	3.8
Tungsten, W	54	1.26	4.8
Vanadium, V	196	1.02	3.9
Zinc, Zn	59.6	1.14	4.2
Zirconium, Zr	440	1.03	4.4

pure metals used as conductors, e.g. Cu, Al, Au, but fails badly for others, such as indium, antimony and, in particular, the magnetic metals, e.g. iron and nickel.

Frequently we are given α_0 at a temperature T_0, and we wish to use some other reference temperature, say $T_0{'}$, that is, we wish to use $\rho_0{'}$ and $\alpha_0{'}$ for ρ_0 and α_0 respectively in (2.29) by changing the reference from T_0 to $T_0{'}$. Then we can find α_1 from α_0,

$$\alpha_0' = \frac{\alpha_0}{1 + \alpha_0(T_0' - T_0)}$$

and $\rho = \rho_0' \left[1 + \alpha_0' \left(T - T_0'\right)\right]$. (2.30)

For example, for Cu $\alpha_0 = 4.31 \times 10^{-3}$ K^{-1} at $T_0 = 0$ °C, but it is $\alpha_0 = 3.96 \times 10^{-3}$ K^{-1} at $T_0 = 20$ °C. Table 2.1 summarizes α_0 for various metals.

2.3.2 Fermi Electrons

The electrical properties of metals depend on the behavior of the electrons at the Fermi surface. The electron states at energies more than a few kT below E_F are almost fully occupied. The Pauli exclusion principle requires that an electron can only be scattered into an empty state, and thus scattering of deep electrons is highly suppressed by the scarcity of empty states (scattering where the energy changes by more than a few kT is unlikely). Only the electrons near E_F undergo scattering. Likewise, under the action of an external field, only the electron occupation near E_F is altered. As a result, the density of states (DOS) near the Fermi level is most important for the metal electrical properties, and only those electrons in a small range ΔE around E_F actually contribute to electrical conduction. The density of these electrons is approximately $g(E_F)\Delta E$ where $g(E_F)$ is the DOS at the Fermi energy. From simple arguments, the overall conductivity can be shown to be [2.5]

$$\sigma = \frac{1}{3} e^2 v_F^2 \tau g(E_F),$$ (2.31)

where v_F is the Fermi speed and τ is the scattering time of these Fermi electrons. According to (2.31), what is important is the density of states at the Fermi energy, $g(E_F)$. For example, Cu and Mg are metals with valencies I and II. Classically, Cu and Mg atoms each contribute 1 and 2 conduction electrons respectively into the crystal. Thus, we would expect Mg to have higher conductivity. However, the Fermi level in Mg is where the top tail of the 3p-band overlaps the bottom tail of the 3s band where the density of states is small. In Cu, on the other hand, E_F is nearly in the middle of the 4s band where the density of states is high. Thus, Mg has a lower conductivity than Cu.

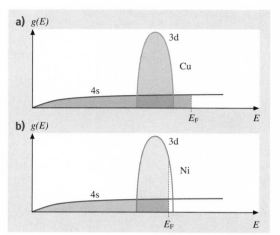

Fig. 2.7a,b Simplified energy band diagrams around E_F for copper (**a**) and nickel (**b**)

for example, in Cu, whose simplified energy band diagram around E_F is shown in Fig. 2.7a. In certain metals, there are two different energy bands that overlap at E_F. For example, in Ni, 3d and 4s bands overlap at E_F as shown in Fig. 2.7b. An electron can be scattered from the 4s to the 3d band and vice versa. Electrons in the 3d band have very low drift mobilities and effectively do not contribute to conduction so that only $g(E_F)$ of the 4s band operates in (2.31). Since 4s to 3d band is an additional scattering mechanism, by virtue of Matthiessen's rule, the effective scattering time τ for the 4s band electrons is shortened and hence σ from (2.31) is smaller. Thus, Ni has poorer conductivity than Cu.

Equation (2.31) does not assume a particular density of states model. If we now apply the *free electron model* for $g(E_F)$, and also relate E_F to the total number of conduction electrons per unit volume n, we would find that the conductivity is the same as that in the Drude model, that is

$$\sigma = \frac{e^2 n \tau}{m_e}. \tag{2.32}$$

The scattering time τ in (2.31) assumes that the scattered electrons at E_F remain in the same energy band as,

2.4 Solid Solutions and Nordheim's Rule

In an *isomorphous alloy* of two metals, i.e. a binary alloy which forms a binary *substitutional solid solution*, an additional mechanism of scattering appears, the *scattering off solute phase atoms*. This scattering contributes to lattice scattering, and therefore increases the overall resistivity. An important semi-empirical equation which can be used to predict the resistivity of an alloy is Nordheim's rule. It relates the impurity part of the resistivity ρ_I to the atomic fraction X of solute atoms in a solid solution via

$$\rho_I = CX(1-X), \tag{2.33}$$

where the constant C is termed the Nordheim coefficient and represents the effectiveness of the solute atom in increasing the resistivity. Nordheim's rule was originally derived for crystals. Combining Nordheim's rule with Matthiessen's rule (2.23), the resistivity of an alloy of composition X should follow

$$\rho = \rho_{\text{matrix}} + CX(1-X), \tag{2.34}$$

where ρ_{matrix} is the resistivity of the matrix due to scattering from thermal vibrations and from other defects, in the absence of alloying elements.

Nordheim's rule assumes that the solid solution has the solute atoms randomly distributed in the lattice. For sufficiently small amounts of impurity, experiments show that the increase in the resistivity ρ_I is nearly always simply proportional to the impurity concentration X, that is, $\rho_I \propto X$. For dilute solutions, Nordheim's rule predicts the same linear behavior, that is, $\rho_I = CX$ for $X \ll 1$.

Originally the theoretical model for ρ_I was developed by *Nordheim* [2.8] by assuming that the solute atoms simply perturb the periodic potential and thereby increase the probability of scattering. Quantum mechanical calculations for electron scattering within a single band, such as the s-band, at E_F show that

$$\rho_I \propto g(E_F) V_{\text{scatter}}^2 X(1-X), \tag{2.35}$$

where $g(E_F)$ is the DOS at E_F, and V_{scatter} is matrix element for scattering from one wavefunction to another at the Fermi surface in the same band, which for an s-band is

$$V_{\text{scatter}} = \langle \psi_s^* | \Delta V | \psi_s \rangle, \tag{2.36}$$

Table 2.2 Nordheim coefficients (at 20 °C) for dilute alloys obtained from $\rho_i = CX$ and $X < 1$ at.%. Note: For many isomorphous alloys, C may be different at higher concentrations; that is, it may depend on the composition of the alloy [2.7, 9]. Maximum solubility data from [2.10]

Solute in solvent (element in matrix)	Nordheim coefficient (nΩ m)	Maximum solubility at 25 °C (at.%)
Au in Cu matrix	5500	100
Mn in Cu matrix	2900	24
Ni in Cu matrix	1250	100
Sn in Cu matrix	2900	0.6
Zn in Cu matrix	300	30
Cu in Au matrix	450	100
Mn in Au matrix	2410	25
Ni in Au matrix	790	100
Sn in Au matrix	3360	5
Zn in Au matrix	950	15

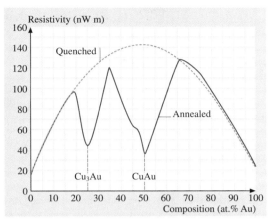

Fig. 2.8 Electrical resistivity versus composition at room temperature in Cu–Au alloys. The quenched sample (*dashed curve*) is obtained by quenching the liquid, and it has the Cu and Au atoms randomly mixed. The resistivity obeys the Nordheim rule. On the other hand, when the quenched sample is annealed or the liquid slowly cooled (*solid curve*), certain compositions (Cu$_3$Au and CuAu) result in an ordered crystalline structure in which Cu and Au atoms are positioned in an ordered fashion in the crystal and the scattering effect is reduced

where ΔV is the difference between the potentials associated with solvent and solute atoms, and ψ_s is the wavefunction of an electron in the s-band at E_F. It is clear that C is only independent of X if $g(E_F)$ and V_{scatter} remain the same for various X which may not be true. For example, if the effective number of free electrons increases with X, E_F will be shifted higher, and C will not be constant.

Table 2.2 lists some typical Nordheim coefficients for various additions to copper and gold. The value of the Nordheim coefficient depends on the type of solute and the solvent. A solute atom that is drastically different in size to the solvent atom will result in a bigger increase in ρ_I and will therefore lead to a larger C. An important assumption in Nordheim's rule in (2.33) is that the alloying does not significantly vary the number of conduction electrons per atom in the alloy. Although this will be true for alloys with the same valency, that is, from the same column in the Periodic Table (e.g., Cu–Au, Ag–Au), it will not be true for alloys of different valency, such as Cu and Zn. In pure copper, there is just one conduction electron per atom, whereas each Zn atom can donate two conduction electrons. As the Zn content in brass is increased, more conduction electrons become available per atom. Consequently, the resistivity predicted by (2.34) at high Zn contents is greater than the actual value because C refers to dilute alloys. To get the correct resistivity from (2.34) we have to lower C, which is equivalent to using an *effective* Nordheim coefficient C_{eff} that decreases as the Zn content increases. In other cases, for example, in Cu–Ni alloys, we have to increase C at high Ni concentrations to account for additional electron scattering mechanisms that develop with Ni addition. Nonetheless, the Nordheim rule is still useful for predicting the resistivities of dilute alloys, particularly in the low-concentration region.

In some solid solutions, at some concentrations of certain binary alloys, such as 75% Cu–25% Au and 50% Cu–50% Au, the annealed solid has an orderly structure; that is, the Cu and Au atoms are not randomly mixed, but occupy regular sites. In fact, these compositions can be viewed as a pure compound – like the solids Cu$_3$Au and CuAu. The resistivities of Cu$_3$Au and CuAu will therefore be less than the same composition random alloy that has been quenched from the melt. As a consequence, the resistivity ρ versus composition X curve does not follow the dashed parabolic curve throughout; rather, it exhibits sharp falls at these special compositions, as illustrated Fig. 2.8. The *effective media approximation* may be used as an effective tool to estimate the resistivities of inhomogeneous media.

2.5 Carrier Scattering in Semiconductors

At low electric fields, ionized impurity scattering and phonon scattering predominate. Other types of scattering include carrier-carrier scattering, inter-valley scattering, and neutral impurity scattering; these may generally be ignored to a first approximation.

For phonon scattering, both *polar* and *non-polar phonon scattering* should be considered. In polar scattering, short wavelength oscillations of atoms on different sub-lattices vibrating out of phase produce an effective dipole moment proportional to the bond polarity. Since such vibrational modes are optically active (since this dipole moment can interact with an incident electromagnetic field), this type of lattice scattering is usually referred to as *polar optical phonon scattering*. Since a sub-lattice is necessary for optical modes, this scattering mechanism is not present in elemental semiconductors such as Si, Ge, or diamond.

Non-polar phonon scattering comes from long wavelength oscillations in the crystal, involving small displacements of tens to thousands of atoms. The wavelength depends on the material and its elastic properties. Such modes are very similar to sound vibrations and are thus referred to as acoustic modes. The associated atomic displacements correspond to an effective built-in strain, with local change in the lattice potential, causing carrier scattering known as *deformation potential acoustic phonon scattering*. Since the change in potential is relatively small, the scattering efficiency is relatively low as compared with polar optical phonon scattering.

Each scattering process contributes to the drift mobility according to *Matthiessen's rule*

$$\frac{1}{\mu_e} = \sum_i \frac{1}{\mu_i} \qquad (2.37)$$

as discussed in Sect. 2.3 above. Figure 2.9 shows the contributions of each scattering process for n-type ZnSe – a material used in optoelectronic devices. At room temperature, polar optical phonon scattering and ionized impurity scattering dominate. These processes depend on the carrier concentration. The curve for ionized impurity scattering decreases markedly with increasing carrier concentration owing to the increasing concentration of ionized donors that supply these carriers. The ionized impurity scattering mobility is roughly inversely proportional to the concentration of ionized impurities. However, as the carrier concentration increases, as in a degenerate semiconductor, the average energy per carrier also increases (i. e., carriers move faster on average) and thus carriers are less susceptible to being scattered from the

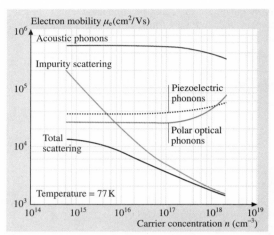

Fig. 2.9 Dependence of electron mobility on carrier concentration for ZnSe at 77 K [2.11]

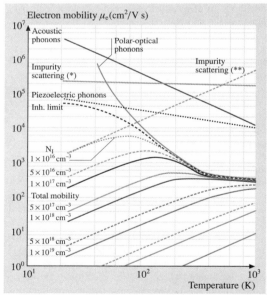

Fig. 2.10 Dependence of electron mobility on temperature and doping for ZnSe [2.11]

ionized impurity centers. In contrast, the polar phonon scattering rate is determined by the number of participating phonons which depends on the thermal energy available to create a given quantized vibrational mode.

The temperature dependence of the mobility may be estimated by considering the effect of temperature on ionized impurity and phonon scattering and combining these mechanisms using Matthiessen's rule. Phonon scattering increases strongly with increasing temperature T due to the increase in the number of phonons resulting in a $T^{-3/2}$ dependence for the polar phonon mobility. For ionized impurity scattering, increasing the temperature increases the average carrier velocity and hence increases the carrier mobility for a set concentration of ionized impurities. Once a temperature is reached such that impurity ionization is complete, the ionized impurity based carrier mobility can be shown to increase with temperature T as approximately, $T^{+3/2}$. At low temperatures, the mobility is basically determined by ionized impurity scattering and at high temperatures by phonon scattering leading to a peaked curve. Invoking the previous discussions for the dependence of the total mobility on carrier concentration, it is clear that the peak mobility will depend on the doping level, and the peak location will shift to higher temperatures with increased doping as shown in Fig. 2.10.

2.6 The Boltzmann Transport Equation

A more rigorous treatment of charge transport requires a discussion of the *Boltzmann Transport Equation*. The electronic system is described by a distribution function $f(\mathbf{k}, \mathbf{r}, t)$ defined in such a way that the number of electrons in a six-dimensional volume element $d^3\mathbf{k}\,d^3\mathbf{r}$ at time t is given by $\frac{1}{4}\pi^{-3} f(\mathbf{k}, \mathbf{r}, t)\,d^3\mathbf{k}\,d^3\mathbf{r}$. In equilibrium, $f(\mathbf{k}, \mathbf{r}, t)$ depends only on energy and reduces to the Fermi distribution f_0 where the probability of occupation of states with momenta $+\mathbf{k}$ equals that for states with $-\mathbf{k}$, and $f_0(\mathbf{k})$ is symmetrical about $\mathbf{k} = 0$, giving no net charge transport. If an external field acts on the system (i.e., non-equilibrium), the occupation function $f(\mathbf{k})$ will become asymmetric in $\mathbf{k}$-space. If this non-equilibrium distribution function $f(\mathbf{k})$ is completely specified and appropriate boundary conditions supplied, the electronic transport properties can be completely determined by solving the *steady state Boltzmann transport equation* [2.12]

$$\mathbf{v} \cdot \nabla_r f + \dot{\mathbf{k}} \cdot \nabla_k f = \left(\frac{\partial f}{\partial t}\right)_c \quad (2.38)$$

where,

1. $\mathbf{v} \cdot \nabla_r f$ represents diffusion through a volume element $d^3 r$ about the point $\mathbf{r}$ in phase space due to a gradient $\nabla_r f$,
2. $\dot{\mathbf{k}} \cdot \nabla_k f$ represents drift through a volume element $d^3 k$ about the point $\mathbf{k}$ in phase space due to a gradient $\nabla_k f$ (for example, $\hbar \dot{\mathbf{k}} = e\left(\mathbf{E} + \frac{1}{c}\mathbf{v} \times \mathbf{B}\right)$ in the presence of electric and magnetic fields)
3. $(\partial f/\partial t)_c$ is the collision term and accounts for the scattering of electrons from a point $\mathbf{k}$ (for example, this may be due to lattice or ionized impurity scattering).

Equation (2.38) may be simplified by using the *relaxation time approximation*

$$\left(\frac{\partial f}{\partial t}\right)_c = \frac{\Delta f}{\tau} = -\frac{f - f_0}{\tau} \quad (2.39)$$

which is based on the assumption that for small changes in f carriers return to equilibrium in a characteristic time τ, dependent on the dominant scattering mechanisms. Further simplifications of (2.38) using (2.39) applicable for low electric fields lead to a simple equation connecting the mobility μ to the *average scattering time* $\langle \tau \rangle$

$$\mu \cong \frac{e\langle \tau \rangle}{m^*} . \quad (2.40)$$

The details of calculations may be found in various advanced textbooks, for example *Bube* [2.13], *Blatt* [2.4]. The average scattering time may be calculated assuming a *Maxwell-Boltzmann* distribution function and a parabolic band

$$\langle \tau \rangle = \frac{2}{3 k_B T} \frac{\int\limits_0^\infty \tau(E) E^{3/2} e^{-E/k_B T}\,dE}{\int\limits_0^\infty E^{1/2} e^{-E/k_B T}\,dE} . \quad (2.41)$$

Quantum mechanical perturbation theory can be used to calculate the carrier scattering rate for different processes i, giving,

$$\tau_i(E) = a E^{-\alpha} , \quad (2.42)$$

where a and α are constants and E is the electron energy. Substituting (2.42) into (2.41) gives

$$<\tau_i> = \frac{4a\Gamma(5/2 - \alpha)}{3\pi^{1/2}(k_B T)^\alpha} \quad (2.43)$$

in terms of the *gamma function* Γ. Using this approach, the expressions for the mobility for the case of lattice and impurity scattering may be easily found

$$\mu_L \propto \frac{4e}{m^*\sqrt{9\pi k_B}} T^{-\frac{3}{2}}, \qquad (2.44)$$

$$\mu_I \propto \frac{8ek_B^{3/2} N_I}{m^*\sqrt{\pi}} T^{+\frac{3}{2}}, \qquad (2.45)$$

where N_I is the concentration of ionized impurities.

2.7 Resistivity of Thin Polycrystalline Films

Two new dominant scattering mechanisms must be considered in polycrystalline thin films – *scattering by grain boundaries* and *scattering at the surface*. The scattering by grain boundaries is schematically shown in Fig. 2.11. As a first approximation, the conduction electron may be considered free within a grain, but becomes scattered at the grain boundary. Its mean free path ℓ_grains is therefore roughly equal to the average grain size d. If $\lambda = \ell_\mathrm{crystal}$ is the mean free path of the conduction electrons in the single crystal, then according to Matthiessen's rule

$$\frac{1}{\ell} = \frac{1}{\ell_\mathrm{crystal}} + \frac{1}{\ell_\mathrm{grains}} = \frac{1}{\lambda} + \frac{1}{d}. \qquad (2.46)$$

The resistivity is inversely proportional to the mean free path which means the resistivity of the single crystal $\rho_\mathrm{crystal} \propto 1/\lambda$ and resistivity of the polycrystalline sample $\rho \propto 1/\ell$. Thus,

$$\frac{\rho}{\rho_\mathrm{crystal}} = 1 + \left(\frac{\lambda}{d}\right). \qquad (2.47)$$

Figure 2.12 clearly demonstrates that even simple considerations agree well with experimental data. However, in a more rigorous theory we have to consider a number of effects. It may take more than one scattering at a grain boundary to totally randomize the velocity so that we need to calculate the effective mean free path that accounts for how many collisions are needed to randomize the velocity. There is a possibility that the electron may be totally reflected back at a grain boundary (bounce back). Let σ be the conductivity of the polycrystalline (grainy) material and σ_crystal be the conductivity of the bulk single crystal. Suppose that the probability of reflection at a grain boundary is R. If d is the average grain size (diameter) then the two conductivities may be related using the *Mayadas* and *Shatzkes*

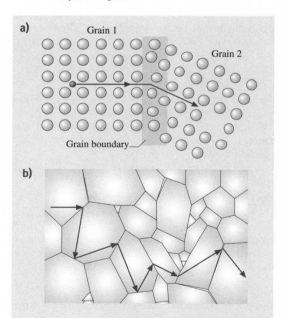

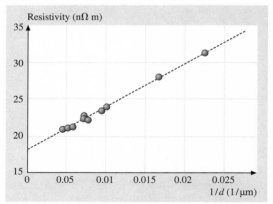

Fig. 2.11 (a) Grain boundaries cause electron scattering and therefore add to the resistivity according to Matthiessen's rule. (b) For a very grainy solid, the electron is scattered from grain boundary to grain boundary, and the mean free path is approximately equal to the mean grain diameter

Fig. 2.12 Resistivity of Cu polycrystalline film vs. reciprocal mean grain size (diameter), $1/d$. Film thickness $D = 250–900$ nm does not affect the resistivity. The straight line $\rho = 17.8\,\mathrm{n\Omega\,m} + (595\,\mathrm{n\Omega\,nm})(1/d)$ [2.14]

formula [2.15]

$$\frac{\sigma}{\sigma_{\text{crystal}}} = 1 - \frac{3}{2}\beta + 3\beta^2 - 3\beta^3 \ln\left(1 + \frac{1}{\beta}\right)$$

$$\text{where} \quad \beta = \frac{\lambda}{d}\left(\frac{R}{1-R}\right), \tag{2.48}$$

which in $0.1 < \beta < 10$, approximates to

$$\frac{\rho}{\rho_{\text{crystal}}} \approx 1 + 1.34\beta. \tag{2.49}$$

For copper typically R values are 0.24–0.40, and somewhat a smaller R for Al. Equation (2.49) for a Cu film with $R \approx 0.3$ predicts $\rho/\rho_{\text{crystal}} \approx 1.20$ for roughly $d \approx 3\lambda$ or a grain size $d \approx 120$ nm since in the bulk crystal $\lambda \approx 40$ nm. *Tellier* et al. [2.17] have given extensive discussions of grain boundary scattering limited resistivity of thin films.

Scattering from the film surfaces must also be included in any resistivity calculation. It is generally assumed that the scattering from a surface is partially *inelastic*, that is the electron loses some of the velocity gained from the field. The inelastic scattering is also called *nonspecular*. (If the electron is elastically reflected from the surface just like a rubber ball bouncing from a wall, then there is no increase in the resistivity.) If the parameter p is the fraction of surface collisions which are specular (elastic) and if the thickness of film D is greater than λ, and $\sigma_{\text{bulk}} = 1/\rho_{\text{bulk}}$, than in accordance with Fuchs-Sondheimer equation [2.18, 19] the conductivity σ of the film is

$$\frac{\sigma}{\sigma_{\text{bulk}}} = 1 - \frac{3\lambda}{8D}(1-p), \quad (\lambda/D > 1). \tag{2.50}$$

If D is much shorter than ℓ,

$$\frac{\sigma}{\sigma_{\text{bulk}}} = \frac{3D}{4\lambda}\left[\ln\left(\frac{\lambda}{D}\right) + 0.423\right]$$
$$\times (1 + 2p), \quad (\lambda/D \ll 1). \tag{2.51}$$

For purely nonspecular (inelastic) scattering, an approximate estimate can be obtained from

$$\frac{\rho}{\rho_{\text{bulk}}} \approx 1 + \frac{3}{8}\left(\frac{\lambda}{D}\right). \tag{2.52}$$

Figure 2.13 shows the resistivity of thin polycrystalline Cu films as a function of film thickness. From (2.50), for sufficiently small thicknesses, ρ is inversely proportional to the thickness D, which is what is observed experimentally in Fig. 2.13. The saturation at higher thicknesses is mostly due to scattering on grain boundaries.

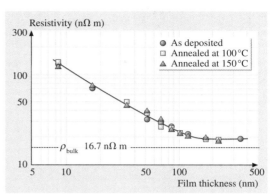

Fig. 2.13 Resistivity of Cu polycrystalline film versus film thickness D. The resistivity is mostly controlled by the surface scattering, and annealing does not significantly affect the resistivity while it reduces the crystallinity [2.16]

For elastic or specular scattering $p = 1$ and there is no change in the conductivity. The value of p depends on the film preparation method (e.g. sputtering, epitaxial growth etc.) and the substrate on which the film has been deposited. Table 2.3 summarizes the resistivities of thin Cu and Au gold films deposited by various preparation techniques. Notice the large difference between the Au films deposited on a noncrystalline glass substrate and on a crystalline mica substrate. Such differences between films are typically attributed to different values of p. The p-value can also change (increase) when the film is annealed. Obviously, the polycrystallinity of the film will also affect the resistivity as discussed above. Typically, most epitaxial thin films, unless very thin ($D \ll \ell$), deposited onto heated crystalline substrates exhibit highly specular scattering with $p = 0.9 - 1$.

It is generally very difficult to separate the effects of surface and grain boundary scattering in thin polycrystalline films; the contribution from grain boundary scattering is likely to exceed that from the surfaces. In any event, both contributions, by Matthiessen's general rule, increase the overall resistivity. Figure 2.12 shows an example in which the resistivity ρ_{film} of thin Cu polycrystalline films is due to grain boundary scattering, and thickness has no effect (D was 250 nm – 900 nm and much greater than λ). The resistivity ρ_{film} is plotted against the reciprocal mean grain size $1/d$, which then follows the expected linear behavior in (2.49). On the other hand, Fig. 2.13 shows the resistivity of Cu films as a function of film thickness D. In this case, annealing (heat treating) the films to reduce the polycrystallinity

Table 2.3 Resistivities of some thin Cu and Au films at room temperature. PC: Polycrystalline film; RT is room temperature; D = film thickness; d = average grain size. At RT for Cu, $\lambda = 38$–40 nm, and for Au $\lambda = 36$–38 nm. Data selectively combined from various sources, including [2.14, 16, 20–22].

Film	D (nm)	d (nm)	ρ (nΩ m)	Comment
Cu films (polycrystalline)				
Cu on TiN, W and TiW [2.14]	> 250	186	21	CVD (chemical vapor deposition).
		45	32	Substrate temperature 200 °C, ρ depends on d not $D = 250$–900 nm
Cu on 500 nm SiO$_2$ [2.20]	20.5		35	Thermal evaporation, substrate at RT
	37		27	
Cu on Si (100) [2.16]	52		38	Sputtered Cu films. Annealing at 150 °C
	100		22	has no effect. $R \approx 0.40$ and $p \approx 0$
Cu on glass [2.21]	40		50	As deposited
	40		29	Annealed at 200 °C
	40		25	Annealed at 250 °C
				All thermally evaporated and PC
Au films				
Au epitaxial film on mica	30		25	Single crystal on mica. $p \approx 0.8$, specular scattering
Au PC film on mica	30		54	PC. Sputtered on mica. p is small
Au film on glass	30		70	PC. Evaporated onto glass. p is small, nonspecular scattering
Au on glass [2.22]	40	8.5	92	PC. Sputtered films. $R = 0.27$–0.33
	40	3.8	189	

does not significantly affect the resistivity because ρ_film is controlled primarily by surface scattering, and is given by (2.52). Gould in Chapt. 29 provides a more advanced treatment of conduction in thin films.

2.8 Inhomogeneous Media. Effective Media Approximation

The *effective media approximation* (EMA) attempts to estimate the properties of inhomogeneous mixture of two or more components using the known physical properties of each component. The general idea of any EMA is to substitute for the original inhomogeneous mixture some imaginary homogeneous substance – the *effective medium* (EM) – whose response to an external excitation is the same as that of the original mixture. The EMA is widely used for investigations of non-uniform objects in a variety of applications such as composite materials [2.23, 24], microcrystalline and amorphous semiconductors [2.25–28], light scattering [2.29], conductivity of dispersed ionic semiconductors [2.30] and many others.

Calculations of the conductivity and dielectric constant of two component mixtures are reviewed by *Reynolds* and *Hough* [2.31] and summarized by *Rossiter* [2.1]. For such a mixture we assume that the two components α and β are randomly distributed in space with volume fractions of χ_α and $\chi_\beta = 1 - \chi_\alpha$. The dielectric properties are described by an *effective permittivity* ε_eff given by the ratio

$$\varepsilon_\text{eff} = \langle D \rangle / \langle E \rangle \;, \tag{2.53}$$

where $\langle E \rangle$ is the average electric field and $\langle D \rangle$ is the average displacement field. The displacement field averaged over a large volume may be calculated from

$$\langle D \rangle = \frac{1}{V} \int_V D \, dv = \frac{1}{V} \left(\int_{V_\alpha} D \, dv + \int_{V_\beta} D \, dv \right)$$
$$= \chi_\alpha \langle D_\alpha \rangle + \chi_\beta \langle D_\beta \rangle \;, \tag{2.54}$$

Table 2.4 Mixture rules for randomly oriented particles

Particle shape	Mixture rule	Factors in (2.58) A	ε^*	References
Spheres	$\dfrac{\varepsilon_{\text{eff}} - \varepsilon_\beta}{\varepsilon_{\text{eff}} + 2\varepsilon_\beta} = \chi_\alpha \dfrac{\varepsilon_\alpha - \varepsilon_\beta}{\varepsilon_\alpha + 2\varepsilon_\beta}$	$\tfrac{1}{3}$	ε_2	[2.32–36]
Spheres	$\dfrac{\varepsilon_{\text{eff}} - \varepsilon_\beta}{3\varepsilon_\beta} = \chi_\alpha \dfrac{\varepsilon_\alpha - \varepsilon_\beta}{\varepsilon_\alpha + 2\varepsilon_\beta}$	$\tfrac{1}{3}$	ε_2	[2.37]
Spheres	$\chi_\alpha \dfrac{\varepsilon_\alpha - \varepsilon_{\text{eff}}}{\varepsilon_\alpha + 2\varepsilon_{\text{eff}}} + \chi_\beta \dfrac{\varepsilon_\beta - \varepsilon_{\text{eff}}}{\varepsilon_{\beta f} + 2\varepsilon_{\text{eff}}} = 0$	$\tfrac{1}{3}$	ε_{eff}	[2.38]
Spheres	$\dfrac{\varepsilon_{\text{eff}} - \varepsilon_\beta}{3\varepsilon_{\text{eff}}} = \chi_\alpha \dfrac{\varepsilon_\alpha - \varepsilon_\beta}{\varepsilon_\alpha + 2\varepsilon_\beta}$	$\tfrac{1}{3}$	ε_{eff}	[2.39]
Spheroids	$\varepsilon_{\text{eff}} = \varepsilon_\beta + \dfrac{\chi_\alpha}{3(1-\chi_\alpha)} \sum_{n=1}^{3} \dfrac{\varepsilon_\alpha - \varepsilon_{\text{eff}}}{1 + A\left(\frac{\varepsilon_\alpha}{\varepsilon_{\text{eff}}} - 1\right)}$	A	ε_2	[2.40]
Spheroids	$\varepsilon_{\text{eff}} = \varepsilon_\beta + \dfrac{\chi_\alpha}{3} \sum_{n=1}^{3} \dfrac{\varepsilon_\alpha - \varepsilon_{\text{eff}}}{1 + A\left(\frac{\varepsilon_\alpha}{\varepsilon_{\text{eff}}} - 1\right)}$	A	ε_{eff}	[2.41]
Lamellae	$\varepsilon_{\text{eff}}^2 = \dfrac{2(\varepsilon_\alpha \chi_\alpha - \varepsilon_\beta \chi_\beta) - \varepsilon_{\text{eff}}}{\frac{\varepsilon_\alpha}{\chi_\alpha} + \frac{\varepsilon_\beta}{\chi_\beta}}$	0	ε_{eff}	[2.38]
Rods	$5\varepsilon_{\text{eff}}^3 + \left(5\varepsilon'_p - 4\varepsilon_p\right)\varepsilon_{\text{eff}}^2 - \left(\chi_\alpha \varepsilon_\alpha^2 + 4\varepsilon_\alpha \varepsilon_\beta + \chi_\beta \varepsilon_\beta^2\right) - \varepsilon_\alpha \varepsilon_\beta \varepsilon_p = 0$ where $\dfrac{1}{\varepsilon'_p} = \dfrac{\chi_\alpha}{\varepsilon_\beta} + \dfrac{\chi_\beta}{\varepsilon_\alpha}$ and $\dfrac{1}{\varepsilon_p} = \dfrac{\chi_\alpha}{\varepsilon_\alpha} + \dfrac{\chi_\beta}{\varepsilon_\beta}$	$\tfrac{1}{2}$	ε_{eff}	[2.42]

where $\langle D_\alpha \rangle$ and $\langle D_\beta \rangle$ are the average displacements fields inside regions of the respective components and V_α and V_β are their volumes. Likewise the electric field is given by

$$\langle E \rangle = \chi_\alpha \langle E_\alpha \rangle + \chi_\beta \langle E_\beta \rangle \,. \tag{2.55}$$

From (2.53) one gets

$$\varepsilon_{\text{eff}} = \varepsilon_\beta + (\varepsilon_\alpha - \varepsilon_\beta) \chi_\alpha f_\alpha \tag{2.56}$$

or

$$(\varepsilon_{\text{eff}} - \varepsilon_\alpha) \chi_\alpha f_\alpha + (\varepsilon_{\text{eff}} - \varepsilon_\beta) \chi_\beta f_\beta = 0 \tag{2.57}$$

where ε_α and ε_β are the permittivities of the components and $f_\alpha = \langle E_\alpha \rangle / \langle E \rangle$ and $f_\beta = \langle E_\beta \rangle / \langle E \rangle$ are so-called *field factors*. The choice between (2.56) and (2.57) depends on particle geometry. Equation (2.56) is better when the particles of component are dispersed in a continuous medium β. Equation (2.57) is preferred when the particle size of the two components is of the same order of magnitude.

The field factors can be calculated analytically only for phase regions with special specific geometries. The field factor for ellipsoids is given by (Stratton [2.43])

$$f_\alpha = \sum_{i=1}^{3} \dfrac{\cos^2 \alpha_i}{1 + A_i \left(\frac{\varepsilon_\alpha}{\varepsilon^*} - 1\right)} \tag{2.58}$$

where α_i are the angles between the ellipsoid axes and the applied field and A_i depends upon the axial ratios of the ellipsoids subject to the condition that

$$\sum_{i=1}^{3} A_i = 1 \,.$$

For a spheroid, $A_2 = A_3 = A$ and $A_1 = 1 - 2A$. For a random orientation of spheroids $\cos^2 \alpha_1 = \cos^2 \alpha_2 = \cos^2 \alpha_3 = \tfrac{1}{3}$. For the case of long particles with aligned axes $\cos^2 \alpha_1 = \cos^2 \alpha_2 = \tfrac{1}{2}$ and $\cos^2 \alpha_3 = 0$. The values of parameters entering (2.58) can be found in Table 2.4 which shows a set of *mixture rules*, i.e.

Table 2.5 Mixture rules for partially oriented particles

Particle shape	Formula	Factors in (2.58) A	ε^*	$\cos\alpha_1 = \cos\alpha_2$	$\cos\alpha_3$	References
Parallel cylinders	$\dfrac{\varepsilon_{\text{eff}} - \varepsilon_\beta}{\varepsilon_{\text{eff}} + \varepsilon_\beta} = \chi_\alpha \dfrac{\varepsilon_\alpha - \varepsilon_\beta}{\varepsilon_\alpha + \varepsilon_\beta}$	$\tfrac{1}{2}$	ε_2	$\tfrac{1}{2}$	0	[2.35, 36]
Parallel cylinders	$\chi_\alpha \dfrac{\varepsilon_\alpha - \varepsilon_{\text{eff}}}{\varepsilon_\alpha + \varepsilon_{\text{eff}}} + \chi_\beta \dfrac{\varepsilon_\beta - \varepsilon_{\text{eff}}}{\varepsilon_\beta + \varepsilon_{\text{eff}}}$	$\tfrac{1}{2}$	ε_{eff}	$\tfrac{1}{2}$	0	[2.38]
Parallel lamellae (with two axes randomly oriented)	$\varepsilon_{\text{eff}}^2 = \dfrac{\varepsilon_\alpha \chi_\alpha + \varepsilon_\beta \chi_\beta}{\dfrac{\varepsilon_\alpha}{\chi_\alpha} + \dfrac{\varepsilon_\beta}{\chi_\beta}}$	0	ε_{eff}	$\tfrac{1}{2}$	0	[2.38]
Lamellae with all axes aligned (current lines are perpendicular to lamellae planes)	$\dfrac{1}{\varepsilon_{\text{eff}}} = \dfrac{\chi_\alpha}{\varepsilon_\alpha} + \dfrac{\chi_\beta}{\varepsilon_\beta}$	0	ε_{eff}	0	1	[2.44]
Lamellae with all axes aligned (current lines are parallel to lamellae planes)	$\varepsilon_{\text{eff}} = \varepsilon_\alpha \chi_\alpha + \varepsilon_\beta \chi_\beta$	0	ε_{eff}	1	0	[2.45, 46]
Spheroids with all axes aligned (current lines are parallel to one of the axes)	$\varepsilon_{\text{eff}} = \varepsilon_\beta + \dfrac{\chi_\alpha(\varepsilon_\alpha - \varepsilon_\beta)}{1 + A\left(\dfrac{\varepsilon_\alpha}{\varepsilon_\beta} - 1\right)}$	A	ε_2	0	1	[2.47]
Spheroids with all axes aligned (current lines are parallel to one of the axes)	$\dfrac{\varepsilon_{\text{eff}}}{\varepsilon_\beta} = 1 + \dfrac{\chi_\alpha}{\left(\dfrac{\varepsilon_\alpha}{\varepsilon_\beta} - 1\right)^{-1} + A\chi_\beta}$	A	ε_2	0	1	[2.48]

Table 2.6 Conductivity / resistivity mixture rules

Particle shape	Formula	Commentary
Lamellae with all axes aligned (current lines are perpendicular to lamellae planes)	$\rho_{\text{eff}} = \chi_\alpha \rho_\alpha + \chi_\beta \rho_\beta$	*Resistivity mixture rule*: ρ_α and ρ_β are the resistivities of two phases and ρ_{eff} is the effective resistivity of mixture
Lamellae with all axes aligned (current lines are parallel to lamellae planes)	$\sigma_{\text{eff}} = \chi_\alpha \sigma_\alpha + \chi_\beta \sigma_\beta$	*Conductivity mixture rule*: σ_α and σ_β are the conductivities of two phases and σ_{eff} is the effective conductivity of mixture
Small spheroids (α-phase) in medium (β-phase)	$\rho_{\text{eff}} = \rho_\beta \dfrac{\left(1 + \tfrac{1}{2}\chi_\alpha\right)}{(1 - \chi_\alpha)}$	$\rho_\alpha > 10\rho_\beta$
Small spheroids (α-phase) in medium (β-phase)	$\rho_{\text{eff}} = \rho_\beta \dfrac{(1 - \chi_\alpha)}{(1 + 2\chi_\alpha)}$	$\rho_\alpha < 0.1\rho_\beta$

a set of formulae allowing one to calculate ε_{eff} for some specific cases (such as spheres, rods, lamellae, etc.). The presence of some degree of orientation somewhat simplifies the calculations as shown in the Table 2.5.

The same formulae can be used to calculate the conductivity of mixtures by substituting the appropriate conductivity σ for ε. For some special cases, the mixture rules of Table 2.5 lead to very simple formulae which allows one to calculate the conductivity of inhomogeneous alloys with those specific geometries. These *mixture rules* are summarized in Table 2.6.

The most general approach to calculating the effective dielectric permittivity comes from

$$\varepsilon_{\text{eff}} = \varepsilon_2 \left(1 - \chi_\alpha \int_0^1 \frac{G(L)}{t - L} \, dL\right) \quad (2.59)$$

where $t = \varepsilon_2/(\varepsilon_2 - \varepsilon_1)$ and $G(L)$ is the *spectral function* which describes the geometry of particles. The advantage of the spectral representation is that it distinguishes between the influence of geometrical quantities and that of the dielectric properties of the components on the effective behavior of the sys-

Table 2.7 Mixture rules and corresponding spectral functions $G(L)$

Mixture rule by *Bruggemann* [2.38]: $\chi_\alpha \dfrac{\varepsilon_\alpha - \varepsilon_{\text{eff}}}{\varepsilon_\alpha + 2\varepsilon_{\text{eff}}} + \chi_\beta \dfrac{\varepsilon_\beta - \varepsilon_{\text{eff}}}{\varepsilon_{\beta f} + 2\varepsilon_{\text{eff}}} = 0$

$G(L) = \dfrac{3\chi_\alpha - 1}{2\chi_\alpha} \delta(L) \Theta(3\chi_\alpha - 1) + \dfrac{3}{4\pi \chi_\alpha L} \sqrt{(L - L^-)(L^+ - L)} \Theta(L - L^-) \Theta(L^+ - L)$

where $L^{+/-} = \dfrac{1}{3}\left(1 + \chi_\alpha \pm 2\sqrt{2\chi_\alpha - 2\chi_\alpha^2}\right)$

Mixture rule by *Maxwell-Garnett* [2.49]: $\dfrac{\varepsilon_{\text{eff}} - \varepsilon_\beta}{\varepsilon_{\text{eff}} + 2\varepsilon_\beta} = \chi_\alpha \dfrac{\varepsilon_\alpha - \varepsilon_\beta}{\varepsilon_\alpha + 2\varepsilon_\beta}$

$G(L) = \delta\left(L - \dfrac{1 - \chi_\alpha}{3}\right)$

Mixture rule by *Looyenga* [2.50]: $\varepsilon_{\text{eff}}^{1/3} = \chi_\alpha \varepsilon_\alpha^{1/3} + \chi_\beta \varepsilon_\beta^{1/3}$

$G(L) = \chi_\alpha^2 \delta(L) + \dfrac{3\sqrt{3}}{2\pi}\left(\chi_\beta^2 \left|\dfrac{L-1}{L}\right|^{1/3} + \chi_\alpha \chi_\beta \left|\dfrac{L-1}{L}\right|^{2/3}\right)$

Mixture rule by *Monecke* [2.51]: $\varepsilon_{\text{eff}} = \dfrac{2(\chi_\alpha \varepsilon_\alpha + \chi_\beta \varepsilon_\beta)^2 + \varepsilon_\alpha \varepsilon_\beta}{(1 + \chi_\alpha)\varepsilon_\alpha + (2 - \chi_\alpha)\varepsilon_\beta}$

$G(L) = \dfrac{2\chi_\alpha}{1 + \chi_\alpha} \delta(L) + \dfrac{1 - \chi_\alpha}{1 + \chi_\alpha} \delta\left(L - \dfrac{1 + \chi_\alpha}{3}\right)$

Mixture rule for hollow sphere equivalent by *Bohren* and *Huffman* [2.52]

$\varepsilon_{\text{eff}} = \varepsilon_\alpha \dfrac{(3 - 2f)\varepsilon_\beta + 2f\varepsilon_\alpha}{f\varepsilon_\beta + (3 - f)\varepsilon_\alpha}$

$G(L) = \dfrac{2}{3 - f} \delta(L) + \dfrac{1 - f}{3 - f} \delta\left(L - \dfrac{3 - f}{3}\right)$

where $f = 1 - \dfrac{r_i^3}{r_o^3}$ and $r_{i/o}$ is the inner/outer radius of the sphere

tem. Although the spectral function $G(L)$ is generally unknown for an arbitrary two-phase composite, it's analytically known or can be numerically derived for any existing mixture rule. Examples of spectral functions and corresponding solutions are shown in Table 2.7.

2.9 The Hall Effect

The Hall effect is closely related to the phenomenon of conductivity and is observed as the occurrence of a voltage appearing across a conductor carrying an electric current in a magnetic field. The schematic of the experiment is shown in the Fig. 2.14. The effect is often characterized by the *Hall coefficient*

$$R_{\text{H}} = \dfrac{E_y}{J_x B_z}, \qquad (2.60)$$

where E_y is the Hall effect electric field in the y-direction, J_x is the current density in the x-direction and B_z is magnetic field in the z-direction.

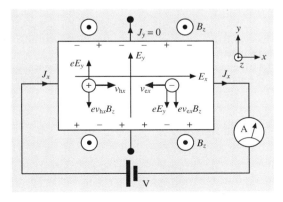

Fig. 2.14 Hall effect for ambipolar conduction. The magnetic field B_z is out of the plane of the paper. Both electrons and holes are deflected toward the bottom surface of the conductor and so the Hall voltage depends on the relative mobilities and concentrations of electrons and holes

The *Hall effect for ambipolar conduction* in a sample where there are both negative and positive charge carriers, e.g. electrons and holes in a semiconductor, involves not only the concentrations of electrons and holes, n and p respectively, but also the electron and hole drift mobilities, μ_e and μ_h. In the first approximation, the Hall coefficient can be calculated in the following way. Both charge carriers experience a Lorentz force in the same direction since they are drifting in the opposite directions but of course have opposite signs as illustrated in Fig. 2.14. Thus, both holes and electrons accumulate near the bottom surface. The magnitude of the Lorentz force, however, will be different since the drift mobilities and hence drift velocities will be different. Once equilibrium is reached, there should be no current flowing in the y-direction as we have an open circuit. The latter physical arguments lead to the following Hall coefficient [2.5],

$$R_H = \frac{p\mu_h^2 - n\mu_e^2}{e(p\mu_h + n\mu_e)^2}$$

or $\quad R_H = \dfrac{p - nb^2}{e(p + nb)^2}$, (2.61)

where $b = \mu_e/\mu_h$.

It is clear that the Hall coefficient depends on both the drift mobility ratio and the concentrations of holes and electrons. For $p > nb^2$, R_H will be positive and for $p < nb^2$, it will be negative. Note that the carrier concentration is not zero when the Hall coefficient is zero but rather $n/p = (\mu_h/\mu_e)^2$. As an example, Fig. 2.15 shows the dependence of Hall coefficient vs. electron concentration for a single crystal silicon. The calculations are based on (2.61) and the law of mass action

$$np = n_i^2, \quad (2.62)$$

where n_i is the electron concentration in the intrinsic semiconductor.

In the case of *monopolar conduction*, e.g. conduction in metals or in doped semiconductors, (2.61) reduces to

$$R_H = -\frac{1}{en}, \quad \text{(for } n \gg p\text{)} \quad (2.63)$$

or

$$R_H = \frac{1}{ep}, \quad \text{(for } p \gg n\text{)}. \quad (2.64)$$

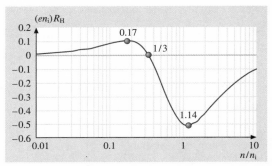

Fig. 2.15 Normalized Hall coefficient versus normalized electron concentration plot for single-crystal silicon. The values 0.17, 1.14 and 0.33 shown are the n/n_i values when the magnitude R reaches maxima and zero respectively. In single-crystal silicon, $n_i = 1.5 \times 10^{10}$ cm^{-3}, $\mu_e = 1350$ cm^{2}V^{-1}s^{-1} and $\mu_h = 450$ cm^{2}V^{-1}s^{-1}

Therefore, (considering as an example a n-type semiconductor where $\sigma = ne\mu_n$) one can write

$$\mu_H = \frac{\sigma}{ne} = -\sigma R_H \quad (2.65)$$

which provides a simple expression for determining the electron mobility known as the *Hall mobility*. Note, however, that the Hall mobility may differ from the drift mobility discussed in the previous sections. The difference arises from the carriers in a semiconductor having a distribution of energies. An average is used to describe the effect of carriers occupying different allowed states. (This is distinct from the earlier discussions of mobility where it was assumed that all the carriers have the same mean free time between collisions.) To include this, it is necessary to use a formal analysis based on the Boltzmann transport equation, as discussed in Sect. 2.6. If we express the averaging of the carriers with energy E and distribution function $f(E)$, we can write the energy-averaged $\tau(E)$ (i.e., $\langle\tau\rangle$) as:

$$\langle\tau\rangle = \frac{\int \tau(E)f(E)\,dE}{\int f(E)\,dE} \quad (2.66)$$

and the energy-averaged τ^2 (i.e., $\langle\tau^2\rangle$) as:

$$\langle\tau\rangle = \frac{\int \tau^2(E)f(E)\,dE}{\int f(E)\,dE}. \quad (2.67)$$

The rigorous analysis [2.12] shows that the Hall mobility μ_H in terms of the drift mobility μ_d is

$$\mu_H = r_H \times \mu_d, \quad (2.68)$$

where r_H is the *Hall factor* given by the ratio

$$r_H = \frac{\langle \tau^2 \rangle}{\langle \tau \rangle^2}. \tag{2.69}$$

2.10 High Electric Field Transport

The previous sections focused on carrier transport in weak electric fields where the energy gained by carriers from the field is lost to the lattice through collisions with phonons or ionized impurities. However at higher fields, the efficiency of collision mechanisms diminishes and the carrier system contains more energy than the lattice. The carriers are then called *hot* with effective temperatures T_e for electrons and T_h for holes. In this case, the drift velocity no longer obeys Ohm's law, and becomes non-linear in the applied field with a clear tendency to saturation due to the appearance of a new dissipation mechanism involving optical phonon generation. Figure 2.16 shows the drift velocity saturation for both electrons and holes in Si and electrons in GaAs; GaAs shows a region of electron velocity overshoot and then negative differential resistivity due to inter-valley scattering, i.e., the transfer of electrons from the Γ minimum to the L conduction band minimum.

Solving the Boltzmann transport equation by analogy with (2.6, 2.7) the mobility may be defined as

$$\mu = \frac{e \langle \tau(T_e) \rangle}{m^*}, \tag{2.70}$$

where e is electron charge and m_e^* is the effective mass and $\langle \tau(T_e) \rangle$ is the mean free time which now strongly de-

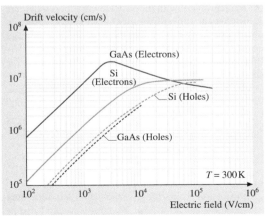

Fig. 2.16 Dependence of carrier drift velocity on electric field for GaAs and Si [2.53]

The magnitude of the *Hall factor* will depend on the magnitude of the scattering mechanisms that contribute to τ, but usually at low magnetic fields, $r_H \approx 1$ and the two mobilities are identical.

pends on T_e. Therefore, the high-field mobility is related to the low field mobility μ_0 by

$$\mu = \mu_0 \left(\frac{T_e}{T}\right)^\beta, \tag{2.71}$$

where β depends on electric field and scattering mechanisms. Thus, in order to determine the dependence of the mobility on electric field, the dependence of the effective carrier temperature on field is required. This may be found by using the time-dependent Boltzmann transport equation. Suppose that F is the field, then for non-degenerate conditions with $T_e \gg T$

$$\frac{T_e}{T} \propto F^{-\frac{2}{2\beta-1}} \tag{2.72}$$

and hence

$$\mu(E) \propto \mu_0 F^{-\frac{2}{2\beta-1}}. \tag{2.73}$$

For acoustic phonon scattering, $\beta = -3/2$ and the drift mobility shows no saturation, increasing with field as $F^{1/4}$. Saturation in the drift velocity may be achieved only when $\beta \to \infty$ due to optical phonon scattering where large energy changes are involved. The saturation velocity v_s (related to the saturation mobility as $v_s = F\mu_s$) may be calculated using the energy and momentum rate equations for optical phonon scattering:

$$\frac{d\langle E \rangle}{dt} = eFv_s - \frac{E_{op}}{\tau_e} \tag{2.74}$$

$$\frac{d\langle m^* v_s \rangle}{dt} = eF - \frac{m^* v_s}{\tau_m} \tag{2.75}$$

where E_{op} is the optical phonon energy, τ_e and τ_m are the energy and momentum relaxation times, respectively. At steady state and for not extremely high fields, one may assume that $\tau_e \approx \tau_m$. (It is worth noting that at the highest electric fields $\tau_e > \tau_m$ and may lead to the appearance of avalanche, as discussed in Sect. 2.11). Therefore, the solution of (2.74, 2.75) is

$$v_s = \left(\frac{E_{op}}{m^*}\right)^{1/2}. \tag{2.76}$$

in agreement with the experimental values shown in Fig. 2.16.

2.11 Avalanche

At very high electric field (in the range 2×10^5 V/cm or larger) a new possibility appears: a carrier may have kinetic energy in excess of the binding energy of a valence electron to its parent atom. In colliding with an atom, such a carrier can break the covalent bond and produce an electron-hole pair. This process is called *impact ionization* and is characterized by the *impact ionization coefficient* α (α_e for electrons and α_h for holes). The released electrons and holes may, in turn, impact ionize more atoms producing new electron-hole pairs. This process is called *avalanche* and is characterized by the *multiplication factor*, M which is the ratio of number of collected carriers to the number of initially injected carriers.

The field dependence of the impact ionization coefficient, at least over the limited fields where avalanche is observed, is usually modeled by experimentalists by using

$$\alpha = A \exp\left[-\left(\frac{B}{F}\right)^n\right], \qquad (2.77)$$

where F is the field A, B, n are constants that depend on the semiconductor material properties such as the E-k electronic structure, phonon energies and spectra, scattering mechanisms and so on. The constant B has been semiquantitatively argued to depend on E_g/τ where E_g is the bandgap and $1/\tau$ is the phonon scattering rate; higher bandgap semiconductors tend to have steeper slopes on log α vs. $1/F$ plots and the log α vs. $1/F$ curve tends to shift to higher fields. Typically, it has been found that $n \approx 1$ at low fields and $n \approx 2$ for high fields. Figure 2.17 shows experimental data for a variety of materials over a wide range of electric fields.

The origin of (2.77) in its simple $n = 1$ form lies in Shockley's [2.64] *lucky electron* model. When a carrier moves a distance z down-stream (along the field) without being scattered, it gains an energy eFz. An unlucky carrier is scattered so frequently that its eFz never reaches the threshold ionization energy E_I for impact ionization. On the other hand, a *lucky electron* is a ballistic electron that avoids scattering for a substantial distance, and hence is able to build its eFz to reach E_I and thereby cause impact ionization. If λ is the mean free path of collisions, then Shockley's model gives

$$\alpha = \frac{eF}{E_I} \exp\left(-\frac{E_I}{e\lambda F}\right). \qquad (2.78)$$

The main problem with the Shockley model is that there are just not enough ballistic electrons to cause sufficient impact ionizations to explain the experiments. A better model was developed by *Baraff* [2.65] who numerically solved the Boltzmann transport equation for a simple parabolic band and energy independent mean free path to provide a relationship between α and F in terms of four parameters, that is, threshold energy for impact ionization, mean free path associated with ionizations, optical phonon energy, and mean free path for optical phonon scattering. Baraff's theory served experimentalists quite well in terms of comparing their results even though the model was not intuitive and was limited in terms of its assumptions and applicability to real semiconductors.

Impact ionization theory in crystalline solids only reached an acceptable level of confidence and understanding in the 1980s and 1990s with the development of the lucky-drift model by *Ridley* [2.66] and its extension

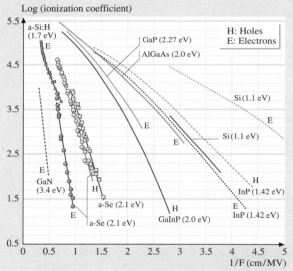

Fig. 2.17 Semilogarithmic plot of the dependence of the impact ionization coefficient on the reciprocal field for not only a-Se and a-Si:H, but also for various crystalline semiconductors for comparison; H indicates holes and E electrons. The y-axis is a base-10 logarithm of α in which α is in 1/cm. Data for a-Si:H from [2.54]; a-Se electrons and holes from [2.55]; a-Se holes from [2.56], from which α_h was obtained by reanalyzing their multiplication data. Data for crystalline semiconductors are for GaP [2.57]; GaInP (Ga$_{0.52}$In$_{0.48}$P) [2.58]; Al$_{0.60}$Ga$_{0.40}$As, [2.59]; InP, [2.60]; Si, [2.61]; GaN (calculation only), [2.62]. *Solid circle*: Electrons in a-Se [2.55]; *Solid square*: electrons in a-Si:H [2.54]; *Open circle*: holes in a-Se [2.55]; *Open square*: holes in a-Se [2.63]

by *Burt* [2.67], and *Mackenzie* and *Burt* [2.68]. The latter major advancement in the theory appeared as the *lucky drift* (LD) *model*, and it was based on the realization that at high fields, hot electrons do not relax momentum and energy at the same rates. Momentum relaxation rate is much faster than the energy relaxation rate. An electron can drift, being scattered by phonons, and have its momentum relaxed, which controls the drift velocity, but it can still gain energy during this drift. Stated differently, the mean free path λ_E for energy relaxation is much longer than the mean free path λ for momentum relaxation.

In the *Mackenzie* and *Burt* [2.68] version of the LD model, the probability $P(E)$ that a carrier attains an energy E is given by,

$$P(E) = \exp\left(-\int_0^E \frac{dE'}{eF\lambda(E')}\right) + \int_0^E \frac{dE_1}{eF\lambda(E_1)}$$
$$\times \exp\left(-\int_0^{E_1} \frac{dE'}{eF\lambda(E')}\right)$$
$$\times \exp\left(-\int_{E_1}^E \frac{dE'}{eF\lambda_E(E')}\right), \quad (2.79)$$

where as mentioned above λ is the mean free path associated with momentum relaxing collisions and λ_E is the mean energy relaxation length associated with the energy relaxing collisions. The first term is the Shockley lucky electron probability, i.e. the electron moves ballistically to energy E. The second term is the lucky drift probability term which is composed of the following: the electron first moves ballistically to some intermediate energy E_1 ($0 < E_1 < E$) from where it begins its lucky drift to energy E_1; hence the integration over all possible E_1. The impact ionization coefficient can then readily be evaluated from

$$\alpha = \frac{eFP(E_I)}{\int_0^{E_I} P(E)dE}. \quad (2.80)$$

The model above is based on a hard threshold ionization energy E_I, that is, when a carrier attains the energy E_I, ionization ensues. The model has been further refined by the inclusion of soft threshold energies which represent the fact the ionization does not occur immediately when the carrier attains the energy E_I, and the carrier drifts further to gain more energy than E_I before impact ionization [2.69–71].

Assuming λ and λ_E are energy independent, which would be the case for a single parabolic band in the crystalline state, (2.79) and (2.80) can be solved analytically to obtain

$$\alpha = \frac{1}{\lambda} \times \frac{\frac{\lambda}{\lambda_E}\exp\left(\frac{-E_I}{eF\lambda_E}\right) + \left(\frac{\lambda}{\lambda_E}\right)^2 \exp\left(\frac{-E_I}{eF\lambda}\right)}{1 - \exp\left(\frac{-E_I}{eF\lambda_E}\right) - \left(\frac{\lambda}{\lambda_E}\right)^2\left[1 - \exp\left(\frac{-E_I}{eF\lambda}\right)\right]}.$$
$$(2.81)$$

For $\lambda_E > \lambda$, and in the "low field region", where typically $(\alpha\lambda) < 10^{-1}$, or $x = E_I/eF\lambda > 10$, (2.81) leads to a simple expression for α,

$$\alpha = \left(\frac{1}{\lambda_E}\right)\exp\left(-\frac{E_I}{eF\lambda_E}\right). \quad (2.82)$$

For crystalline semiconductors, one typically also assumes that λ_E depends on the field F, λ and the optical phonon energy $\hbar\omega$ as

$$\lambda_E = \frac{eF\lambda^2}{2\hbar\omega}\coth\left(\frac{\hbar\omega}{2kT}\right). \quad (2.83)$$

As the field increases, λ_E eventually exceeds λ, and allows lucky drift to operate and the LD carriers to reach the ionization energy.

It is worth noting that the model of lucky drift is successfully used not only for crystalline semiconductors but to amorphous semiconductors [2.72].

2.12 Two-Dimensional Electron Gas

Heterostructures offer the ability to spatially engineer the potential in which carriers move. In such structures having layers deposited in the z-direction, when the width of a region with confining potential $t_z < \lambda_{dB}$, the de Broglie electron wavelength, electron states become stationary states in that direction, retaining Bloch wave character in the other two directions (i. e., x- and y-directions), and is hence termed a *2-D electron gas* (2DEG). These structures are notable for their extremely high carrier mobility.

High mobility structures are formed by selectively doping the wide bandgap material behind an initially undoped spacer region of width d as shown in Fig. 2.18a. Ionization and charge transfer leads to carrier build-up

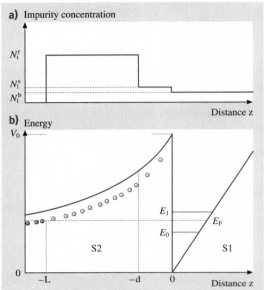

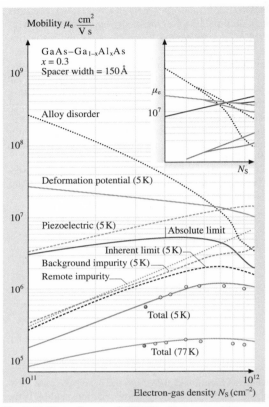

Fig. 2.18 (**a**) Doping profile for a selectively doped 2DEG heterostructure. The S1 is a narrow bandgap material adjacent to the heterointerface. The S2 is selectively doped wide-bandgap material including an initially undoped spacer region of width d. (**b**) Conduction energy band structure for a selectively doped 2DEG heterostructure. The equilibrium band bending (through Poisson's equation) in the well region results from the equalization of the ionized donor concentration in the wide-bandgap material and the 2DEG concentration adjacent to the heterointerface. When the associated interfacial field is sufficiently strong, carriers are confined within λ_{dB} and electron states are quantized into the sub-bands E_0 and E_1 [2.73]

in the low potential region of narrow bandgap material adjacent to the hetero-interface. The equilibrium band bending (i.e., through Poisson's equation) in the well region, as shown in Fig. 2.18b, results from the equalization of ionized donor concentration in the wide bandgap material and 2DEG concentration adjacent to the heterointerface. When the associated interfacial field is sufficiently strong, carriers are confined within λ_{dB} and electron states are quantized into sub-bands (i.e., E_0 and E_1), as shown in Fig. 2.18b.

Figure 2.19 shows the contributions of component scattering mechanisms to the low temperature mobility of a 2DEG formed at a $Ga_{0.70}Al_{0.3}As$-GaAs heterointerface, as a function of the electron gas density. As for bulk samples, the most important mechanism limiting

Fig. 2.19 Dependence of 2DEG electron mobility on carrier concentration for a $Ga_{0.70}Al_{0.3}As$-GaAs heterostructure [2.73]

the low temperature mobility is ionized impurity scattering, except at high electron densities, where so-called *alloy disorder scattering* is significant. Ionized impurity scattering may be further broken down into scattering from ionized impurities that are with the GaAs quantum well, known as *background impurities*, those beyond the spacer region, termed *remote impurity scattering*. For high purity growth, the unintentional background impurity concentration can be kept to very low limits and impurity scattering based mobility values are then dictated by remote impurity scattering. Since carriers in the well are only weakly scattered by the tail field of these remote Coulomb centers, the mobility of such 2DEG systems can be orders of magnitude higher than bulk samples. The temperature dependence of the electron mobility for such a system is shown in Fig. 2.20. Notice how, similar to bulk sample, increasing temperature

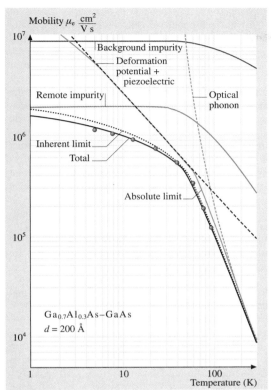
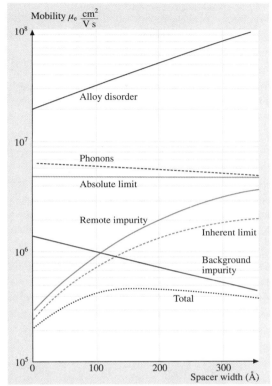

Fig. 2.20 Dependence of 2DEG electron mobility on temperature for a $Ga_{0.70}Al_{0.3}As$-GaAs heterostructure [2.73]

Fig. 2.21 Dependence of 2DEG electron mobility on spacer width for a $Ga_{0.70}Al_{0.3}As$-GaAs heterostructure [2.73]

increases the phonon population such that for the example shown, above about 100 K, polar optical phonon scattering controls the mobility.

Figure 2.21 shows the dependence of the 2DEG mobility on the spacer width. Two competing factors are active – for narrow spacer widths, the transfer efficiency of carriers to the GaAs well is high and so a lower remote doping concentration is sufficient to provide for a given constant 2DEG concentration, but since the Coulomb scatters are so close to the 2DEG, they scatter very efficiently and limit the mobility. On the other hand, for large spacer widths, carrier transfer efficiency is quite poor requiring higher remote doping to supply the given 2DEG concentration; however, being relatively far away each of the scattering centers are less effective at lowering the mobility, but given their high concentration, the net effect is still a decrease in the mobility at large spacer widths as seen in the figure.

2.13 One Dimensional Conductance

In the case where carriers are confined within regions of width $L_x, L_y < \lambda_{dB}$ in two directions x and y, respectively, the electron energy in those two directions become quantized with quantum numbers n_x and n_y, respectively. In the third direction z, electrons travel as Bloch waves with energy that may be approximated by $\hbar^2 k_z^2 / 2m^*$ giving an expression for the total energy E of the so-called *1-D electron system* or *quantum wire* as:

$$E(n_x, n_y, n_z) = \frac{\hbar^2 \pi^2}{2m^*}\left(\frac{n_x^2}{L_x^2} + \frac{n_y^2}{L_y^2}\right) + \frac{\hbar^2 k_z^2}{2m^*}. \quad (2.84)$$

The associated electron wavefunctions are:

$$\Psi(x,y,z) = \frac{1}{2\sqrt{L_x L_y L_z}} \sin\left(\frac{n_x \pi x}{L_x}\right)$$
$$\times \sin\left(\frac{n_y \pi y}{L_y}\right) e^{ik_z z}. \quad (2.85)$$

Using these equations, one can readily derive an expression for the density of states per unit energy range:

$$\text{DOS} = 2 \times 2 \left(\frac{L_z}{2\pi}\right) (\nabla_{k_z} E)^{-1}$$
$$= \frac{2L_z}{h} \sqrt{\frac{m^*}{2(E - E_{n_x,n_y})}}. \quad (2.86)$$

In order to evaluate the conductance of this quantum wire, consider the influence of a weak applied potential V. Similar to the case for bulk transport the applied field displaces the *Fermi surface* and results in a change in the electron wave-vector from k_0 (i.e., with no applied potential) to k_V (i.e., when the potential is applied). When V is small compared with the electron energy:

$$k_0 = \sqrt{\frac{2m^*(E - E_{n_x,n_y})}{\hbar^2}}, \quad (2.87)$$

$$k_V = k_0 \sqrt{1 + \frac{eV}{E - E_{n_x,n_y}}}$$
$$\approx k_0 \left(1 + \frac{1}{2} \frac{eV}{E - E_{n_x,n_y}}\right). \quad (2.88)$$

This leads to establishing a current density J in the wire

$$J = \frac{2e^2 \, (\text{DOS}) \sqrt{(E_F - E_{n_x,n_y})}}{\sqrt{2m^*}}. \quad (2.89)$$

Which may be simplified to the following expressions for J and the current flowing in the wire for a given quantum state $E\{n_x,n_y\}$, I

$$J_{n_x,n_y} = \frac{2e^2 V L_z}{h} \quad \text{and} \quad I_{n_x,n_y} = \frac{2e^2 V}{h}. \quad (2.90)$$

The expression for the conductance through one channel corresponding to a given quantum state $\{n_x, n_y\}$ is then given by

$$G_{n_x,n_y} = \frac{I_{n_x,n_y}}{V} = \frac{2e^2}{h}. \quad (2.91)$$

Notice how the conductance is quantized in units of e^2/h with each populated channel contributing equally to the conductance – moreover, this is a fundamental result, being independent of the material considered. In practice, deviations from this equation can occur (although generally less than 1%) owing to the finiteness of real nanowires and impurities in or near the channel, influencing the conductivity and even resulting in weak localization. Generally, unlike both bulk and 2DEG systems, ionized impurity scattering is suppressed in nanowires. The main reason for this is that an incident electron in a quantum state $\{n_x,n_y\}$ traveling along the wire with wave-vector $k_z\{n_x,n_y\}$, can not be elastically scattered into any states except those in a small region of k-space in the vicinity of $-k_z\{n_x,n_y\}$. Such a scattering event involves a large change in momentum of $\approx 2k_z\{n_x,n_y\}$ and thus, the probability of such events is very small. As a result, the mean free path and mobility of carriers in such quantum wires are substantially increased.

The nature of carrier transport in quantum wires depends on the wire dimensions (i.e., length L_Wire and diameter d_Wire) as compared with the carrier mean free path, l_Carrier. When $l_\text{Carrier} \gg L_\text{Wire}, d_\text{Wire}$ the only potential seen by the carriers is that associated with the wire walls, and carriers exhibit wavelike behavior, being guided through the wire as if it were a waveguide without any internal scattering. Conversely, if $d_\text{Wire} \ll \lambda_\text{DeBroglie}$, only a few energy states in the wire are active, and in the limit of an extremely small waveguide, only one state or channel is active, analogous to a single mode waveguide cavity – this case is termed *quantum ballistic transport*. In the limit, $l_\text{Carrier} \ll L_\text{Wire}, d_\text{Wire}$, scattering dominates transport throughout the wire – with numerous scattering events occurring before a carrier can traverse the wire or move far along its length. In such a case the transport is said to be *diffusive*. As discussed previously, ionized impurity and lattice scattering contribute to l_Carrier, with l_Carrier decreasing with increasing temperature due to phonon scattering. For strong impurity scattering, this may not occur until relatively high temperatures. In the intermediate case of $L_\text{Wire} \gg l_\text{Carrier} \gg d_\text{Wire}$ and where $d_\text{Wire} \ll \lambda_\text{DeBroglie}$ scattering is termed "mixed mode" and is often called *quasi-ballistic*.

2.14 The Quantum Hall Effect

The observation of, and first explanation for the Hall Effect in a 2DEG by von *Klitzing* et al. [2.74], won them a Nobel Prize. As shown in Fig. 2.22 the Hall resistivity exhibits plateaus for integer values of h/e^2, independent of any material dependent parameters. This discovery was later shown to be correct to a precision of at least one part in 10^7, enabling extremely accurate determinations of the fine structure constant (i. e., $\alpha = (\mu_0 c e^2/2h) \approx 1/137$) and a fundamental resistance standard to be established.

For the six point Hall geometry used (as shown in the insert to Fig. 2.18), one can define the *Hall resistivity* $\rho_{xy} = g_1 V_{H,y}/I_x$ and *longitudinal resistivity* $\rho_{xx} = g_2 V_{L,x}/I_x$ where g_1 and g_2 are geometric constants related to the sample geometry. These resistivities are related to corresponding conductivities through the conductivity and resistivity tensors

$$\rho_{xy} = \frac{\sigma_{xy}}{\sigma_{xx}^2 + \sigma_{xy}^2},$$
$$\rho_{xx} = \frac{\sigma_{xx}}{\sigma_{xx}^2 + \sigma_{xy}^2},$$
$$\rho_{yx} = -\rho_{xy},$$
$$\rho_{yy} = \rho_{xx}. \qquad (2.92)$$

Starting from a classical equation of motion for electrons in an electric field E_x, magnetic field B_z, and defining the *cyclotron frequency* $\omega_c = eB_z/m^*$, the velocities perpendicular to the applied magnetic field can be deduced as

$$v_x = \frac{E_x}{B_x} \sin \omega_c t \quad \text{and} \quad v_y = \frac{E_x}{B_x}(\cos \omega_c t - 1) \qquad (2.93)$$

with the time averaged velocities

$$\langle v_x \rangle = \frac{E_x}{B_x} \quad \text{and} \quad \langle v_y \rangle = \frac{E_x}{B_x} \qquad (2.94)$$

leading to Hall and longitudinal resistivities of

$$\sigma_{xx} = \sigma_{yy} = 0 \quad \text{and} \quad \sigma_{xy} = -\sigma_{yx} = \frac{N_s e}{B_z}, \qquad (2.95)$$

where N_S is the areal electron concentration. Below, a quantum approach is used to establish a relationship for the electron concentration N_s. Note that the motion of the electrons in the crossed fields are quantized with allowed levels termed *Landau levels* E_n:

$$E_n = \left(n + \frac{1}{2}\right)\hbar\omega_c + g^*\mu_B B_z + \varepsilon_z, \qquad (2.96)$$

where n is the quantum number describing the particular Landau Level, g^* is the *Landé factor*, μ_B is the Bohr magneton and $g^*\mu_B B_z$ is *the spin magnetic energy*, and E_z is the energy associated with the z-motion of the carriers. xy-plane carrier motion is characterized by the cyclotron energy term E_{xy},

$$E_{xy} = \frac{\hbar^2 k_{xy}^2}{2m_e^*} = \left(n + \frac{1}{2}\right)\hbar\omega_c. \qquad (2.97)$$

Following this description and noting that the motion of electrons in the xy-plane may be expressed in terms of wavefunctions of the *harmonic oscillator* using the *Landau gauge* of the *vector potential* $[0, xB_z, 0]$, we may write the density of states per unit area, DOS_A as:

$$DOS_A = \frac{m_e^* \omega_c L_x L_y}{2\pi\hbar}. \qquad (2.98)$$

Since the degeneracy of each Landau level is one (i. e., since they are single spin states), this enables one to find N_s assuming Landau state filling up to the pth level (for integer p):

$$N_s = \frac{m_e^* \omega_c p}{2\pi\hbar}. \qquad (2.99)$$

Using the definition of the cyclotron frequency gives the final form

$$N_s = \frac{peB_z}{\hbar} \qquad (2.100)$$

which may be used to rewrite the previous expressions for Hall and longitudinal resistivity

$$\rho_{xx} = \rho_{yy} = 0 \quad \text{and} \quad \rho_{xy} = -\rho_{yx} = \frac{h}{pe^2}. \qquad (2.101)$$

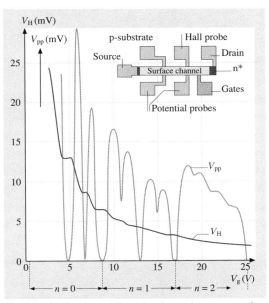

Fig. 2.22 Hall voltage V_H and voltage drop across electrodes V_{PP} as a function of gate voltage V_g at 1.5 K, when $B = 18$ T. Source-drain current is 1 μA. *Insert* shows plan view of a device with length 400 μm, width 50 μm and an interprobe separation of 130 μm [2.74]

Equation (2.101) shows that Hall resistivity is quantized in units of h/pe^2 whenever the Fermi energy lies between filled Landau levels. Consistent with observation, the result is independent of the semiconductor being studied. Although this model provides an excellent basis for understanding experiments, understanding the details of the results (i. e., in particular the existence of a finite width for the Hall effect plateaus and zero longitudinal resistance dips) requires a more complete treatment involving so-called localized states.

References

2.1 P. L. Rossiter: *The Electrical Resisitivity of Metals and Alloys* (Cambridge Univ. Press, Cambridge 1987)
2.2 J. S. Dugdale: *The Electrical Properties of Metals and Alloys* (Arnold, London 1977)
2.3 B. R. Nag: *Theory of Electrical Transport in Semiconductors* (Pergamon, Oxford 1972)
2.4 F. J. Blatt: *Physics of Electronic Conduction in Solids* (McGraw-Hill, New York 1968) Chap. 5, 6
2.5 S. O. Kasap: *Principles of Electronic Materials and Devices*, 3 edn. (McGraw-Hill, New York 2005)
2.6 G. T. Dyos, T. Farrell (Eds.): *Electrical Resistivity Handbook* (Peregrinus, London 1992)
2.7 D. G. Fink, D. Christiansen (Eds.): *Electronics Engineers' Handbook*, 2 edn. (McGraw-Hill, New York 1982) Section 6
2.8 L. Nordheim: Ann. Phys. **9**, 664 (1931)
2.9 J. K. Stanley: *Electrical and Magnetic Properties of Metals* (American Society for Metals, Metals Park 1963)
2.10 M. Hansen, K. Anderko: *Constitution of Binary Alloys*, 2 edn. (McGraw-Hill, New York 1985)
2.11 H. E. Ruda: J. Appl. Phys. **59**, 1220 (1986)
2.12 M. Lundstrom: *Fundamentals of Carrier Transport* (Cambridge Univ. Press, Cambridge 2000)
2.13 R. H. Bube: *Electronic Properties of Crystalline Solids* (Academic, New York 1974) Chap. 7
2.14 S. Riedel, J. Röber, T. Geßner: Microelectron. Eng., **33**, 165 (1997)
2.15 A. F. Mayadas, M. Shatzkes: Phys. Rev. B, **1**, 1382(1970)
2.16 J.-W. Lim, K. Mimura, M. Isshiki: Appl. Surf. Sci. **217**, 95 (2003)
2.17 C. R. Tellier, C. R. Pichard, A. J. Tosser: J. Phys. F, **9**, 2377 (1979) (and references therein)
2.18 K. Fuchs: Proc. Camb. Philos. Soc., **34**, 100 (1938)
2.19 E. H. Sondheimer: Adv. Phys., **1**, 1 (1952)
2.20 H.-D. Liu, Y.-P. Zhao, G. Ramanath, S. P. Murarka, G.-C. Wang: Thin Solid Films **384**, 151 (2001)
2.21 R. Suri, A. P. Thakoor, K. L. Chopra: J. Appl. Phys., **46**, 2574 (1975)
2.22 R. H. Cornely, T. A. Ali: J. Appl. Phys., **49**, 4094(1978)
2.23 J. S. Ahn, K. H. Kim, T. W. Noh, D. H. Riu, K. H. Boo, H. E. Kim: Phys. Rev. B, **52**, 15244 (1995)
2.24 R. J. Gehr, G. L. Fisher, R. W. Boyd: J. Opt. Soc. Am. B, **14**, 2310 (1997)
2.25 D. E. Aspnes, J. B. Theeten, F. Hottier: Phys. Rev. B, **20**, 3292 (1979)
2.26 Z. Yin, F. W. Smith: Phys. Rev. B, **42**, 3666 (1990)
2.27 M. F. MacMillan, R. P. Devaty, W. J. Choyke, D. R. Goldstein, J. E. Spanier, A. D. Kurtz: J. Appl. Phys., **80**, 2412 (1996)
2.28 C. Ganter, W. Schirmacher: Phys. Status Solidi B, **218**, 71 (2000)
2.29 R. Stognienko, Th. Henning, V. Ossenkopf.: Astron. Astrophys. **296**, 797 (1995)
2.30 A. G. Rojo, H. E. Roman: Phys. Rev. B, **37**, 3696 (1988)
2.31 J. A. Reynolds, J. M. Hough: Proc. Phys. Soc., **70**, 769 (1957)
2.32 R. Clausius: *Die Mechanische Wärmetheorie*, Vol. 2 (Wieveg, Braunschweig 1879)
2.33 L. Lorenz: Ann. Phys. Lpz., **11**, 70 (1880)
2.34 O. F. Mosotti: Mem. Math. Fisica Modena II, **24**, 49 (1850)
2.35 V. I. Odelevskii: Zh. Tekh. Fiz., **6**, 667 (1950)
2.36 Lord Rayleigh: Philos. Mag., **34**, 481 (1892)
2.37 K. W. Wagner: Arch. Electrochem., **2**, 371 (1914)
2.38 D. A. G. Bruggeman: Ann. Phys. Lpz. **24**, 636 (1935)
2.39 C. J. F. Bottcher: Rec. Trav. Chim. Pays-Bas **64**, 47 (1945)
2.40 H. Fricke: Phys. Rev. **24**, 575 (1924)
2.41 D. Polder, J. M. Van Santen: Physica **12**, 257 (1946)
2.42 W. Niesel: Ann. Phys. Lpz. **10**, 336 (1952)
2.43 J. A. Stratton: *Electromagnetic Theory* (McGraw-Hill, New York 1941)
2.44 O. Wiener: Abh. Sachs. Ges. Akad. Wiss. Math. Phys. **32**, 509 (1912)
2.45 L. Silberstein: Ann. Phys. Lpz. **56**, 661 (1895)
2.46 O. Wiener: Abh. Sachs. Ges. Akad. Wiss. Math. Phys. **32**, 509 (1912)
2.47 R. W. Sillars: J. Inst. Elect. Eng. **80**, 378 (1937)
2.48 F. Ollendorf: Arch. Electrochem. **25**, 436 (1931)
2.49 J. C. M. Maxwell-Garnett: Phil. Trans. R. Soc. Lond. **203**, ,385 (1904)
2.50 H. Looyenga: Physica **31**, 401 (1965)
2.51 J. Monecke: J. Phys. Condens. Mat. **6**, 907 (1994)
2.52 C. F. Bohren, D. R. Huffman: *Absorption and Scattering of Light by Small Particles* (Wiley, New York 1983)
2.53 P. Y. Yu, M. Cardona: *Fundamentals of Semiconductors* (Springer, Berlin, Heidelberg 1996)

2.54 M. Akiyama, M. Hanada, H. Takao, K. Sawada, M. Ishida: Jpn. J. Appl. Phys **41**, 2552 (2002)
2.55 K. Tsuji, Y. Takasaki, T. Hirai, K. Taketoshi: J. Non-Cryst. Solids **14**, 94 (1989)
2.56 G. Juska, K. Arlauskas: Phys. Status Solidi **77**, 387 (1983)
2.57 R. A. Logan, H. G. White: J. Appl. Phys. **36**, 3945 (1965)
2.58 R. Ghin, J. P. R. David, S. A. Plimmer, M. Hopkinson, G. J. Rees, D. C. Herbert, D. R. Wight: IEEE Trans. Electron Dev. **ED45**, 2096 (1998)
2.59 S. A. Plimmer, J. P. R. David, R. Grey, G. J. Rees: IEEE Trans. Electron Dev. **ED47**, 21089 (2000)
2.60 L. W. Cook, G. E. Bulman, G. E. Stillma: Appl. Phys. Lett. **40**, 589 (1982)
2.61 C. A. Lee, R. A. Logan, R. L. Batdorf, J. J. Kleimack, W. Wiegmann: Phys. Rev. **134**, B766 (1964)
2.62 C. Bulutay: Semicond. Sci. Technol. **17**, L59 (2002)
2.63 G. Juska, K. Arlauskas: Phys. Status Solidi **59**, 389 (1980)
2.64 W. Shockley: Solid State Electron. **2**, 35 (1961)
2.65 G. A. Baraff: Phys. Rev. **128**, 2507 (1962)
2.66 B. K. Ridley: J. Phys. C **16**, 4733 (1983)
2.67 M. G. Burt: J. Phys. C **18**, L477 (1985)
2.68 S. MacKenzie, M. G. Burt: Semicond. Sci. Technol. **2**, 275 (1987)
2.69 B. K. Ridley: Semicond. Sci. Technol. **2**, 116 (1987)
2.70 J. S. Marsland: Solid State Electron. **30**, 125 (1987)
2.71 J. S. Marsland: Semicond. Sci. Technol. **5**, 177 (1990)
2.72 S. O. Kasap, J. A. Rowlands, S. D. Baranovskii, K. Tanioka: J. Appl. Phys. **96**, 2037 (2004)
2.73 W. Walukiewicz, H. E. Ruda, J. Lagowski, H. C. Gatos: Phys. Rev. B **30**, 4571 (1984)
2.74 K. V. Klitzing, G. Dorda, M. Pepper: Phys. Rev. Lett. **45**, 494 (1980)

3. Optical Properties of Electronic Materials: Fundamentals and Characterization

Light interacts with materials in a variety of ways; this chapter focuses on refraction and absorption. Refraction is characterized by a material's refractive index. We discuss some of the most useful models for the frequency dependence of the refractive index, such as those due to Cauchy, Sellmeier, Gladstone–Dale, and Wemple–Di Dominico. Examples are given of the applicability of the models to actual materials. We present various mechanisms of light absorption, including absorption by free carriers, phonons, excitons and impurities. Special attention is paid to fundamental and excitonic absorption in disordered semiconductors and to absorption by rare-earth, trivalent ions due to their importance to modern photonics. We also discuss the effect of an external electric field on absorption, and the Faraday effect. Practical techniques for determining the optical parameters of thin films are outlined. Finally, we present a short technical classification of optical glasses and materials.

3.1 **Optical Constants** 47
    3.1.1 Refractive Index and Extinction Coefficient 47
    3.1.2 Kramers–Kronig Relations 49

3.2 **Refractive Index** 50
    3.2.1 Cauchy Dispersion Equation 50
    3.2.2 Sellmeier Dispersion Equation 51
    3.2.3 Gladstone–Dale Formula 51
    3.2.4 Wemple–Di Dominico Dispersion Relation 52
    3.2.5 Group Index (N) 53

3.3 **Optical Absorption** 53
    3.3.1 Lattice or Reststrahlen Absorption and Infrared Reflection 54
    3.3.2 Free Carrier Absorption (FCA) 55
    3.3.3 Band-to-Band or Fundamental Absorption 57
    3.3.4 Exciton Absorption 63
    3.3.5 Impurity Absorption 66
    3.3.6 Effects of External Fields 69

3.4 **Thin Film Optics** 70
    3.4.1 Swanepoel's Analysis of Optical Transmission Spectra ... 71
    3.4.2 Ellipsometry 72

3.5 **Optical Materials** 74
    3.5.1 Abbe Number or Constringence ... 74
    3.5.2 Optical Materials 74
    3.5.3 Optical Glasses 76

References 76

3.1 Optical Constants

The changes that light undergoes upon interacting with a particular substance are known as the optical properties of that substance. These optical properties are influenced by the macroscopic and microscopic properties of the substance, such as the nature of its surface and its electronic structure. Since it is usually far easier to detect the way a substance modifies light than to investigate its macroscopic and microscopic properties directly, the optical properties of a substance are often used to probe other properties of the material. There are many optical properties, including the most well known: reflection, refraction, transmission and absorption. Many of these optical properties are associated with important optical constants, such as the refractive index and the extinction coefficient. In this section we review these optical constants, such as the refractive index and the extinction coefficient. Books by *Adachi* [3.1], *Fox* [3.2] and *Simmons* and *Porter* [3.3] are highly recommended. In addition, *Adachi* also discusses the optical properties of III–V compounds in this handbook.

3.1.1 Refractive Index and Extinction Coefficient

The refractive index n of an optical or dielectric medium is the ratio of the velocity of light c in vacuum to its velocity v in the medium; $n = c/v$. Using this and Maxwell's equations, one obtains the well-known formula for the refractive index of a substance as $n = \sqrt{\varepsilon_r \mu_r}$, where ε_r is the static constant or relative permittivity and μ_r is the magnetic permeability. As $\mu_r = 1$ for nonmagnetic substances, one gets $n = \sqrt{\varepsilon_r}$, which is very useful for relating the dielectric properties of a material to its optical properties at any particular frequency of interest. Since ε_r depends on the wavelength of the light, the refractive index depends on it too, which is called dispersion. In addition to dispersion, an electromagnetic wave propagating through a lossy medium (one that absorbs or scatters radiation passing through it) experiences attenuation, which means that it loses its energy due to various loss mechanisms such as the generation of phonons (lattice waves), photogeneration, free carrier absorption and scattering. In such materials, the refractive index is a complex function of the frequency of the light wave.

The complex refractive index N, with real part n and imaginary part K (called the extinction coefficient), is related to the complex relative permittivity, $\varepsilon_r = \varepsilon_r' - i\varepsilon_r''$ by:

$$N = n - iK = \sqrt{\varepsilon_r} = \sqrt{\varepsilon_r' - i\varepsilon_r''} \quad (3.1)$$

where ε_r' and ε_r'' are, respectively, the real and imaginary parts of ε_r. Equation (3.1) gives:

$$n^2 - K^2 = \varepsilon_r' \quad \text{and} \quad 2nK = \varepsilon_r'' . \quad (3.2)$$

The optical constants n and K can be determined by measuring the reflectance from the surface of a material as a function of polarization and the angle of incidence. For normal incidence, the reflection coefficient r is obtained as:

$$r = \frac{1-N}{1+N} = \frac{1-n+iK}{1+n-iK} . \quad (3.3)$$

The reflectance R is then defined by:

$$R = |r|^2 = \left|\frac{1-n+iK}{1+n-iK}\right|^2 = \frac{(1-n)^2 + K^2}{(1+n)^2 + K^2} . \quad (3.4)$$

Notice that whenever K is large, for example over a range of wavelengths, the absorption is strong, and the reflectance is almost unity. The light is then reflected, and any light in the medium is highly attenuated.

Optical properties of materials are typically presented by showing either the frequency dependence (dispersion relation) of n and K or ε_r' and ε_r''. An intuitive guide to explaining dispersion in insulators is based on a single oscillator model in which the electric field in the light induces forced dipole oscillations in the material (displaces the electron shells to oscillate about the nucleus) with a single resonant frequency ω_0 [3.1]. The frequency dependences of ε_r' and ε_r'' are then obtained as:

$$\varepsilon_r' = 1 + \frac{N_{at}}{\varepsilon_0}\alpha_e' \quad \text{and} \quad \varepsilon_r'' = 1 + \frac{N_{at}}{\varepsilon_0}\alpha_e'' , \quad (3.5)$$

where N_{at} is the number of atoms per unit volume, ε_0 is the vacuum permittivity, and α_e' and α_e'' are the real and imaginary parts of the electronic polarizability, given respectively by:

$$\alpha_e' = \alpha_{e0} \frac{1 - (\omega/\omega_0)^2}{[1-(\omega/\omega_0)^2]^2 + (\gamma/\omega_0)^2(\omega/\omega_0)^2} ,$$
(3.6a)

and

$$\alpha_e'' = \alpha_{e0} \frac{(\gamma/\omega_0)(\omega/\omega_0)}{[1-(\omega/\omega_0)^2]^2 + (\gamma/\omega_0)^2(\omega/\omega_0)^2} ,$$
(3.6b)

where α_{e0} is the dc polarizability corresponding to $\omega = 0$ and γ is the loss coefficient. Using (3.1), (3.2) and (3.5), it is then possible to study (3.6a, 6b), the frequency dependence of n and K. Figure 3.1a shows the dependence of n and K on the normalized frequency ω/ω_0 for a simple single electronic dipole oscillator of resonance frequency ω_0. It is seen that n and K peak close to $\omega = \omega_0$. If a material has a $\varepsilon_r'' \gg \varepsilon_r'$, then $\varepsilon_r \approx -i\varepsilon_r''$ and $n = K \approx \sqrt{\varepsilon_r''/2}$ is obtained from (3.2). Figure 3.1b shows the dependence of the reflectance R on the frequency. It is observed that R reaches its

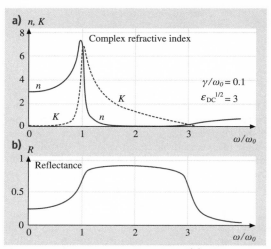

Fig. 3.1a,b The dipole oscillator model. (**a**) Refractive index and extinction coefficient versus normalized frequency. (**b**) Reflectance versus normalized frequency

maximum value at a frequency slightly above $\omega = \omega_0$, and then remains high until ω reaches nearly $3\omega_0$; so the reflectance is substantial while absorption is strong. The normal dispersion region is the frequency range below ω_0 where n falls as the frequency decreases; that is, n decreases as the wavelength λ increases. The anomalous dispersion region is the frequency range above ω_0, where n decreases as ω increases. Below ω_0, K is small and if ε_{dc} is $\varepsilon_r(0)$, the refractive index becomes:

$$n^2 \approx 1 + (\varepsilon_{dc} - 1)\frac{\omega_0^2}{\omega_0^2 - \omega^2}; \quad \omega < \omega_0 . \tag{3.7}$$

Since, $\lambda = 2\pi c/\omega$, defining $\lambda_0 = 2\pi c/\omega_0$ as the resonance wavelength, one gets:

$$n^2 \approx 1 + (\varepsilon_{dc} - 1)\frac{\lambda^2}{\lambda^2 - \lambda_0^2}; \quad \lambda > \lambda_0 . \tag{3.8}$$

While intuitively useful, the dispersion relation in (3.8) is far too simple. More rigorously, we have to consider the dipole oscillator quantum-mechanically, which means that a photon excites the oscillator to a higher energy level; see, for example, [3.2, 3]. The result is that we have a series of $\lambda^2/(\lambda^2 - \lambda_i^2)$ terms with various weighting factors A_i that add to unity, where the λ_i represent different resonance wavelengths. The weighting factors A_i involve quantum-mechanical matrix elements.

Figure 3.2 shows the complex relative permittivity and the complex refractive index of crystalline silicon in terms of photon energy $h\nu$. For photon energies below the bandgap energy (1.1 eV), both ε_r'' and K are negligible and n is close to 3.7. Both ε_r'' and K increase and change strongly as the photon energy becomes greater than 3 eV, far beyond the bandgap energy. Notice that both ε_r' and n peak at $h\nu \approx 3.5$ eV, which corresponds to a direct photoexcitation process – the excitation of electrons from the valence band to the conduction band – as discussed later.

3.1.2 Kramers–Kronig Relations

If we know the frequency dependence of the real part, ε_r', of the relative permittivity of a material, then we can use the Kramers–Kronig relations to determine the frequency dependence of the imaginary part ε_r'', and vice versa. The transform requires that we know the frequency dependence of either the real or imaginary part over a frequency range that is as wide as possible, ideally from dc to infinity, and that the material is linear; in

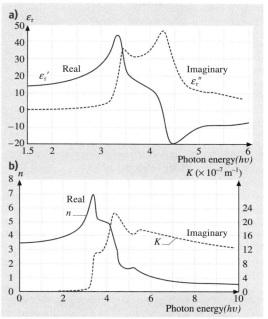

Fig. 3.2 (a) Complex relative permittivity of a silicon crystal as a function of photon energy plotted in terms of real (ε_r') and imaginary (ε_r'') parts. (b) Optical properties of a silicon crystal versus photon energy in terms of real (n) and imaginary (K) parts of the complex refractive index. After [3.4]

other words it has a relative permittivity that is independent of the applied field. The Kramers–Kronig relations for $\varepsilon_r = \varepsilon_r' + i\varepsilon_r''$ are given by [3.4]:

$$\varepsilon_r'(\omega) = 1 + \frac{2}{\pi} P \int_0^\infty \frac{\omega' \varepsilon_r''(\omega')}{\omega'^2 - \omega^2} d\omega' , \tag{3.9a}$$

and

$$\varepsilon_r''(\omega) = -\frac{2\omega}{\pi} P \int_0^\infty \frac{\varepsilon_r'(\omega')}{\omega'^2 - \omega^2} d\omega' , \tag{3.9b}$$

where ω' is the integration variable, P represents the Cauchy principal value of the integral and the singularity at $\omega = \omega'$ is avoided.

Similarly, one can relate the real and imaginary parts of the polarizability, $\alpha'(\omega)$ and $\alpha''(\omega)$, and those of the complex refractive index, $n(\omega)$ and $K(\omega)$ as well. For $\alpha'(\omega)$ and $\alpha''(\omega)$, one can analogously write:

$$\alpha'(\omega) = \frac{2}{\pi} P \int_0^\infty \frac{\omega' \alpha''(\omega')}{\omega'^2 - \omega^2} d\omega' \quad (3.10a)$$

and

$$\alpha''(\omega') = -\frac{2\omega}{\pi} P \int_0^\infty \frac{\alpha'(\omega')}{\omega'^2 - \omega^2} d\omega'. \quad (3.10b)$$

3.2 Refractive Index

There are several simplified models describing the spectral dependence of the refractive index n.

3.2.1 Cauchy Dispersion Equation

The dispersion relationship for the refractive index (n) versus the wavelength of light (λ) is stated in the following form:

$$n = A + \frac{B}{\lambda^2} + \frac{C}{\lambda^4}, \quad (3.11)$$

where A, B and C are material-dependent specific constants. The Cauchy equation (3.11) is typically used in the visible region of the spectrum for various optical glasses, and is applied to normal dispersion. The third term is sometimes dropped for a simpler representation of n versus λ behavior. The original expression was a series in terms of the wavelength, λ, or frequency, ω, of light as:

$$n = a_0 + a_2\lambda^{-2} + a_4\lambda^{-4} + a_6\lambda^{-6} + \ldots \lambda > \lambda_h, \quad (3.12a)$$

or

$$n = n_0 + n_2\omega^2 + n_4\omega^4 + n_6\omega^6 + \ldots \omega < \omega_h, \quad (3.12b)$$

where $\hbar\omega$ is the photon energy, $\hbar\omega_h = hc/\lambda_h$ is the optical excitation threshold (the bandgap energy), while a_0, $a_2 \ldots$, and $n_0, n_2, \ldots$ are constants. A Cauchy relation of the following form:

$$n = n_{-2}(\hbar\omega)^{-2} + n_0 + n_2(\hbar\omega)^2 + n_4(\hbar\omega)^4, \quad (3.13)$$

can be used satisfactorily over a wide range of photon energies. The dispersion parameters, calculated from (3.13), of a few materials are listed in Table 3.1.

Table 3.1 Cauchy's dispersion parameters (obtained from (3.11)) for a few materials

Material	$\hbar\omega$ (eV)	n_{-2} (eV^2)	n_0	n_2 (eV^{-2})	n_4 (eV^{-4})
Diamond	0.05 to 5.47	-1.07×10^{-5}	2.378	8.01×10^{-3}	1.04×10^{-4}
Si	0.002 to 1.08	-2.04×10^{-8}	3.4189	8.15×10^{-2}	1.25×10^{-2}
Ge	0.002 to 0.75	-1.00×10^{-8}	4.0030	2.20×10^{-1}	1.40×10^{-1}

Table 3.2 Sellmeier coefficients of a few materials (λ_1, λ_2, λ_3 are in μm)

Material	A_1	A_2	A_3	λ_1	λ_2	λ_3
SiO$_2$ (fused silica)	0.696749	0.408218	0.890815	0.0690660	0.115662	9.900559
86.5%SiO$_2$-13.5%GeO$_2$	0.711040	0.451885	0.704048	0.0642700	0.129408	9.425478
GeO$_2$	0.80686642	0.71815848	0.85416831	0.068972606	0.15396605	11.841931
Barium fluoride	0.63356	0.506762	3.8261	0.057789	0.109681	46.38642
Sapphire	1.023798	1.058264	5.280792	0.0614482	0.110700	17.92656
Diamond	0.3306	4.3356		0.175	0.106	
Quartz, n_0	1.35400	0.010	0.9994	0.092612	10.700	9.8500
Quartz, n_e	1.38100	0.0100	0.9992	0.093505	11.310	9.5280
KDP, n_0	1.2540	0.0100	0.0992	0.09646	6.9777	5.9848
KDP, n_e	1.13000	0.0001	0.9999	0.09351	7.6710	12.170

Cauchy's dispersion relation, given in (3.13), was originally called the elastic-ether theory of the refractive index [3.5–7]. It has been widely used for many materials, although in recent years it has been largely replaced by the Sellmeier equation, which we consider next.

3.2.2 Sellmeier Dispersion Equation

The dispersion relationship can be quite complicated in practice. An example of this is the Sellmeier equation, which is an empirical relation between the refractive index n of a substance and the wavelength λ of light in the form of a series of $\lambda^2/(\lambda^2 - \lambda_i^2)$ terms, given by:

$$n^2 = 1 + \frac{A_1 \lambda^2}{\lambda^2 - \lambda_1^2} + \frac{A_2 \lambda^2}{\lambda^2 - \lambda_2^2} + \frac{A_3 \lambda^2}{\lambda^2 - \lambda_3^2} + \cdots, \quad (3.14)$$

where λ_i is a constant and A_1, A_2, A_3, λ_1, λ_2 and λ_3 are called Sellmeier coefficients, which are determined by fitting this expression to the experimental data. The full Sellmeier formula has more terms of similar form, such as $A_i \lambda^2/(\lambda^2 - \lambda_i^2)$, where $i = 4, 5, \ldots$ but these can generally be neglected when considering n versus λ behavior over typical wavelengths of interest and by ensuring that the three terms included in the Sellmeier equation correspond to the most important or relevant terms in the summation. Examples of Sellmeier coefficients for some materials, including pure silica (SiO_2) and 86.5 mol.%SiO_2-13.5 mol.% GeO_2, are given in Table 3.2. Two methods are used to find the refractive index of silica-germania glass $(SiO_2)_{1-x}(GeO_2)_x$: (a) a simple, but approximate, linear interpolation of the refractive index between known compositions, for example $n(x) - n(0.135) = (x - 0.135)[n(0.135) - n(0)]/0.135$ for $(SiO_2)_{1-x}(GeO_2)_x$, so $n(0.135)$ is used for 86.5 mol.%SiO_2-13.5 mol.% GeO_2 and $n(0)$ is used for SiO_2; (b) an interpolation for the coefficients A_i and λ_i between SiO_2 and GeO_2:

$$n^2 - 1 = \frac{\{A_1(S) + X[A_1(G) - A_1(S)]\}\lambda^2}{\lambda^2 - \{\lambda_1(S) + X[\lambda_1(G) - \lambda_1(S)]\}^2} + \cdots, \quad (3.15)$$

where X is the atomic fraction of germania, S and G in parentheses refer to silica and germania [3.10]. The theoretical basis of the Sellmeier equation is that the solid is represented as a sum of N lossless (frictionless) Lorentz oscillators such that each takes the form of $\lambda^2/(\lambda^2 - \lambda_i^2)$ with different λ_i, and each has different strengths, with weighting factors (A_i, $i = 1$ to N) [3.11, 12]. Knowledge of appropriate dispersion relationships is essential when designing photonic devices, such as waveguides.

There are other dispersion relationships that inherently take account of various contributions to optical properties, such as the electronic and ionic polarization and the interactions of photons with free electrons. For example, for many semiconductors and ionic crystals, two useful dispersion relations are:

$$n^2 = A + \frac{B\lambda^2}{\lambda^2 - C} + \frac{D\lambda^2}{\lambda^2 - E}, \quad (3.16)$$

and

$$n^2 = A + \frac{B}{\lambda^2 - \lambda_0^2} + \frac{C}{(\lambda^2 - \lambda_0^2)^2} + D\lambda^2 + E\lambda^4, \quad (3.17)$$

where A, B, C, D, E and λ_0 are constants particular to a given material. Table 3.3 provides a few examples.

The refractive index of a semiconductor material typically decreases with increasing bandgap energy E_g. There are various empirical and semi-empirical rules and expressions that relate n to E_g. In *Moss' rule*, n and E_g are related by $n^4 E_g = K =$ constant (≈ 100 eV). In the *Hervé–Vandamme* relationship [3.13],

$$n^2 = 1 + \left(\frac{A}{E_g + B}\right)^2, \quad (3.18)$$

where A and B are constants ($A \approx 13.6$ eV and $B \approx 3.4$ eV and $dB/dT \approx 2.5 \times 10^{-5}$ eV/K). The refractive index typically increases with increasing temperature. The temperature coefficient of the refractive index (TCRI) of a semiconductor can be found from the Hervé–Vandamme relationship [3.13]:

$$\text{TCRI} = \frac{1}{n} \cdot \frac{dn}{dT} = \frac{(n^2 - 1)^{3/2}}{13.6 n^2}\left(\frac{dE_g}{dT} + \frac{dB}{dT}\right). \quad (3.19)$$

Table 3.3 Parameters from Eqs. (3.16) and (3.17) for some selected materials (Si data from [3.8]; others from [3.9])

Material	λ_0(μm)	A	B(μm)^2	C(μm)$^{-4}$	D(μm)$^{-2}$	E(μm)$^{-4}$
Silicon	0.167	3.41983	0.159906	−0.123109	1.269×10^{-6}	-1.951×10^{-9}
MgO	0.11951	2.95636	0.021958	0	-1.0624×10^{-2}	-2.05×10^{-5}
LiF	0.16733	1.38761	0.001796	-4.1×10^{-3}	-2.3045×10^{-3}	-5.57×10^{-6}
AgCl	0.21413	4.00804	0.079009	0	-8.5111×10^{-4}	-1.976×10^{-7}

Table 3.4 Examples of parameters for the Wemple–DiDomenico dispersion relationship (3.23), for various materials

Material	N_c	Z_a	N_e	E_0(eV)	E_d(eV)	β(eV)	β	Comment
NaCl	6	1	8	10.3	13.6	0.28	β_i	Halides, LiF, NaF, etc.
CsCl	8	1	8	10.6	17.1	0.27	β_i	CsBr, CsI, etc.
TlCl	8	1	10	5.8	20.6	0.26	β_i	TlBr
CaF$_2$	8	1	8	15.7	15.9	0.25	β_i	BaF$_2$, etc.
CaO	6	2	8	9.9	22.6	0.24	β_i	Oxides, MgO, TeO$_2$, etc.
Al$_2$O$_3$	6	2	8	13.4	27.5	0.29	β_i	
LiNbO$_3$	6	2	8	6.65	25.9	0.27	β_i	
TiO$_2$	6	2	8	5.24	25.7	0.27	β_i	
ZnO	4	2	8	6.4	17.1	0.27	β_i	
ZnSe	4	2	8	5.54	27	0.42	β_c	II–VI, Zinc blende, ZnS, ZnTe, CdTe
GaAs	4	3	8	3.55	33.5	0.35	β_c	III–V, Zinc blende, GaP, etc.
Si (crystal)	4	4	8	4.0	44.4	0.35	β_c	Diamond, covalent bonding; C (diamond), Ge, β-SiC etc.
SiO$_2$ (crystal)	4	2	8	13.33	18.10	0.28	β_i	Average crystalline form
SiO$_2$ (amorphous)	4	2	8	13.38	14.71	0.23	β_i	Fused silica
CdSe	4	2	8	4.0	20.6	0.32	$\beta_i - \beta_c$	Wurtzite

TCRI is typically in the range 10^{-6} to 10^{-4}.

3.2.3 Gladstone–Dale Formula

The *Gladstone–Dale formula* is an empirical equation that allows the average refractive index n of an oxide glass to be calculated from its density ρ and its constituents as:

$$\frac{n-1}{\rho} = p_1 k_1 + p_2 k_2 + \cdots = \sum_{i=1}^{N} p_i k_i = C_{GD}, \quad (3.20)$$

where the summation is for various oxide components (each a simple oxide), p_i is the weight fraction of the i-th oxide in the compound, and k_i is the refraction coefficient that represents the polarizability of the i-th oxide. The right hand side of (3.20) is called the Gladstone–Dale coefficient C_{GD}. In more general terms, as a mixture rule for the overall refractive index, the Gladstone–Dale formula is frequently written as:

$$\frac{n-1}{\rho} = \frac{n_1 - 1}{\rho_1} w_1 + \frac{n_2 - 1}{\rho_2} w_2 + \cdots, \quad (3.21)$$

where n and ρ are the effective refractive index and effective density of the whole mixture, n_1, n_2, ... are the refractive indices of the constituents, and ρ_1, ρ_2, ... represent the densities of each constituent. Gladstone–Dale equations for the polymorphs of SiO$_2$ and TiO$_2$ give the average n values as [3.14]:

$$n(\text{SiO}_2) = 1 + 0.21\rho \quad \text{and} \quad n(\text{TiO}_2) = 1 + 0.40\rho. \quad (3.22)$$

3.2.4 Wemple–Di Dominico Dispersion Relation

The *Wemple–Di Dominico dispersion relation* is a semi-empirical single oscillator-based relationship used to find the refractive indices of a variety of materials for photon energies below the interband absorption edge, given by

$$n^2 = 1 + \frac{E_0 E_d}{E_0^2 - (h\nu)^2}, \quad (3.23)$$

where ν is the frequency, h is the Planck constant, E_0 is the single oscillator energy, E_d is the dispersion energy, which is a measure of the average strength of interband optical transitions; $E_d = \beta N_c Z_a N_e$ (eV), where N_c is the effective coordination number of the cation nearest-neighbor to the anion ($N_c = 6$ in NaCl, $N_c = 4$

in Ge), Z_a is the formal chemical valency of the anion ($Z_a = 1$ in NaCl; 2 in Te; 3 in GaP), N_e is the effective number of valence electrons per anion excluding the cores ($N_e = 8$ in NaCl, Ge; 10 in TlCl; 12 in Te; $9\frac{1}{3}$ in As$_2$Se$_3$), and β is a constant that depends on whether the interatomic bond is ionic (β_i) or covalent (β_c): $\beta_i = 0.26 \pm 0.04$ eV (this applies to halides such as NaCl and ThBr and most oxides, including Al$_2$O$_3$, for example), while $\beta_c = 0.37 \pm 0.05$ eV (applies to tetrahedrally bonded $A^N B^{8-N}$ zinc blende- and diamond-type structures such as GaP, ZnS, for instance). (Note that wurtzite crystals have a β that is intermediate between β_i and β_c.). Also, empirically, $E_0 = C E_g(D)$, where $E_g(D)$ is the lowest direct bandgap and C is a constant, typically ≈ 1.5. E_0 has been associated with the main peak in the $\varepsilon_r''(h\nu)$ versus $h\nu$ spectrum. The parameters required to calculate n from (3.23) are listed in Table 3.4 [3.15].

3.2.5 Group Index (N)

The group index represents the factor by which the group velocity of a group of waves in a dielectric medium is reduced with respect to propagation in free space,

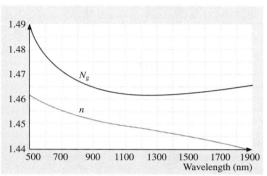

Fig. 3.3 Refractive index n and the group index N_g of pure SiO$_2$ (silica) glass as a function of wavelength

$N_g = c/v_g$, where v_g is the group velocity. The group index can be determined from the ordinary refractive index n via

$$N_g = n - \lambda \frac{dn}{d\lambda}, \qquad (3.24)$$

where λ is the wavelength of light. The relation between N_g and n for SiO$_2$ is illustrated in Fig. 3.3.

3.3 Optical Absorption

The main optical properties of a semiconductor are typically its refractive index n and its extinction coefficient K or absorption coefficient α (or equivalently the real and imaginary parts of the relative permittivity), as well as their dispersion relations (their dependence on the electromagnetic radiation wavelength λ or photon energy $h\nu$) and the changes in the dispersion relations with temperature, pressure, alloying, impurities, and so on. A typical relationship between the absorption coefficient and the photon energy observed in a crystalline semiconductor is shown in Fig. 3.4, where various possible absorption processes are illustrated. The important features of the behavior of the α versus $h\nu$ can be summarized as follows. (a) Free carrier absorption due to the presence of free electrons and holes, an effect that decreases with increasing photon energy. (b) An impurity absorption band (usually narrow) due the various dopants. (c) Reststrahlen or lattice absorption in which the radiation is absorbed by vibrations of the crystal ions. (d) Exciton absorption peaks that are usually observed at low temperatures and are close the fundamental absorption edge. (e) Band-to-band or fundamental absorption of photons, which excites an electron from the valence to the conduction band. Type (e) absorption has a large absorption coefficient and occurs when the photon energy reaches the bandgap energy

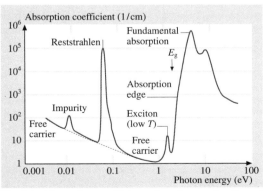

Fig. 3.4 Absorption coefficient plotted as a function of the photon energy in a typical semiconductor, illustrating various possible absorption processes

Table 3.5 Crystal structure, lattice parameter a, bandgap energy E_g at 300 K, type of bandgap (D=direct and I=indirect), change in E_g per unit temperature change (dE_g/dT) at 300 K, bandgap wavelength λ_g and refractive index n close to λ_g (A=amorphous, D=diamond, W=wurtzite, ZB=zinc blende). Approximate data from various sources

Semiconductors	Crystal	a (nm)	E_g (eV)	Type	dE_g/dT (meV/K)	λ_g (μm)	$n(\lambda_g)$	dn/dT (10^{-5} K^{-1})
Group IV								
Diamond	D	0.3567	5.48	I	−0.05	0.226	2.74	1.1
Ge	D	0.5658	0.66	I	−0.37	1.87	4	27.6
								42.4 (4 μm)
Si	D	0.5431	1.12	I	−0.25	1.11	3.45	13.8
								16 (5 μm)
a-Si:H	A		1.7–1.8			0.73		
SiC(α)	W	0.3081a 1.5120c	2.9	I	−0.33	0.42	2.7	9
III–V Compounds								
AlAs	ZB	0.5661	2.16	I	−0.50	0.57	3.2	15
AlP	ZB	0.5451	2.45	I	−0.35	0.52	3	11
AlSb	ZB	0.6135	1.58	I	−0.3	0.75	3.7	
GaAs	ZB	0.5653	1.42	D	−0.45	0.87	3.6	15
GaAs$_{0.88}$Sb$_{0.12}$	ZB		1.15	D		1.08		
GaN	W	0.3190 a 0.5190 c	3.44	D	−0.45	0.36	2.6	6.8
GaP	ZB	0.5451	2.26	I	−0.54	0.40	3.4	
GaSb	ZB	0.6096	0.73	D	−0.35	1.7	4	33
In$_{0.53}$Ga$_{0.47}$As on InP	ZB	0.5869	0.75	D		1.65		
In$_{0.58}$Ga$_{0.42}$As$_{0.9}$P$_{0.1}$ on InP	ZB	0.5870	0.80	D		1.55		
In$_{0.72}$Ga$_{0.28}$As$_{0.62}$P$_{0.38}$ on InP	ZB	0.5870	0.95	D		1.3		
InP	ZB	0.5869	1.35	D	−0.40	0.91	3.4–3.5	9.5
InAs	ZB	0.6058	0.36	D	−0.28	3.5	3.8	2.7
InSb	ZB	0.6479	0.18	D	−0.3	7	4.2	29
II–VI Compounds								
ZnSe	ZB	0.5668	2.7	D	−0.50	0.46	2.3	6.3
ZnTe	ZB	0.6101	2.3	D	−0.45	0.55	2.7	

E_g. It is probably the most important absorption effect; its characteristics for $h\nu > E_g$ can be predicted using the results of Sect. 3.3.3. The values of E_g, and its temperature shift, dE_g/dT, are therefore important factors in semiconductor-based optoelectronic devices. In nearly all semiconductors E_g decreases with temperature, hence shifting the fundamental absorption to longer wavelengths. The refractive index n also changes with temperature. dn/dT depends on the wavelength, but for many semiconductors $(dn/dT)/n \approx 5 \times 10^{-5}$ K^{-1}; for example, for GaAs, $(dn/dT)/n = 4 \times 10^{-4}$ K^{-1} at $\lambda = 2$ μm. There is a good correlation between the refractive indices and the bandgaps of semiconductors in which, typically, n decreases as E_g increases; semiconductors with wider bandgaps have lower refractive indices. The refractive index n and the extinction coefficient K (or α) are related by the Kramers–Kronig relations. Thus, large increases in the absorption coefficient for $h\nu$ near and above the bandgap energy E_g also result in increases in the refractive index n versus $h\nu$ in this region. Optical (and some structural) properties of various semiconductors are listed in Table 3.5.

from the EM wave to optical phonons, and ω_0 is a resonance frequency that is related to the "spring" constant between the ions. By definition, the frequency ω_T is

$$\omega_T^2 = \omega_0^2 \left(\frac{\varepsilon_{r\infty} + 2}{\varepsilon_{r0} + 2} \right). \qquad (3.26)$$

The loss ε_r'' and the absorption are maxima when $\omega = \omega_T$, and the wave is attenuated by the transfer of energy to the transverse optical phonons, so the EM wave couples to the transverse optical phonons. At $\omega = \omega_L$, the wave couples to the longitudinal optical (LO) phonons. Figure 3.5 shows the optical properties of AlSb [3.16] in terms of n, K and R versus wavelength. The peaks in the extinction coefficient K and reflectance R occur over roughly the same wavelength region, corresponding to the coupling of the EM wave to the transverse optical phonons. At wavelengths close to $\lambda_T = 2\pi/\omega_T$, n and K peak, and there is strong absorption of light (which corresponds to the EM wave resonating with the TO lattice vibrations), and then R rises sharply.

3.3.2 Free Carrier Absorption (FCA)

An electromagnetic wave with sufficiently low frequency oscillations can interact with free carriers in a material and thereby drift the carriers. This interaction results in an energy loss from the EM wave to the lattice vibrations through the carrier scattering processes. Based on the Drude model, the relative permittivity $\varepsilon_r(\omega)$ due to N free electrons per unit volume is given by

$$\varepsilon_r = \varepsilon_r' - i\varepsilon_r'' = 1 - \frac{\omega_p^2}{\omega^2 - i\omega/\tau} ;$$

$$\omega_p^2 = \frac{Ne^2}{\varepsilon_0 m_e}, \qquad (3.27)$$

where ω_p is a plasma frequency which depends on the electron concentration, while τ is the relaxation time of the scattering process (the mean scattering time). For metals where the electron concentration is very large, ω_p is of the order of $\approx 10^{16}$ rad/s, at UV frequencies, and for $\omega > \omega_p$, $\varepsilon_r \approx 1$, and the reflectance becomes very small. Metals lose their free-electron reflectance in the UV range, thus becoming UV-transparent. The reflectance does not fall to zero because there are other absorption processes such as interband electron excitations or excitations from core levels to energy bands. Plasma edge transparency, where the reflectance almost vanishes, can also be observed in doped semiconductors. For example, the reflectance of doped InSb has a plasma edge wavelength that decreases with increasing free carrier concentration [3.17]. Equation (3.27) can be written

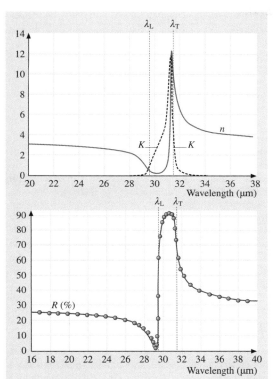

Fig. 3.5 Infrared refractive index n, extinction coefficient K (*top*), and reflectance R (*bottom*) of AlSb. Note that the wavelength axes are not identical, and the wavelengths λ_T and λ_L, corresponding to ω_T and ω_L, respectively, are shown as *dashed vertical lines*. After [3.16]

3.3.1 Lattice or Reststrahlen Absorption and Infrared Reflection

In the infrared wavelength region, ionic crystals reflect and absorb light strongly due to the resonance interaction of the electromagnetic (EM) wave field with the transverse optical (TO) phonons. The dipole oscillator model based on ions driven by an EM wave results in

$$\varepsilon_r = \varepsilon_r' - i\varepsilon_r'' = \varepsilon_{r\infty} + \frac{\varepsilon_{r\infty} - \varepsilon_{r0}}{\left(\frac{\omega}{\omega_T}\right)^2 - 1 + i\frac{\gamma}{\omega_T}\left(\frac{\omega}{\omega_T}\right)}, \qquad (3.25)$$

where ε_{r0} and $\varepsilon_{r\infty}$ are the relative permittivity at $\omega = 0$ (very low frequencies) and $\omega = \infty$ (very high frequencies) respectively, γ is the loss coefficient per unit reduced mass, representing the rate of energy transfer

At low frequencies where $\omega < 1/\tau$, we have $\alpha(\lambda) \propto \sigma_0/n(\lambda)$ so that α should be controlled by the dc conductivity, and hence the amount of doping. Furthermore, α will exhibit the frequency dependence of the refractive index n ($\alpha(\lambda) \propto 1/n(\lambda)$), in which n will typically be determined by the electronic polarization of the crystal.

At high frequencies where $\omega > 1/\tau$,

$$\alpha \propto \sigma_0/\omega^2 \propto N\lambda^2, \tag{3.32}$$

where α is proportional to N, the free carrier concentration, and λ^2. Experimental observations on FCA in doped semiconductors are in general agreement with these predictions. For example, α increases with N, whether N is increased by doping or by carrier injection [3.23]. However, not all semiconductors show the simple $\alpha \propto \lambda^2$ behavior. A proper account of the field-driven electron motion and scattering must consider the fact that τ will depend on the electron energy. The correct approach is to use the Boltzmann transport equation [3.24] with the appropriate scattering mechanism. FCA can be calculated using a quantum-mechanical approach based on second-order time-dependent perturbation theory with Fermi–Dirac statistics [3.25].

Absorption due to free carriers is commonly written as $\alpha \propto \lambda^p$, where the index p depends primarily on the scattering mechanism, although it is also influenced by the compensation doping ratio, if the semiconductor has been doped by compensation, and the free carrier concentration. In the case of lattice scattering, one must consider scattering from acoustic and optical phonons. For acoustic phonon scattering $p \approx 1.5$, for optical phonon scattering $p \approx 2.5$, and for impurity scattering $p \approx 3.5$. The observed free carrier absorption coefficient will then have all three contributions:

$$\alpha = A_{\text{acoustic}}\lambda^{1.5} + A_{\text{optical}}\lambda^{2.5} + A_{\text{impurity}}\lambda^{3.5}. \tag{3.33}$$

Inasmuch as α for FCA depends on the free carrier concentration N, it is possible to evaluate the latter from the experimentally measured α, given its wavelength dependence and p as discussed by [3.26]. The free carrier absorption coefficients $\alpha(\text{mm}^{-1})$ for GaP, n-type PbTe and n-type ZnO are shown in Fig. 3.6.

Free carrier absorption in p-type Ge demonstrates how the FCA coefficient α can be dramatically different than what is expected from (3.33). Figure 3.7a shows the wavelength dependence of the absorption coefficient for p-Ge over the wavelength range from about 2 to 30 μm [3.27]. The observed absorption is due to excitations of electrons from the spin-off band to the heavy

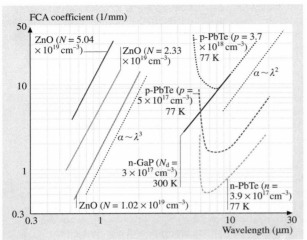

Fig. 3.6 Free carrier absorption in n-GaP at 300 K [3.18], p- and n-type PbTe [3.19] at 77 K and In doped n-type ZnO at room temperature. After [3.20]

in terms of the conductivity σ_0 at low frequencies (dc) as

$$\varepsilon_r = \varepsilon_r' - i\varepsilon_r'' = 1 - \frac{\tau\sigma_0}{\varepsilon_0[(\omega\tau)^2 + 1]} - i\frac{\sigma_0}{\varepsilon_0\omega[(\omega\tau)^2 + 1]}. \tag{3.28}$$

In metals, σ_0 is high. At frequencies where $\omega < 1/\tau$, the imaginary part $\varepsilon_r'' = \sigma_0/\varepsilon_0\omega$ is normally much more than 1, and $n = K \approx \sqrt{(\varepsilon_r''/2)}$, so that the free carrier attenuation coefficient α is then given by

$$\alpha = 2k_0 K \approx \frac{2\omega}{c}\left(\frac{\varepsilon_r''}{2}\right)^{1/2} \approx (2\sigma_0\mu_0)^{1/2}\omega^{1/2}. \tag{3.29}$$

Furthermore, the reflectance can also be calculated using $n = K \approx \sqrt{(\varepsilon_r''/2)}$, which leads to the well-known Hagen–Rubens relationship [3.21]

$$R \approx 1 - 2\left(\frac{2\omega\varepsilon_0}{\sigma_0}\right)^{1/2}. \tag{3.30}$$

In semiconductors, one typically encounters $\sigma_0/\varepsilon_0\omega < 1$, since the free electron concentration is small, and we can treat n as constant due to various other polarization mechanisms such as electronic polarization. Since $2nK = \varepsilon_r''$, the absorption coefficient becomes [3.22]:

$$\alpha = 2k_0 K \approx \frac{2\omega}{c}\left(\frac{\varepsilon_r''}{2n}\right) = \frac{\sigma_0}{nc\varepsilon_0[(\omega\tau)^2 + 1]}. \tag{3.31}$$

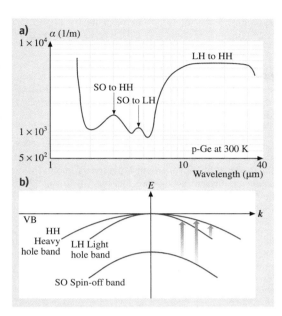

Fig. 3.7 (a) Free carrier absorption due to holes in p-Ge [3.27]. (b) The valence band of Ge has three bands; heavy hole, light hole and a spin-off bands

two types of band-to-band absorptions, corresponding to direct and indirect transitions.

A direct transition is a photoexcitation process in which no phonons are involved. The photon momentum is negligible compared with the electron momentum, so that when the photon is absorbed, exciting an electron from the valence band (VB) to the conduction band (CB), the electron's k-vector does not change. A direct transition on an $E - k$ diagram is a vertical transition from an initial energy E and wavevector k in the VB to a final energy E' and a wavevector k' in the CB where $k' = k$, as shown in Fig. 3.8a. The energy $(E' - E_c)$ is the kinetic energy $(\hbar k)^2/(2m_e^*)$ of the electron with an effective mass m_e^*, and $(E_v - E)$ is the kinetic energy $(\hbar k)^2/(2m_h^*)$ of the hole left behind in the VB. The ratio of the kinetic energies of the photogenerated electron and hole depends inversely on the ratio of their effective masses.

The absorption coefficient α is derived from the quantum-mechanical probability of transition from E to E', the occupied density of states at E in the VB from which electrons are excited, and the unoccupied density of states in the CB at $E + h\nu$. Thus, α depends on the joint density of states at E and $E + h\nu$, and we have to suitably integrate this joint density of states. Near the band edges, the density of states can be approximated by a parabolic band, and α rises with the photon energy following

$$\alpha h\nu = A(h\nu - E_g)^{1/2}, \qquad (3.34)$$

hole band, from the spin-off band to the light hole band, and from the light hole band to the heavy hole band, as marked in the Fig. 3.7b.

3.3.3 Band-to-Band or Fundamental Absorption

Crystalline Solids

Band-to-band absorption or fundamental absorption of radiation is due to the photoexcitation of an electron from the valence band to the conduction band. There are

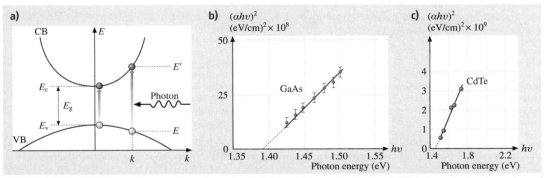

Fig. 3.8 (a) A direct transition from the valence band (VB) to the conduction band (CB) through the absorption of a photon. Absorption behavior represented as $(\alpha h\nu)^2$ versus photon energy $h\nu$ near the band edge for single crystals of (b) p-type GaAs, from [3.28] and (c) CdTe. After [3.29]

where the constant $A \approx [e^2/(nch^2 m_e^*)](2\mu^*)^{3/2}$ in which μ^* is a reduced electron and hole effective mass, n is the refractive index, and E_g is the direct bandgap, with minimum $E_c - E_v$ at the same k value. Experiments indeed show this type of behavior for photon energies above E_g and close to E_g, as shown in Fig. 3.8b for a GaAs crystal [3.28] and in (c) for a CdTe crystal [3.29]. The extrapolation to zero photon energy gives the direct bandgap E_g, which is about 1.40 eV for GaAs and 1.46–1.49 eV for CdTe. For photon energies very close to the bandgap energy, the absorption is usually due to exciton absorption, especially at low temperatures, and is discussed later in this chapter.

In indirect bandgap semiconductors, such as Si and Ge, the photon absorption for photon energies near E_g requires the absorption and emission of phonons during the absorption process, as illustrated in Fig. 3.9a. The absorption onset corresponds to a photon energy of $(E_g - h\vartheta)$, which represents the absorption of a phonon with energy $h\vartheta$. For the latter, α is proportional to $[h\nu - (E_g - h\vartheta)]^2$. Once the photon energy reaches $(E_g + h\vartheta)$, then the photon absorption process can also occur by phonon emission, for which the absorption coefficient is larger than that for phonon absorption. The absorption coefficients for the phonon absorption and emission processes are given by [3.30]

$$\alpha_{\text{absorption}} = A[f_{\text{BE}}(h\vartheta)][h\nu - (E_g - h\vartheta)]^2 \; ;$$
$$h\nu > (E_g - h\vartheta) \, , \quad (3.35)$$

and

$$\alpha_{\text{emission}} = A[(1 - f_{\text{BE}}(h\vartheta))][h\nu - (E_g + h\vartheta)]^2 \; ;$$
$$h\nu > (E_g + h\vartheta) \, , \quad (3.36)$$

where A is a constant and $f_{\text{BE}}(h\vartheta)$ is the Bose–Einstein distribution function at the phonon energy $h\vartheta$, $f_{\text{BE}}(h\vartheta) = [(\exp(h\vartheta/k_B T) - 1]^{-1}$, where k_B is the Boltzmann constant and T is the temperature. As we increase the photon energy in the range $(E_g - h\vartheta) < h\nu < (E_g + h\vartheta)$, the absorption is controlled by $\alpha_{\text{absorption}}$ and the plot of $\alpha^{1/2}$ versus $h\nu$ has an intercept of $(E_g - h\vartheta)$.

On the other hand, for $h\nu > (E_g + h\vartheta)$, the overall absorption coefficient is $\alpha_{\text{absorption}} + \alpha_{\text{emission}}$, but at slightly higher photon energies than $(E_g + h\vartheta)$, α_{emission} quickly dominates over $\alpha_{\text{absorption}}$ since $[f_{\text{BE}}(h\vartheta)] \gg [(1 - f_{\text{BE}}(h\vartheta))]$. Figure 3.9b shows the behavior of $\alpha^{1/2}$ versus photon energy for Si at two temperatures for $h\nu$ near band edge absorption. At low temperatures, $f_{\text{BE}}(h\vartheta)$ is small and $\alpha_{\text{absorption}}$ decreases with decreasing temperature, as apparent from Fig. 3.9b.

Equations (3.35) and (3.36) intersect the photon energy axis at $(E_g - h\vartheta)$ and $(E_g + h\vartheta)$, which can be used to obtain E_g.

An examination of the extinction coefficient K or ε_r'' versus photon energy for Si in Fig. 3.2 shows that absorption peaks at certain photon energies, $h\nu \approx 3.5$ and 4.3 eV. These peaks are due to the fact that the joint density of states function peaks at these energies. The absorption coefficient peaks whenever there is a direct transition in which the E versus k curve in the VB is parallel to the E versus k curve in the CB, as schematically illustrated in Fig. 3.9a, where a photon of energy $h\nu_{12}$ excites an electron from state 1 in the VB to state 2 in the CB in a direct transition, i.e. $k_1 = k_2$. Such transitions where E versus k curves are parallel at a photon energy $h\nu_{12}$ result in a peak in the absorption versus photon energy behavior, and can be represented by the condition that

$$(\nabla_k E)_{\text{CB}} - (\nabla_k E)_{\text{VB}} = 0 \, . \quad (3.37)$$

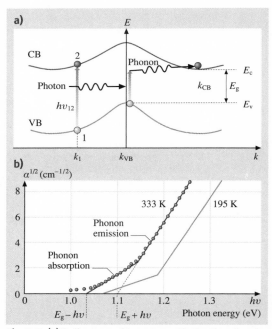

Fig. 3.9 (a) Indirect transitions across the bandgap involve phonons. Direct transitions in which dE/dk in the CB is parallel to dE/dk in the VB lead to peaks in the absorption coefficient. (b) Fundamental absorption in Si at two temperatures. The overall behavior is well described by (3.35) and (3.36)

The above condition is normally interpreted as the joint density of states reaching a peak value at certain points in the Brillouin zone called van Hove singularities. Identification of peaks in K versus $h\nu$ involves the examination of all E versus $\mathbf{k}$ curves of a given crystal that can participate in a direct transition. The silicon ε_r'' peaks at $h\nu \approx 3.5$ eV and 4.3 eV correspond to (3.37) being satisfied at points L, along $\langle 111 \rangle$ in $\mathbf{k}$-space, and X along $\langle 100 \rangle$ in $\mathbf{k}$-space, at the edges of the Brillouin zone.

In degenerate semiconductors, the Fermi level E_F is in a band; for example, in the CB for a degenerate n-type semiconductor. Electrons in the VB can only be excited to states above E_F in the CB rather than to the bottom of the CB. The absorption coefficient then depends on the free carrier concentration since the latter determines E_F. Fundamental absorption is then said to depend on band filling, and there is an apparent shift in the absorption edge, called the *Burstein–Moss shift*. Furthermore, in degenerate indirect semiconductors, the indirect transition may involve a non-phonon scattering process, such as impurity or electron–electron scattering, which can change the electron's wavevector $\mathbf{k}$. Thus, in degenerate indirect bandgap semiconductors, absorption can occur without phonon assistance and the absorption coefficient becomes:

$$\alpha \propto [h\nu - (E_g + \Delta E_F)]^2 , \quad (3.38)$$

where ΔE_F is the energy depth of E_F into the band measured from the band edge.

Heavy doping of degenerate semiconductors normally leads to a phenomenon called bandgap narrowing and bandtailing. Bandtailing means that the band edges at E_v and E_c are no longer well defined cut-off energies, and there are electronic states above E_v and below E_c where the density of states falls sharply with energy away from the band edges. Consider a degenerate direct band gap p-type semiconductor. One can excite electrons from states below E_F in the VB, where the band is nearly parabolic, to tail states below E_c, where the density of states decreases exponentially with energy into the bandgap, away from E_c. Such excitations lead to α depending exponentially on $h\nu$, a dependence that is called the *Urbach rule* [3.31, 32], given by:

$$\alpha = \alpha_0 \exp[(h\nu - E_0)/\Delta E] \quad (3.39)$$

where α_0 and E_0 are material-dependent constants, and ΔE, called the Urbach width, is also a material-dependent constant. The Urbach rule was originally reported for alkali halides. It has been observed for many ionic crystals, degenerately doped crystalline semiconductors, and almost all amorphous semiconductors. While exponential bandtailing can explain the observed Urbach tail of the absorption coefficient versus photon energy, it is also possible to attribute the absorption tail behavior to strong internal fields arising, for example, from ionized dopants or defects.

Amorphous Solids

In a defect-free crystalline semiconductor, a well-defined energy gap exists between the valence and conduction bands. In contrast, in an amorphous semiconductor, the distributions of conduction band and valence band electronic states do not terminate abruptly at the band edges. Instead, some electronic states called the tail states encroach into the gap region [3.33]. In addition to tail states, there are also other localized states deep within the gap region [3.34]. These localized tail states in amorphous semiconductors are contributed by defects. The defects in amorphous semiconductors are considered to be all cases of departure from the normal nearest-neighbor coordination (or normal valence requirement). Examples of defects are: broken and dangling bonds (typical for amorphous silicon), over- and under-coordinated atoms (such as "valence alternation pairs" in chalcogenide glasses), voids, pores, cracks and other macroscopic defects. Mobility edges exist, which separate these localized states from their extended counterparts; tail and deep defect states are localized [3.35–37]. These localized tail and deep defect states are responsible for many of the unique properties exhibited by amorphous semiconductors.

Despite years of intensive investigation, the exact form of the distribution of electronic states associated with amorphous semiconductors remains a subject of some debate. While there are still some unresolved theoretical issues, there is general consensus that the tail states arise as a consequence of the disorder present within amorphous networks, and that the width of these tails reflects the amount of disorder present [3.38]. Experimental results (from, for example, [3.39, 40]) suggest exponential distributions for the valence and conduction band tail states in a-Si:H, although other possible functional forms [3.41] cannot be ruled out. *Singh* and *Shimakawa* [3.37] have derived separate effective masses of charge carriers in their extended and tail states. That means the density of states (DOS) of extended and tail states can be represented in two different parabolic forms. The relationship between the absorption coefficient and the distribution of electronic states for the case of a-Si:H may be found in [3.37, 42–44].

The existence of tail states in amorphous solids has a profound impact upon the band-to-band optical absorption. Unlike in a crystalline solid, the absorption of photons in an intrinsic amorphous solid can also occur for photon energies $\hbar\omega \leq E_0$ due to the presence of tail states in the forbidden gap. E_0 is the energy of the optical gap, which is usually close to the mobility gap – the energy difference between the conduction band and valence band mobility edges.

In a crystalline semiconductor, the energy and crystal momentum of an electron involved in an optical transition must be conserved. In an amorphous semiconductor, however, only the energy needs to be conserved [3.36, 37]. As a result, for optical transitions caused by photons of energy $\hbar\omega \geq E_0$ in amorphous semiconductors, the approach of a joint density of states is not applicable [3.37, 45]. One has to consider the product of the densities of both conduction and valence electronic states when calculating the absorption coefficient [3.37, 46]. Assuming that both the valence band and conduction band DOS functions have square-root dependencies on energy, one can derive the absorption coefficient α as [3.37]:

$$(\alpha\hbar\omega)^{1/2} = B^{1/2}(\hbar\omega - E_0), \qquad (3.40)$$

where, if one assumes that the transition matrix element is constant,

$$B = \frac{1}{nc\varepsilon_0}\left(\frac{e}{m_e^*}\right)^2 \left(\frac{L(m_e^* m_h^*)^{3/2}}{2^2 \hbar^3}\right) \qquad (3.41)$$

where m_e^* and m_h^* are the effective masses of an electron and a hole, respectively, L denotes the average interatomic separation in the sample, and n is the refractive index. Plotting $(\alpha\hbar\omega)^{1/2}$ as a function of $\hbar\omega$ yields a straight line that is usually referred to as Tauc's plot, from which one can determine the optical gap energy E_0 experimentally. Equation (3.40) is also known as Tauc's relation and B as Tauc's constant [3.47]. Experimental data from many, but not all, amorphous semiconductors and chalcogenides fit to (3.40) very well. Deviations from Tauc's relation have been observed. For instance, some experimental data in a-Si:H fit much better to a cubic root relation given by *Mott* and *Davis* [3.36].

$$(\alpha\hbar\omega)^{1/3} = C(\hbar\omega - E_0) \qquad (3.42)$$

and therefore the cubic root has been used to determine the optical gap E_0. Here C is another constant.

If one considers that the optical transition matrix element is independent of photon energy [3.37, 48], one finds that the cubic root dependence on photon energy can be obtained only when the valence band and conduction band DOS depend linearly on energy. Using such DOS functions, the cubic root dependence has been explained by *Mott* and *Davis* [3.36]. Another way to obtain the cubic root dependence has been suggested in [3.49]. Using (3.40), *Sokolov* et al. [3.49] have modeled the cubic root dependence on photon energy by considering the fluctuations in the optical band gap due to structural disorder. Although their approach gives a way of getting the cubic root dependence, Sokolov et al.'s model is hardly different from the linear density of states model suggested by *Mott* and *Davis* [3.36]. *Cody* [3.50] has shown an alternative approach using a photon energy-dependent transition matrix element. Thus, the absorption coefficient is obtained as [3.37]:

$$(\alpha\hbar\omega) = B'(\hbar\omega)^2(\hbar\omega - E_0)^2, \qquad (3.43)$$

where

$$B' = \frac{e^2}{nc\varepsilon_0}\left(\frac{(m_e^* m_h^*)^{3/2}}{2\pi^2\hbar^7 \nu \rho_A}\right) Q_a^2, \qquad (3.44)$$

and Q_a is the average separation between the excited electron and hole pair in an amorphous semiconductor, ν denotes the number of valence electrons per atom and ρ_A represents the atomic density per unit volume.

Equation (3.43) suggests that $(\alpha\hbar\omega)$ depends on the photon energy in the form of a polynomial of order 4. Then, depending on which term of the polynomial may be more significant in which material, one can get square, cubic or fourth root dependences of $(\alpha\hbar\omega)$ on

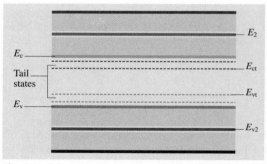

Fig. 3.10 Schematic illustration of the electronic energy states E_2, E_c, E_{ct}, E_{vt}, E_v and E_{v2} in amorphous semiconductors. The *shaded region* represents the extended states. Energies E_2 and E_{v2} correspond to the centers of the conduction and valence extended states and E_{ct} and E_{vt} represent the ends of the conduction and valence tail states, respectively

the photon energy. In this case, (3.43) may be expressed as:

$$(\alpha\hbar\omega)^x \propto (\hbar\omega - E_0),\qquad(3.45)$$

where $x \le 1/2$. Thus, in a way, any deviation from the square root or Tauc's plot may be attributed to the energy-dependent matrix element [3.37, 46]. Another possible explanation has been recently discussed by *Shimakawa* and coworkers on the base of fractal theory [3.54].

Another problem is how to determine the constants B (3.41) and B' (3.44), which involve the effective masses of an electron and a hole. In other words, how do we determine the effective masses of charge carriers in amorphous solids? Recently, a simple approach [3.37, 46] has been developed to calculate the effective masses of charge carriers in amorphous solids. Different effective masses of charge carriers are obtained in the extended and tail states. The approach applies the concepts of tunneling and effective medium, and one obtains the effective mass of an electron in the conduction extended states, denoted by m^*_{ex}, and in the tail states, denoted by m^*_{et}, as:

$$m^*_{\text{ex}} \approx \frac{E_L}{2(E_2 - E_c)a^{1/3}} m_e,\qquad(3.46)$$

and

$$m^*_{\text{et}} \approx \frac{E_L}{(E_c - E_{ct})b^{1/3}} m_e,\qquad(3.47)$$

where:

$$E_L = \frac{\hbar^2}{m_e L^2}.\qquad(3.48)$$

Here $a = \frac{N_1}{N} < 1$, N_1 is the number of atoms contributing to the extended states, $b = \frac{N_2}{N} < 1$, N_2 is the number of atoms contributing to the tail states, such that $a + b = 1 (N = N_1 + N_2)$, and m_e is the free electron mass. The energy E_2 in (3.46) corresponds to the energy of the middle of the extended conduction states, at which the imaginary part of the dielectric constant becomes maximum (Fig. 3.10; see also Fig. 3.2).

Likewise, the effective masses of the hole m^*_{hx} and m^*_{ht} in the valence extended and tail states are obtained, respectively, as:

$$m^*_{\text{hx}} \approx \frac{E_L}{2(E_v - E_{v2})a^{1/3}} m_e,\qquad(3.49)$$

and

$$m^*_{\text{ht}} \approx \frac{E_L}{(E_{vt} - E_v)b^{1/3}} m_e,\qquad(3.50)$$

where E_{v2} and E_{vt} are energies corresponding to the half-width of the valence extended states and the end of the valence tail states, respectively; see Fig. 3.10.

Using (3.46) and (3.47) and the values of the parameters involved, different effective masses of an electron are obtained in the extended and tail states. Taking, for example, the density of weak bonds contributing to the tail states as 1 at. %, so $b = 0.01$ and $a = 0.99$, the effective masses and energies E_L calculated for hydrogenated amorphous silicon (a-Si:H) and germanium (a-Ge:H) are given in Table 3.6.

According to (3.46), (3.47), (3.49) and (3.50), for sp^3 hybrid amorphous semiconductors such as a-Si:H and a-Ge:H, effective masses of the electron and hole are expected to be the same. In these semiconductors, since the conduction and valence bands are two equal halves of the same electronic band, their widths are the same and that gives equal effective masses for the electron and the hole [3.37, 55]. This is one of the reasons for using $E_{ct} = E_{vt} = E_c/2$, which gives equal effective masses for electrons and holes in the tail states as well. This is different from crystalline solids where m^*_e and m^*_h are usually not the same. This difference between amorphous and crystalline solids is similar to, for example, having direct and indirect crystalline semiconductors but only direct amorphous semiconductors.

Using the effective masses from Table 3.6 and (3.41), B can be calculated for a-Si:H and a-Ge:H. The values thus obtained with the refractive index $n = 4$ for a-Si:H and a-Ge:H are $B = 6.0 \times 10^6$ cm^{-1} eV^{-1} for a-Si:H and $B = 4.1 \times 10^6$ cm^{-1} eV^{-1} for a-Ge:H, which are an order of magnitude higher than those estimated from

Table 3.6 Effective mass of electrons in the extended and tail states of a-Si:H and a-Ge:H calculated using Eqs. (3.46) and (3.47) for $a = 0.99$, $b = 0.01$ and $E_{ct} = E_{vt} = E_c/2$. E_L is calculated from (3.48). All energies are given in eV. Note that since the absorption coefficient is measured in cm^{-1}, the value used for the speed of light is in cm/s (a [3.51]; b [3.52]; c [3.33]; d [3.53])

	L(nm)	E_2	E_c	E_L	$E_c - E_{ct}$	m^*_{ex}	m^*_{et}
a-Si:H	0.235[a]	3.6[b]	1.80[c]	1.23	0.9	0.34m_e	6.3m_e
a-Ge:H	0.245[a]	3.6	1.05[d]	1.14	0.53	0.22m_e	10.0m_e

experiments [3.36]. However, considering the quantities involved in B (3.41), this can be regarded as a reasonable agreement.

In a recent paper, *Malik* and *O'Leary* [3.56] studied the distributions of conduction and valence band electronic states associated with a-Si:H. They noted that the effective masses associated with a-Si:H are material parameters that are yet to be experimentally determined. In order to remedy this deficiency, they fitted square-root DOS functions to experimental DOS data and found that $m_h^* = 2.34 m_e$ and $m_e^* = 2.78 m_e$.

The value of the constant B' in (3.44) can also be calculated theoretically, provided that Q_a is known. Using the atomic density of crystalline silicon and four valence electrons per atom, *Cody* [3.50] estimated $Q_a^2 = 0.9 \text{ Å}^2$, which gives $Q_a \approx 0.095$ nm, less than half the interatomic separation of 0.235 nm in a-Si:H, but of the same order of magnitude. Using $\nu = 4$, $\rho_A = 5 \times 10^{28}$ m^{-3}, $Q_a^2 = 0.9 \text{ Å}^2$, and extended state effective masses, we get $B' = 4.6 \times 10^3$ cm^{-1} eV^{-3} for a-Si:H and 1.3×10^3 cm^{-1} eV^{-3} for a-Ge:H. Cody estimated an optical gap, $E_0 = 1.64$ eV, for a-Si:H, which, using in (3.43), gives $\alpha = 1.2 \times 10^3$ cm^{-1} at a photon energy of $\hbar\omega = 2$ eV. This agrees reasonably well with the $\alpha = 6.0 \times 10^2$ cm^{-1} used by Cody. If we use interatomic spacing in place of Q_a in (3.44), we get $B' = 2.8 \times 10^4$ cm^{-1} eV^{-3}, and then the corresponding absorption coefficient becomes 3.3×10^3 cm^{-1}. This suggests that, in order to get an estimate, one may use the interatomic spacing in place of Q_a if the latter is unknown. Thus, both B and B' can be determined theoretically, a task not possible before due to a lack of knowledge of the effective masses in amorphous semiconductors.

The absorption of photons of energy less than the band gap energy, $\hbar\omega < E_0$ in amorphous solids involves the localized tail states and hence follows neither (3.40) nor (3.42). Instead, the absorption coefficient depends on the photon energy exponentially, as given in (3.39), giving rise to Urbach's tail. *Abe* and *Toyozawa* [3.57] have calculated the interband absorption spectra in crystalline solids, introducing the Gaussian site diagonal disorder and applying the coherent potential approximation. They have shown that Urbach's tail occurs due to static disorder (structural disorder). However, the current stage of understanding is that Urbach's tail in amorphous solids occurs due to both thermal effects and static disorder [3.50]. More recent issues in this area have been addressed by *Orapunt* and *O'Leary* [3.48].

Keeping the above discussion in mind, the optical absorption in amorphous semiconductors near the absorption edge is usually characterized by three types of optical transitions corresponding to transitions between tail and tail states, tail and extended states, and extended and extended states. The first two types correspond to $\hbar\omega \leq E_0$, and the third one corresponds to $\hbar\omega \geq E_0$. Thus, the plot of absorption coefficient versus photon energy (α versus $\hbar\omega$) has three different regions, A, B and C, respectively, that correspond to these three characteristic optical transitions shown in Fig. 3.11.

In the small absorption coefficient range A (also called the weak absorption tail, WAT), where $\alpha < 10^{-1}$ cm^{-1}, the absorption is controlled by optical transitions from tail-to-tail states. As stated above, the localized tail states in amorphous semiconductors are contributed by defects. To some extent, the absolute value of absorption in region A may be used to estimate the density of defects in the material. In region B, where typically $10^{-1} < \alpha < 10^4$ cm^{-1}, the absorption is related to transitions from the localized tail states above the valence band edge to extended states in the conduction band and/or from extended states in the valence band to localized tail states below the conduction band. The spectral dependence of α usually follows the so-called Urbach rule, given in (3.39). For many amorphous semiconductors, ΔE has been related to the width of the valence (or conduction) band tail states, and may be used to compare the "widths" of such localized tail states in

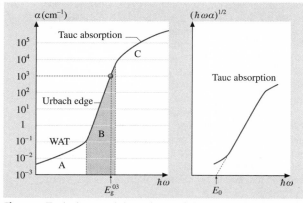

Fig. 3.11 Typical spectral dependence of the optical absorption coefficient in amorphous semiconductors. In the A and B regions, the absorption is controlled by optical transitions between tail and tail states and tail and extended states, respectively, and in the C region it is dominated by transitions from extended to extended states. In domain B, the absorption coefficient follows the Urbach rule (3.39). In region C, the absorption coefficient follows Tauc's relation (3.40) in a-Si:H, as shown in the figure on the *right-hand side*

Table 3.7 Optical bandgaps of a-Si$_{1-x}$C$_x$:H films obtained from Tauc's (3.40), Sokolov et al.'s (3.42) and Cody's (3.43) relations [3.58]

	E_g at $\alpha = 10^3$ cm^{-1}	E_g at $\alpha = 10^4$ cm^{-1}	$E_g = E_0$ (Tauc)	$E_g = E_0$ (Cody)	$E_g = E_0$ (Sokolov)	ΔE (meV)
a-Si:H	1.76	1.96	1.73	1.68	1.60	46
a-Si$_{0.88}$C$_{0.18}$:H	2.02	2.27	2.07	2.03	1.86	89

different materials; ΔE is typically 0.05–0.1 eV. In region C, the absorption is controlled by transitions from extended to extended states. For many amorphous semiconductors, the α versus $\hbar\omega$ behavior follows the Tauc relation given in (3.40). The optical bandgap, E_0, determined for a given material from the α versus $\hbar\omega$ relations obtained in (3.40), (3.42) and (3.43), can vary as shown in Table 3.7 for a-Si:H alloys.

3.3.4 Exciton Absorption

Excitons in Crystalline Semiconductors

Optical absorption in crystalline semiconductors and insulators can create an exciton, which is an electron and hole pair excited by a photon and bound together through their attractive Coulomb interaction. This means that the absorbed optical energy remains held within the solid for the lifetime of the exciton. There are two types of excitons that can be formed in nonmetallic solids: Wannier or Wannier–Mott excitons and Frenkel excitons. The concept of Wannier–Mott excitons is valid for inorganic semiconductors like Si, Ge and GaAs, because in these materials the large overlap of interatomic electronic wavefunctions enables the electrons and holes to be far apart but bound in an excitonic state. For this reason, these excitons are also called large-radii orbital excitons. Excitons formed in organic crystals are called Frenkel excitons. In organic semiconductors/insulators or molecular crystals, the intermolecular separation is large and hence the overlap of intermolecular electronic wavefunctions is very small and electrons remain tightly bound to individual molecules. Therefore, the electronic energy bands are very narrow and closely related to individual molecular electronic energy levels. In these solids, the absorption of photons occurs close to the individual molecular electronic states and excitons are also formed within the molecular energy levels (see [3.59]). Such excitons are therefore also called molecular excitons. For a more detailed look at the theory of Wannier and Frenkel excitons, readers may like to refer to *Singh* [3.59].

In Wannier–Mott excitons, the Coulomb interaction between the hole and electron can be viewed as an effective hydrogen atom with, for example, the hole establishing the coordinate reference frame about which the reduced mass electron moves. If the effective masses of the isolated electron and hole are m_e^* and m_h^*, respectively, their reduced mass μ_x is given by:

$$\mu_x^{-1} = (m_e^*)^{-1} + (m_h^*)^{-1} . \tag{3.51}$$

Note that in the case of so-called hydrogenic impurities in semiconductors (both shallow donor and acceptor impurities), the reduced mass of the nucleus takes the place of one of the terms in (3.51) and hence the reduced mass is given to good approximation by the effective mass of the appropriate carrier. When the exciton is the carrier, the effective masses are comparable and hence the reduced mass is markedly lower; accordingly, the exciton binding energy is markedly lower than that for hydrogenic impurities. The energy of a Wannier–Mott exciton is given by (e.g. [3.59]):

$$E_x = E_g + \frac{\hbar^2 K^2}{2M} - E_n , \tag{3.52}$$

where E_g is the band gap energy, $\hbar K$ is the linear momentum, $M (= m_e^* + m_h^*)$ is the effective mass associated with the center of mass of an exciton, and E_n is the exciton binding energy given by

$$E_n = \frac{\mu_x e^4 \kappa^2}{2\hbar^2 \varepsilon^2} \frac{1}{n^2} = \frac{R_y^x}{n^2} , \tag{3.53}$$

where e is the electronic charge, $\kappa = \frac{1}{4\pi\varepsilon_0}$, ε is the static dielectric constant of the solid, and n is the principal quantum number associated with the internal excitonic states $n = 1(s), 2(p), \ldots$ (see Fig. 3.12). According to (3.53), the excitonic states are formed within the band gap below the conduction band edge. However, as the exciton binding energy is very small (a few meV in bulk Si and Ge crystals), exciton absorption peaks can only be observed at very low temperatures. R_y^x is the so-called effective Rydberg of the exciton given by $Ry(\mu_x/m_e)/\varepsilon^2$, where $Ry = 13.6$ eV. For bulk GaAs, the exciton binding energy ($n = 1$) corresponds to about 5 meV. Extrapolating from the hydrogen atom model, the extension of the excitonic wavefunction can be found from an effective Bohr radius a_B^* given in terms of the Bohr radius

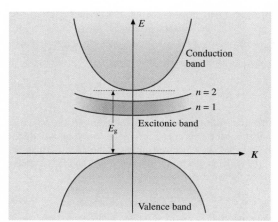

Fig. 3.12 Schematic illustration of excitonic bands for $n = 1$ and 2 in semiconductors. E_g represents the energy gap

as $a_B = h^2/\pi e^2$; that is, $a_B^* = a_B(\varepsilon/\mu_x)$. For GaAs this corresponds to about 12 nm or about 21 lattice constants – in other words the spherical volume of the exciton radius contains $\approx (a_B^*/a_0)^3$ or ≈ 9000 unit cells, where a_0 is the lattice constant of GaAs. As $Ry^* \ll E_g$ and $a_B^* \gg a_0$, it is clear from this example that the excitons in GaAs are large-radii orbital excitons, as stated above. It should be noted that the binding energies of excitons in semiconductors tend to be a strong function of the bandgap. The dependencies of R_y^x (exciton binding energy) and a_B^*/a_0 on the semiconductor bandgap are shown in Fig. 3.13a and b, respectively. For excitons with large binding energies and correspondingly small radii (approaching the size of a lattice parameter), the excitons become localized on a lattice site, as observed in most organic semiconductors. As stated above, such excitons are commonly referred to as Frenkel excitons or molecular excitons. Unlike Wannier-based excitons which are typically dissociated at room temperature, these excitations are stable at room temperature. For the binding energies of Frenkel excitons, one can refer to *Singh* for example [3.59].

Excitons can recombine radiatively, emitting a series of hydrogen-like spectral lines, as described by (3.52). In bulk (3-D) semiconductors such as GaAs, exciton lines can only be observed at low temperatures: they are easily dissociated by thermal fluctuations. On the other hand, in quantum wells and other structures of reduced dimensionality, the spatial confinement of both the electron and hole wavefunctions in the same layer ensures strong excitonic transitions of a few meV below the bandgap, even at room temperature. Excitonic absorption is well-located spectrally and very sensitive to optical saturation. For this reason, it plays an important role in nonlinear semiconductor devices (nonlinear Fabry–Perot resonators, nonlinear mirrors, saturable absorbers, and so on). If the valence hole is a heavy hole, the exciton is called a heavy exciton; conversely, if the valence hole is light, the exciton is a light exciton. For practical purposes, the excitonic contribution to the overall susceptibility around the resonance frequency ν_{ex} can be written as:

$$\chi_{exc} = -A_0 \frac{(\nu - \nu_{ex}) + i\Gamma_{ex}}{(\nu - \nu_{ex})^2 + \Gamma_{ex}^2 (1 + S)}, \quad (3.54)$$

where Γ_{ex} is the linewidth and $S = I/I_S$ the saturation parameter of the transition. For instance, in GaAs MQW, the saturation intensity I_S is as low as $1\,\text{kW/cm}^2$, and Γ_{ex} ($\approx 3.55\,\text{meV}$ at room temperature) varies with the temperature according to

$$\Gamma_{ex} = \Gamma_0 + \frac{\Gamma_1}{\exp(\hbar\omega_{LO}/kT) - 1}, \quad (3.55)$$

where $\hbar\Gamma_0$ is the inhomogeneous broadening (≈ 2 meV), $\hbar\Gamma_1$ is the homogeneous broadening (≈ 5 meV), and

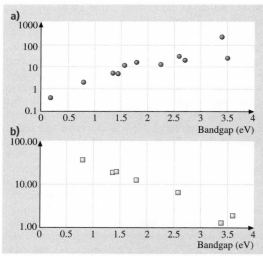

Fig. 3.13a,b Dependence of the (**a**) exciton binding energy (R_y^x) (3.29) and (**b**) size (in terms of the ratio of the excitonic Bohr radius to lattice constant (a_B^*/a_0)) as a function of the semiconductor bandgap. Exciton binding energy increases along with a marked drop in exciton spreading as bandgap increases. The Wannier-based description is not appropriate above a bandgap of about 2 eV

$\hbar\omega_{LO}$ is the longitudinal optical phonon energy (≈ 36 meV). At high carrier concentrations (provided either by electrical pumping or by optical injection), the screening of the Coulomb attractive potential by free electrons and holes provides an efficient mechanism for saturating the excitonic line.

The above discussion refers to so-called free excitons formed between conduction band electrons and valence band holes. According to (3.52), such an excitation is able to move throughout a material with a given center-of-mass kinetic energy (second term on the right hand side). It should be noted, however, that free electrons and holes move with a velocity $\hbar(dE/dk)$ where the derivative is taken for the appropriate band edge. To move through a crystal, both the electron and the hole must have identical translational velocities, restricting the regions in k-space where these excitations can occur to those with $(dE/dk)_{\text{electron}} = (dE/dk)_{\text{hole}}$, commonly referred to as critical points.

A number of more complex pairings of carriers can also occur, which may also include fixed charges or ions. For example, for the case of three charged entities with one being an ionized donor impurity (D+), the following possibilities can occur: (D+)(+)(−), (D+)(−)(−) and (+)(+)(−) as excitonic ions, and (+)(+)(−)(−) and (D+)(+)(−)(−) as biexcitons or even bigger excitonic molecules (see [3.60]). Complexity abounds in these systems, as each electronic level possesses a fine structure corresponding to allowed rotational and vibrational levels. Moreover, the effective mass is often anisotropic. Note that when the exciton or exciton complex is bound to a fixed charge, such as an ionized donor or acceptor center in the material, the exciton or exciton complex is referred to as a bound exciton. Indeed, bound excitons may also involve neutral fixed impurities. It is usual to relate the exciton in these cases to the center binding them; thus, if an exciton is bound to a donor impurity, it is usually termed a donor-bound exciton.

Excitons in Amorphous Semiconductors

The concept of excitons is traditionally valid only for crystalline solids. However, several observations in the photoluminescence spectra of amorphous semiconductors have revealed the occurrence of photoluminescence associated with singlet and triplet excitons [3.37]. Applying the effective mass approach, a theory for the Wannier–Mott excitons in amorphous semiconductors has recently been developed in real coordinate space [3.37, 46, 55, 61]. The energy of an exciton thus derived is obtained as:

$$W_x = E_0 + \frac{P^2}{2M} - E_n(S), \quad (3.56)$$

where P is the linear momentum associated with the exciton's center of motion and $E_n(S)$ is the binding energy of the exciton, given by

$$E_n(S) = \frac{\mu_x e^4 \kappa^2}{2\hbar^2 \varepsilon'(S)^2 n^2}, \quad (3.57)$$

where

$$\varepsilon'(S) = \varepsilon \left[1 - \frac{(1-S)}{A} \right]^{-1}, \quad (3.58)$$

where S is the spin (S being $=0$ for singlet and $=1$ for triplet) of an exciton and A is a material-dependent constant representing the ratio of the magnitude of the Coulomb and exchange interactions between the electron and the hole of an exciton. Equation (3.57) is analogous to (3.53) obtained for excitons in crystalline solids for $S=1$. This is because (3.53) is derived within the large-radii orbital approximation, which neglects the exchange interaction and hence is valid only for triplet excitons [3.59, 62]. As amorphous solids lack long-range order, the exciton binding energy is found to be larger in amorphous solids than in their crystalline counterparts; for example, the binding energy is higher in hydrogenated amorphous silicon (a-Si:H) than in crystalline silicon (c-Si). This is the reason that it is possible to observe both singlet and triplet excitons in a-Si:H [3.63] but not in c-Si.

Excitonic Absorption

Since exciton states lie below the edge of the conduction band in a crystalline solid, absorption to excitonic states is observed below this edge. According to (3.53), the difference in energy in the bandgap and the excitonic absorption gives the binding energy. As the exciton–photon interaction operator and excited electron and hole pair and photon interaction operator depend only on their relative motion, the these interactions take the same form for band-to-band and excitonic absorption. Therefore, to calculate the excitonic absorption coefficient, one can use the same form of interaction as that used for band-to-band absorption, but one must use the joint density of states. Using the joint density of states, the absorption coefficient associated with the excitonic states in crystalline semiconductors is obtained as ([3.37]):

$$\alpha \hbar \omega = A_x (\hbar \omega - E_x)^{1/2} \quad (3.59)$$

with the constant $A_x = 4\sqrt{2}\,e^2|p_{xv}|^2/nc\sqrt{\mu_x}\,\hbar^2$, where p_{xv} is the transition matrix element between the valence and excitonic bands. Equation (3.59) is similar to that seen for direct band-to-band transitions, discussed above ((3.60)), and is only valid for the photon energies $\hbar\omega \geq E_x$. There is no absorption below the excitonic ground state in pure crystalline solids. Absorption of photons to excitonic energy levels is possible through either the excitation of electrons to higher energy levels in the conduction band and then nonradiative relaxation to the excitonic energy level, or through the excitation of an electron directly to the exciton energy level. Excitonic absorption occurs in both direct and indirect semiconductors.

In amorphous semiconductors, the excitonic absorption and photoluminescence can be quite complicated. According to (3.56), the excitonic energy level is below the optical band gap by an energy equal to the binding energy given in (3.57). However, there are four transition possibilities: (i) extended valence to extended conduction states, (ii) valence to extended conduction states, (iii) valence extended to conduction tail states, and (iv) valence tail to conduction tail states. These possibilities will have different optical gap energies, E_0, and different binding energies. Transition (i) will give rise to absorption in the free exciton states, transitions (ii) and (iii) will give absorption in the bound exciton states, because one of the charge carriers is localized in the tail states, and absorption through transition (iv) will create localized excitons, which are also called geminate pairs. This can be visualized as follows: if an electron–hole pair is excited by a high-energy photon through transition (i) and forms an exciton, initially its excitonic energy level and the corresponding Bohr radius will have a reduced mass corresponding to both charge carriers being in extended states. As such an exciton relaxes downward nonradiatively, its binding energy and excitonic Bohr radius will change because its effective mass changes in the tail states. When both charge carriers reach the tail

Fig. 3.14 Energy level diagram of the low-lying $4f^N$ states of trivalent ions doped in LaCl$_3$. After [3.64–66]. The pendant semicircles indicate fluorescent levels ▶

states (transition (iv)), their excitonic Bohr radius will be maintained although they are localized.

The excitonic absorption coefficient in amorphous semiconductors can be calculated using the same approach as presented in Sect. 3.3.3, and similar expressions to (3.40) and (3.43) are obtained. This is because the concept of the joint density of states is not applicable in amorphous solids. Therefore, by replacing the effective masses of the charge carriers by the excitonic reduced mass and the distance between the excited electron and hole by the excitonic Bohr radius, one can use (3.40) and (3.43) to calculate the excitonic absorption coefficients for the four possible transitions above in amorphous semiconductors. However, such a detailed calculation of the excitonic transitions in amorphous semiconductors is yet to be performed.

3.3.5 Impurity Absorption

Impurity absorption can be observed as the absorption coefficient peaks lying below fundamental (band-to-band) and excitonic absorption. It is usually related to the presence of ionized impurities or, simply, ions. The peaks occur due to electronic transitions between ionic electronic states and the conduction/valence band or due to intra-ionic transitions (within d or f shells, between s and d shells, and so on). The first case leads to intense and broad lines, while the characteristics of the features arising from the latter case depend on whether or not these transitions are allowed by parity selection rules. For allowed transitions, the absorption peaks are quite intense and broad, while forbidden transitions produce weak and narrow peaks. General reviews of this topic may be found in *Blasse* and *Grabmaier* [3.67], *Hen-*

Table 3.8 Occupation of outer electronic shells for rare earth elements

57	La	$4s^2$	$4p^2$	$4d^{10}$	–	$5s^2$	$5p^6$	$5d^1$	$6s^2$
58	Ce	$4s^2$	$4p^2$	$4d^{10}$	$4f^1$	$5s^2$	$5p^6$	$5d^1$	$6s^2$
59	Pr	$4s^2$	$4p^2$	$4d^{10}$	$4f^3$	$5s^2$	$5p^6$	–	$6s^2$
60	Nd	$4s^2$	$4p^2$	$4d^{10}$	$4f^4$	$5s^2$	$5p^6$	–	$6s^2$
...									
68	Er	$4s^2$	$4p^2$	$4d^{10}$	$4f^{12}$	$5s^2$	$5p^6$	–	$6s^2$
...									
70	Yb	$4s^2$	$4p^2$	$4d^{10}$	$4f^{14}$	$5s^2$	$5p^6$	–	$6s^2$
71	Lu	$4s^2$	$4p^2$	$4d^{10}$	$4f^{14}$	$5s^2$	$5p^6$	$5d^1$	$6s^2$

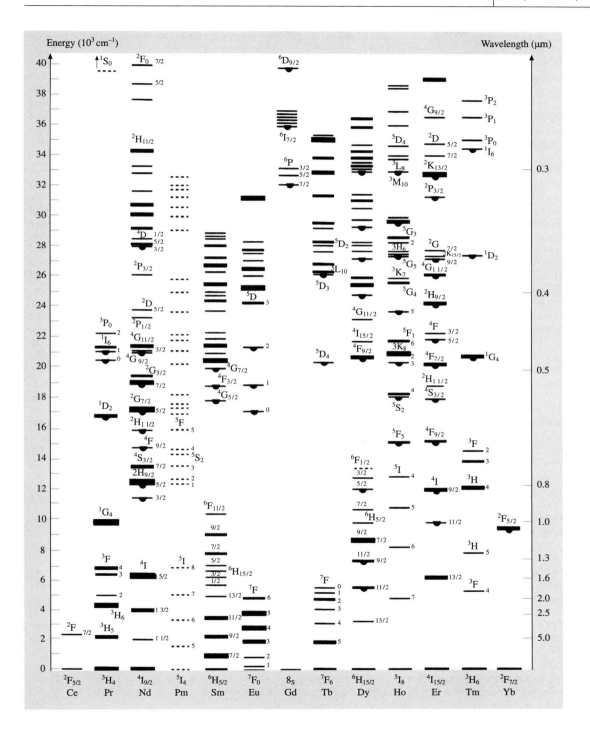

derson and *Imbusch* [3.65] and *DiBartolo* [3.68]. In the following section, we concentrate primarily on the properties of rare earth ions, which are of great importance in modern optoelectronics.

Optical Absorption of Trivalent Rare Earth Ions: Judd–Ofelt Analysis

Rare earths (REs) is the common name used for the elements from Lanthanum (La) to Lutetium (Lu). They have atomic numbers of 57 to 71 and form a separate group in Periodic Table. The most notable feature of these elements is an incompletely filled 4f shell. The electronic configurations of REs are listed in Table 3.8. The RE may be embedded in different host materials in the form of divalent or trivalent ions. As divalent ions, REs exhibit broad absorption–emission lines related to allowed 4f→5d transitions. In trivalent form, REs lose two 6s electrons and one 4f or 5d electron. The Coulomb interaction of a 4f electron with a positively charged core means that the 4f level gets split into complicated set of manifolds with energies, to a first approximation, that are virtually independent of the host matrix because the 4f level is well screened from external influences by the 5s and 5p shells [3.69]. The Fig. 3.14 shows an energy level diagram for the low-lying $4f^N$ states of the trivalent ions embedded in $LaCl_3$. To a second approximation, the exact construction and precise energies of the manifolds depend on the host material, via crystal field and via covalent interactions with the ligands surrounding the RE ion. A ligand is an atom (or molecule or radical or ion) with one or more unshared pairs of electrons that can attach to a central metallic ion (or atom) to form a coordination complex. Examples of ligands include ions (F^-, Cl^-, Br^-, I^-, S^{2-}, CN^-, NCS^-, OH^-, NH_2^-) and molecules (NH_3, H_2O, NO, CO) that donate a pair of electrons to a metal atom or ion. Some ligands that share electrons with metals form very stable complexes.

Optical transitions between 4f manifold levels are forbidden by a parity selection rule which states that the wavefunctions of the initial and final states of an atomic (ionic) transition must have different parities for it to be permitted. Parity is a property of any function (or quantum mechanical state) that describes the function after mirror reflection. Even functions (states) are symmetric (identical after reflection, for example a cosine function), while odd functions (states) are antisymmetric (for example a sine function). The parity selection rule may be partially removed for an ion (or atom) embedded in host material due to the action of the crystal field, which gives rise to "forbidden lines". The crystal field is the electric field created by a host material at the position of the ion.

The parity selection rule is slightly removed by the admixture of 5d states with 4f states and by the disturbed RE ion symmetry due to the influence of the host, which increases with the covalency. Higher covalency implies stronger sharing of electrons between the RE ion and the ligands. This effect is known as the nephelauxetic effect. The resulting absorption–emission lines are characteristic of individual RE ions and quite narrow because they are related to forbidden inner shell 4f transitions.

Judd–Ofelt (JO) analysis allows the oscillator strength of an electric dipole (ED) transition between two states of a trivalent rare earth (RE) ion embedded in a particular host lattice to be calculated. The possible states of an RE ion are often referred to as $^{2S+1}L_J$, where $L = 0, 1, 2, 3, 4, 5, 6 \ldots$ determines the electron's total angular momentum, and is conventionally represented by the letters S, P, D, F, G, I. The term $(2S+1)$ is called the spin multiplicity and represents the number of spin configurations, while J is the total angular momentum, which is the vector sum of the overall (total) angular momentum and the overall spin ($J = L + S$). The value $(2J+1)$ is called the multiplicity and corresponds to the number of possible combinations of overall angular momentum and overall spin that yield the same J. Thus, the notation $^4I_{15/2}$ for the ground state of Er^{3+} corresponds to the term $(J, L, S) = (15/2, 6, 3/2)$, which has a multiplicity of $2J+1 = 16$ and a spin multiplicity of $2S+1 = 4$. If the wavefunctions $|\psi_i\rangle$ and $|\psi_f\rangle$ correspond to the initial ($^{2S+1}L_J$) and final ($^{2S'+1}L_{J'}$) states of an electric dipole transition of an RE ion, the line strength of this transition, according to JO theory, can be calculated using:

$$S_{ed} = |\langle \psi_f | H_{ed} | \psi_i \rangle|^2$$
$$= \sum_{k=2,4,6} \Omega_k \left| \left\langle f_\gamma^N S'L'J' \left| U^{(k)} \right| f_\gamma^N SLJ \right\rangle \right|^2, \quad (3.60)$$

where H_{ed} is the ED interaction Hamiltonian, Ω_k are coefficients reflecting the influence of the host material, and $U^{(k)}$ are reduced tensor operator components, which are virtually independent of the host material, and their values are calculated using the so-called intermediate coupling approximation (see [3.70]). The theoretical values of S_{ed} calculated from this are compared with the values derived from experimental data using

$$S_{exp} = \frac{3hcn}{8\pi^3 e^2 \langle \lambda \rangle} \frac{2J+1}{\chi_{ed}} \int_{band} \frac{\alpha(\lambda)}{\rho} d\lambda, \quad (3.61)$$

where $\langle\lambda\rangle$ is the mean wavelength of the transition, h is the Plank constant, c is the speed of light, e is the elementary electronic charge, $\alpha(\lambda)$ is the absorption coefficient, ρ is the RE ion concentration, n is the refraction index and the factor $\chi_{\text{ed}} = (n^2 + 2)^2/9$ is the so-called local field correction. The key idea of JO analysis is to minimize the discrepancy between experimental and calculated values of line strength, S_{ed} and S_{exp}, by choosing the coefficients Ω_k, which are used to characterize and compare materials, appropriately. The complete analysis should also include the magnetic dipole transistions [3.71]. The value of Ω_2 is of prime importance because it is the most sensitive to the local structure and material composition and is correlated with the degree of covalence. The values of Ω_k are used to calculate radiative transition probabilities and appropriate radiative lifetimes of excited states, which are very useful for numerous optical applications. More detailed analysis may be found in, for example, [3.71]. Ω_k values for different ions and host materials can be found in *Gschneidner, Jr.* and *Eyring* [3.72].

3.3.6 Effects of External Fields

Electroabsorption and the Franz–Keldysh Effect

Electroabsorption is the absorption of light in a device where the absorption is induced by an applied (or changing) electric field within the device. Such a device is an electroabsorption modulator. There are three fundamental types of electroabsorption processes. In the Franz–Keldysh process, a strong applied field modifies the photon-assisted probability of an electron tunneling from the valence band to the conduction band, and thus it corresponds to an effective reduction in the "bandgap energy", inducing the absorption of light with photon energies of slightly less than the bandgap. It was first observed for CdS, in which the absorption edge was observed to shift to lower energies with the applied field; that is, photon absorption shifts to longer wavelengths with the applied field. The effect is normally quite small but is nonetheless observable. In this type of electroabsorption modulation, the wavelength is typically chosen to be slightly smaller than the bandgap wavelength so that absorption is negligible. When a field is applied, the absorption is enhanced by the Franz–Keldysh effect. In free carrier absorption, the concentration of free carriers N in a given band is changed (modulated), for example, by an applied voltage, changing the extent of photon absorption. The absorption coefficient is proportional to N and to the wavelength λ of the light raised to some power, typically 2–3. In the confined Stark effect, the applied electric field modifies the energy levels in a quantum well. The energy levels are reduced by the field by an amount proportional to the square of the applied field. A multiple quantum well (MQW) pin-type device has MQWs in its intrinsic layer. Without any applied bias, light with photon energy just less than the QW exciton excitation energy will not be significantly absorbed. When a field is applied, the energy levels are lowered and the incident photon energy is then sufficient to excite an electron and hole pair in the QWs. The relative transmission decreases with the reverse bias V_r applied to the pin device. Such MQW pin devices are usually not very useful in the transmission mode because the substrate material often absorbs the light (for example a GaAs/AlGaAs MQW pin would be grown on a GaAs substrate, which would absorb the radiation that excites the QWs). Thus, a reflector would be needed to reflect the light back before it reaches the substrate; such devices have indeed been fabricated.

The Faraday Effect

The Faraday effect, originally observed by Michael Faraday in 1845, is the rotation of the plane of polarization of a light wave as it propagates through a medium subjected to a magnetic field parallel to the direction of propagation of the light. When an optically inactive material such as glass is placed in a strong magnetic field and plane-polarized light is sent along the direction of the magnetic field, the emerging light's plane of polarization is rotated. The magnetic field can be applied, for example, by inserting the material into the core of a magnetic coil – a solenoid. The specific rotatory power induced, given by θ/L, has been found to be proportional to the magnitude of the applied magnetic field B, which gives the amount of rotation as:

$$\theta = \vartheta BL, \qquad (3.62)$$

where L is the length of the medium, and ϑ is the so-called Verdet constant, which depends on the material and the wavelength of the light. The Faraday effect is typically small. For example, a magnetic field of ≈ 0.1 T causes a rotation of about $1°$ through a glass rod of length 20 mm. It appears that "optical activity" is induced by the application of a strong magnetic field to an otherwise optically inactive material. There is, however, an important distinction between natural optical activity and the Faraday effect. The sense of rotation θ observed in the

Table 3.9 Verdet constants for some materials

Material	Quartz 589.3nm	Flint glass 632nm	Tb–Ga garnet 632nm	Tb–Ga garnet 1064nm	ZnS 589.3nm	Crown glass 589.3	NaCl 589.3
ϑ (rad m^{-1} T^{-1})	4.0	4.0	134	40	82	6.4	9.6

Faraday effect for a given material (Verdet constant) depends only on the direction of the magnetic field B. If ϑ is positive, for light propagating parallel to B, the optical field E rotates in the same sense as an advancing right-handed screw pointing in the direction of B. The direction of light propagation does not change the absolute sense of rotation of θ. If we reflect the wave to pass through the medium again, the rotation increases to 2θ. The Verdet constant depends not only on the wavelength λ but also on the charge-to-mass ratio of the electron and the refractive index $n(\lambda)$ of the medium through:

$$\vartheta = -\frac{(e/m_e)}{2c}\lambda\frac{dn}{d\lambda}. \quad (3.63)$$

Verdet constants for some glasses are listed in Table 3.9.

3.4 Thin Film Optics

Thin film optics involves multiple reflections of light entering a thin film dielectric (typically on a substrate) so that the reflection and transmission coefficients are determined by multiple wave interference phenomena, as shown in Fig. 3.15.

Consider a thin film coated on a substrate. Assuming that the incident wave has an amplitude of A_0, then there are various transmitted and reflected waves, as shown in Fig. 3.15. We then have the following amplitudes based on the definitions of the reflection and transmission coefficients:

$A_1 = A_0 r_{12}$
$A_2 = A_0 t_{12} r_{23} t_{21}$
$A_3 = A_0 t_{12} r_{23} r_{21} r_{23} t_{21}$
$B_1 = A_0 t_{12}$
$B_2 = A_0 t_{12} r_{23}$
$B_3 = A_0 t_{12} r_{23} r_{21}$
$B_4 = A_0 t_{12} r_{23} r_{21} r_{23}$
$B_5 = A_0 t_{12} r_{23} r_{21} r_{23} r_{21}$
$B_6 = A_0 t_{12} r_{23} r_{21} r_{23} r_{21} r_{23}$
$C_1 = A_0 t_{12} t_{23}$
$C_2 = A_0 t_{12} r_{23} r_{21} t_{23}$
$C_3 = A_0 t_{12} r_{23} r_{21} r_{23} r_{21} t_{23}$ (3.64)

and so on, where r_{12} is the reflection coefficient of a wave in medium 1 incident on medium 2, and t_{12} is the transmission coefficient from medium 1 into 2. For simplicity, we will assume normal incidence. The phase change upon traversing the thin film thickness d is $\phi = (2\pi/\lambda)n_2 d$, where λ is the free space wavelength. The wave must be multiplied by $\exp(i\phi)$ to account for this phase difference.

The reflection and transmission coefficients are given by

$$r_1 = r_{12} = \frac{n_1 - n_2}{n_1 + n_2} = -r_{21},$$
$$r_2 = r_{23} = \frac{n_2 - n_3}{n_2 + n_3}, \quad (3.65a)$$

and

$$t_1 = t_{12} = \frac{2n_1}{n_1 + n_2}, \quad t_2 = t_{21} = \frac{2n_2}{n_1 + n_2},$$
$$t_3 = t_{23} = \frac{2n_3}{n_2 + n_3}, \quad (3.65b)$$

where

$$1 - t_1 t_2 = r_1^2. \quad (3.66)$$

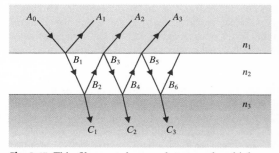

Fig. 3.15 Thin film coated on a substrate and multiple reflections of incident light, where n_1, n_2 and n_3 are the refractive indices of the medium above the thin film, the thin film, and the substrate, respectively

The total reflection coefficient is then obtained as:

$$r = \frac{A_{\text{reflected}}}{A_0} = r_1 - \frac{t_1 t_2}{r_1} \sum_{k=1}^{\infty} \left(-r_1 r_2 e^{-i2\phi}\right)^k, \quad (3.67)$$

which can be summed to be:

$$r = \frac{r_1 + r_2 e^{-i2\phi}}{1 + r_1 r_2 e^{-i2\phi}}. \quad (3.68)$$

The total transmission coefficient is obtained as:

$$t = \frac{C_{\text{transmitted}}}{A_0} = -\frac{t_1 t_{23} e^{i\phi}}{r_1 r_2} \sum_{k=1}^{\infty} \left(-r_1 r_2 e^{-i2\phi}\right)^k$$

$$= \left(\frac{t_1 t_3 e^{i\phi}}{r_1 r_2}\right) \frac{r_1 r_2 e^{-i2\phi}}{1 + r_1 r_2 e^{-i2\phi}}, \quad (3.69)$$

which can be summed to be:

$$t = \frac{t_1 t_3 e^{-i\phi}}{1 + r_1 r_2 e^{-i2\phi}}. \quad (3.70)$$

Equations (3.68) and (3.70) are very useful when studying the optical properties of thin films coated on a substrate as discussed by *Gould* in this handbook. (The reader is referred to the chapter on Thin Films by Gould.) In practice, the two most popular approaches are either to analyze optical transmission spectra, which may observed using a standard spectrophotometer, or ellipsometric investigations of refection. Both approaches are briefly explained below.

3.4.1 Swanepoel's Analysis of Optical Transmission Spectra

One of the simplest and most practically realizable approaches to the problem was developed by *Swanepoel* [3.73]. He showed that for a uniform film with a thickness d, a refractive index n and an absorption coefficient α deposited on the substrate with a refractive index s, the transmittance can be expressed as

$$T(\lambda) = \frac{Ax}{B - Cx\cos\varphi + Dx^2}, \quad (3.71a)$$

where

$$A = 16n^2 s, \quad (3.71b)$$
$$B = (n+1)^3(n+s^2), \quad (3.71c)$$
$$C = 2(n^2-1)(n^2-s^2), \quad (3.71d)$$
$$D = (n-1)^3(n-s^2), \quad (3.71e)$$
$$\varphi = 4\pi nd/\lambda, \quad (3.71f)$$
$$x = \exp(-\alpha d), \quad (3.71g)$$

and n is a function of λ.

For a nonuniform film with a wedge-like cross-section, (3.71a) must be integrated over the thickness of the film, giving

$$T_{\Delta d}(\lambda) = \frac{1}{\varphi_2 - \varphi_1} \int_{\varphi_1}^{\varphi_2} \frac{Ax}{B - Cx\cos\varphi + Dx^2} d x, \quad (3.72)$$

where

$$\varphi_1 = 4\pi n(d - \Delta d)/\lambda$$

and

$$\varphi_2 = 4\pi n(d + \Delta d)/\lambda, \quad (3.73)$$

where d is the average thickness of the film and Δd is the variation of the thickness throughout the illumination (testing) area. The concept behind Swanepoel's method is to construct two envelopes $T_M(\lambda)$ and $T_m(\lambda)$ that pass through the maxima and minima of $T(\lambda)$ and to split the entire spectral range into three regions: negligible absorption, weak absorption and strong absorption regions, as shown in Fig. 3.16.

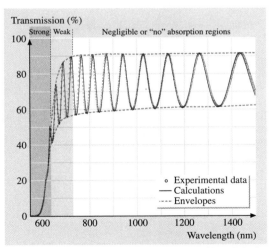

Fig. 3.16 Optical transmission of a-Se thin film. Calculations are done using (3.71) with the $n(\lambda)$ and $\alpha(\lambda)$ relations shown in Fig. 3.17. The film was prepared by the thermal evaporation of photoreceptor-grade selenium pellets. Film thickness was 2 µm. Tentative regions of strong, weak and negligible absorption are also shown

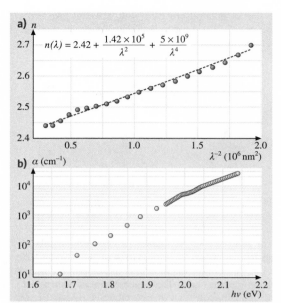

Fig. 3.17 (a) The spectral dependence of the refractive index of the a-Se thin film from Fig. 3.16. The *line* corresponds to the Cauchy approximation with the parameters shown on the figure. (b) The spectral dependence of the absorption coefficient of the same a-Se thin film

In the region of negligible absorption, (3.72) yields

$$T_{M/m} = \frac{\lambda}{2\pi n \Delta d} \frac{a}{\sqrt{1-b^2}} \tan^{-1} \left[\frac{1 \pm b}{\sqrt{1-b^2}} \tan\left(\frac{2\pi n \Delta d}{\lambda}\right) \right], \quad (3.74)$$

where $a = A/(B+D)$ and $b = C/(B+D)$. The "plus" sign in (3.74) corresponds to transmission maxima and "minus" to minima. Equation (3.74) is used to find Δd and n.

In the region of weak absorption,

$$T_{M/m} = \frac{\lambda}{2\pi n \Delta d} \frac{a_x}{\sqrt{1-b_x^2}} \tan^{-1} \left[\frac{1 \pm b_x}{\sqrt{1-b_x^2}} \tan\left(\frac{2\pi n \Delta d}{\lambda}\right) \right], \quad (3.75)$$

where $a_x = Ax/(B+Dx^2)$ and $b_x = Cx/(B+Dx^2)$. Equation (3.75) allows us to find x and n using the previously found Δd.

The previous results are used to construct $n(\lambda)$ and to create approximations using the Cauchy, Sellmeier or Wemple–Di Dominico dispersion equations. An example of a Cauchy approximation is shown in Fig. 3.17.

The positions of the extrema are given by the equation

$$2nd = m\lambda, \quad (3.76)$$

where $m = 1, 2, 3, \ldots$ for maxima and $m = 1/2, 3/2, \ldots$ for minima. Therefore, d can be found from the slope of (n/λ) versus m.

In the region of strong absorption, the absorption coefficient is calculated as

$$x = \frac{A - \sqrt{A^2 - 4T_i^2 BD}}{2T_i D}, \quad (3.77)$$

where

$$T_i = \frac{2T_M T_m}{T_M + T_m} \quad (3.78)$$

and A, B and D are calculated using the above-mentioned $n(\lambda)$ approximation.

It is worth noting that the division into negligible absorption, weak absorption and strong absorption regions is quite arbitrary and should be checked using trial-and-error methods.

3.4.2 Ellipsometry

Ellipsometry measures changes in the polarization of light incident on a sample in order to sensitively characterize surfaces and thin films, for example. Interactions of the polarized light with the sample cause a change in the polarization of the light, which may then be measured by analyzing the light reflected from the sample. The plane-polarized light incident on the sample can be resolved into two components – the component parallel to the plane of incidence (the p-component), and the component perpendicular to it (the s-component). That is, the incident light can be resolved into two components E_p (incident) and E_s (incident), where E refers to the electric field in the light wave, and similarly, the reflected light can be resolved into two components E_p (reflected) and E_s (reflected), enabling so-called Fresnel reflection coefficients, $R_p = E_p$ (reflected)$/E_p$ (incident) and $R_s = E_s$ (reflected)$/E_s$ (incident), to be obtained for the p- and s-components, respectively. Ellipsometry measures ρ, the ratio of these two coefficients, which is given by:

$$\rho = (R_p/R_s) = e^{i\Delta} \tan(\Psi), \quad (3.79)$$

where ρ is expressed in terms of the so-called ellipsometric angles Ψ ($0° \leq \Psi \leq 90°$) and Δ ($0° \leq \Delta \leq 360°$). These angles are defined as $\Psi = \tan^{-1}|\rho|$ and the differential phase change, $\Delta = \Delta_p - \Delta_s$. Thus, ellipsometry measures a change in the polarization, expressed as Ψ and Δ, in order to characterize materials. Because ellipsometry measures the ratio of two values, it can be made to be highly accurate and reproducible. The ratio ρ is a complex number; it contains the "phase" information Δ, which makes the measurement very sensitive. However, establishing values for Ψ and Δ is not particularly useful in itself for sample characterization. What one really wants to determine are the parameters of the sample, including, for example, the film thickness, optical constants, and the refractive index. These characteristics can be found by using the measured values of Ψ and Δ in an appropriate model describing the interaction of light with the sample. As an example, consider light reflected off an optically absorbing sample in air (in other words, with a refractive index of unity). The sample can be characterized by a complex refractive index $n - ik$, where n and k are the sample's refractive index and extinction coefficients at a particular wavelength. From Fresnel's equations, assuming that the light is incident at an angle ϕ to the sample normal, one gets

$$n^2 - k^2 = \sin^2(\phi)[1 + \{\tan^2(\phi) \\ \times [\cos^2(2\Psi) - \sin^2(2\Psi)\sin^2(\Delta)] \\ /[1 + \sin(2\Psi)\cos(\Delta)]^2\}] \quad (3.80a)$$

and

$$2nk = \sin^2(\phi)\tan^2(\phi)\sin(4\Psi)\sin(\Delta) \\ /[1 + \sin(2\Psi)\cos(\Delta)]^2 \,. \quad (3.80b)$$

Since the angle ϕ is set in the experiment, the two parameters measured from the experiment (Ψ and Δ) can be used to deduce the two remaining unknown variables in the equation above – namely, n and k. For a given model (a given set of equations used to describe the sample), the mean squared error between the model and the measured Ψ and Δ values is minimized, typically using the Marquardt–Levenberg algorithm, in order to quickly determine the minimum or best fit within some predetermined confidence limits. Thus, the n and k values are established using this procedure. In practice, ellipsometers consist of a source of linearly polarized light, polarization optics, and a detector. There are a number of different approaches to conducting an ellipsometry experiment, including null ellipsometers, rotating analyzer/compensator ellipsometers, and spectroscopic ellipsometers. Each of these approaches is discussed briefly below.

Historically, the first ellipsometers that were developed were null ellipsometers. In this configuration, the orientation of the polarizer, compensator and analyzer are adjusted such that the light incident on the detector is extinguished or "nulled". It should be noted that there are 32 combinations of polarizer, compensator and analyzer angles that can result in a given pair of Ψ and Δ values. However, because any two angles of the polarizer, compensator and analyzer that are 180° apart are optically equivalent, only 16 combinations of angles need to be considered if all angles are restricted to values below 180°. However, even when automated, this approach is thus inherently slow and spectroscopic measurements are difficult to make. However, this configuration can be very accurate and has low systematic errors. In order to speed up measurements, rotating analyzer/polarizer ellipsometers were developed. In these systems, either the analyzer or polarizer is continuously rotated at a constant angular velocity (typically about 100 Hz) about the optical axis. The operating characteristics of both of these configurations are similar. However, the rotating polarizer system requires the light source to be totally unpolarized. Any residual polarization in the source results in a source of measurement error unless corrected. Similarly, a rotating analyzer system is susceptible to the polarization sensitivity of the detector. However, solid state semiconductor photodetectors have extremely high polarization sensitivities. Thus, commercial systems tend to use rotating analyzer systems where residual polarization in the source is not an issue. In this case, the (sinusoidal) variation in the amplitude of the detector signal can be directly related to the ellipticity of the reflected light – Fourier analysis of this output provides values for Ψ and Δ. Such systems can provide high-speed, accurate measurements and the lack of a compensator actually improves the measurement by eliminating any errors associated with these components. Spectroscopic ellipsometers extend the concepts developed for measurement at a single wavelength to measurements at multiple wavelengths – typically as many as 40 wavelengths. Being able to measure the dispersion in optical constants with wavelength adds another dimension to the analysis, permitting unambiguous determinations of material and structure parameters.

3.5 Optical Materials

3.5.1 Abbe Number or Constringence

In an optical medium, the Abbe number is defined as the inverse of its dispersive power; that is, it represents the relative importance of refraction and dispersion. There are two common definitions based on the use of different standard wavelengths. The Abbe number ν_d is defined by:

$$\nu_d = (n_d - 1)/(n_F - n_C), \quad (3.81)$$

where n_F, n_d and n_C are the refractive indices of the medium at the Fraunhofer standard wavelengths corresponding to the helium d-line ($\lambda_d = 587.6$ nm, yellow), the hydrogen F-line ($\lambda_F = 486.1$ nm, blue) and the hydrogen C-line ($\lambda_C = 656.3$ nm, red), respectively. The Abbe number of a few glasses are listed in Table 3.10. The Abbe number ν_e, on the other hand, is defined by

$$\nu_e = (n_e - 1)/(n_{F'} - n_{C'}) : \quad (3.82)$$

where n_e, $n_{F'}$ and $n_{C'}$ are the refractive indices at wavelengths of the e-line (546.07 nm), the F'-line (479.99 nm) and the C'-line (643.85 nm), respectively.

The relationship between the refractive index n_d and the Abbe number ν_d is given by the Abbe diagram, where n_d is plotted on the y-axis against its corresponding ν_d value for different glasses, usually with the Abbe numbers decreasing on the x-axis, as shown in Fig. 3.18.

3.5.2 Optical Materials

Optical materials are those crystalline solids and glasses that are commonly used in the construction of optical components, such as lenses, prisms, window, mirrors and polarizers. In addition to having the required optical properties, such as a well defined refractive index n with a known dependence on wavelength $n = n(\lambda)$ and temperature, optical materials should also have various other desirable material properties, such as good homogeneity (including negligible macroscopic variations in the refractive index), negligible thermal expansion (small α_L, thermal expansion coefficient), resistance to mechanical damage (resistance to scratching), and resistance to chemical degradation (chemical corrosion, staining). Also, in the case of optical glasses, it is desirable to have negligible air (or gas) bubbles incorporated into the glass structure during fabrication (during shaping for example), negligible stress-induced birefringence, negligible nonlinear properties (unless used specifically due to its nonlinearity), and negligible fluorescence. In all of these cases, negligible implies less than a tolerable quantity within the context of the particular optical application. For example, at a given wavelength and for many glass materials used in optics, n should not vary by more than 10^{-5}, while 10^{-6} is required for certain optics applications in astronomy. In addition, they should have good or reliable man-

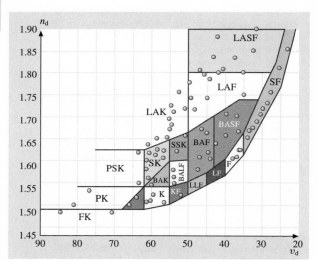

Fig. 3.18 The Abbe diagram is a diagram in which the refractive indices n_d of glasses are plotted against their Abbe numbers in a linear n_d versus ν_d plot and, usually, with the Abbe number decreasing along the x-axis, rather than increasing. A last letter of F or K represents flint or crown glass. Other symbols are as follows: S, dense; L, light; LL, extra light; B, borosilicate; P, phosphate; LA or La, lanthanum; BA or Ba, barium. Examples: BK, dense flint; LF, light flint; LLF, extra light flint; SSK, extra dense crown; PK, phosphate crown; BAK, barium crown; LAF, lanthanum flint

Table 3.10 Abbe numbers for a few glasses. PC denotes polycarbonate, PMMA is polymethylmethacrylate, and PS represents polystyrene

Optical glass →	SF11	F2	BaK1	Crown glass	Fused silica	PC	PMMA	PS
$\nu_d \rightarrow$	25.76	36.37	57.55	58.55	67.80	34	57	31

Table 3.11 The refractive indices, n_d, and Abbe numbers, v_d, (3.81) of selected optical materials (compiled from the websites of Oriel, Newport and Melles-Griot); n_d at $\lambda_d = 587.6$ nm, α_L is the linear thermal expansion coefficient

Glasses	Transmission (typical, nm)	n_d	v_d	Applications	Comment
Fused silica	175–2000	1.45846	67.8	Lenses, windows, prisms, interferometric FT-IR components. UV lithography	Synthetic. Has UV properties; transmittance and excellent thermal low α_L. Resistant to scratching
SF 11, flint	380–2350	1.78472	25.76	Lenses, prisms	Flint glasses have $v_d < 50$
LaSFN9, flint	420–2300	1.85025	32.17	Lenses, prisms	High refractive index. More lens power for less curvature
BK7, borosilicate crown	380–2100	1.51680	64.17	Visible and near IR optics. Lenses, windows, prisms, interferometric components	All around excellent optical lens material. Not recommended for temperature-sensitive applications
BaK1, barium crown	380–2100	1.57250	57.55	Visible and near IR optics. Lenses, windows, prisms, interferometric components	All around excellent optical lens material. Not recommended for temperature-sensitive applications
Optical crown	380–2100	1.52288	58.5	Lenses, windows, prisms, interferometric components	Lower quality than BK7
Pyrex, borosilicate glass		1.43385	66	Mirrors	Low thermal expansion
Crystals					
CaF$_2$ crystal	170–7000	1.43385	94.96	Lenses, windows for UV optics, especially for excimer laser optics	Sensitive to thermal shock
MgF$_2$ crystal	150–7000	$n_0 = 1.37774$ $n_e = 1.38956$		Lenses, windows, polarizers, UV transmittance	Positive birefringent crystal. Resistant to thermal and mechanical shock
Quartz, SiO$_2$ crystal	150–2500	$n_0 = 1.54431$ $n_e = 1.55343$		UV optics. Wave plates. Polarizers	Positive uniaxial birefringent crystal
Sapphire, Al$_2$O$_3$ crystal	150–6000	1.7708 (546.1 nm)		UV-Far IR windows, high power laser optics	High surface hardness, Scratch resistant. Chemically inert
Auxiliary optical materials					
ULE SiO$_2$-TiO$_2$ glass				Optical spacers	Very small thermal expansion
Zerodur, glass ceramic composite		1.5424	56–66	Mirror substrates. Not suitable for transmission optics due to internal scattering	Ultra-low α_L. Fine mixture of glass and ceramic crystals (very small size)

ufacturability at an affordable cost. There are various useful optical materials which encompass not only single crystals (such as CaF_2, MgF_2, quartz, sapphire) but also a vast range of glasses (which are supercooled liquids with high viscosity, such as flint and crown glasses as well as fused silica). Higher refractive index materials have more refractive power and allow lens designs that need less curvature to focus light, and hence tend to give fewer aberrations. Flint glasses have a larger refractive index than crown glasses. On the other hand, crown glasses are chemically more stable, and can be produced more to specification. While most optical materials are used for their optical properties (such as in optical transmission), certain "optical" materials (auxilary materials) are used in optical applications such as mirror substrates and optical spacers for their nonoptical properties, such as their negligible thermal expansion coefficients. Some optical properties of selected optical materials and their applications are listed in Table 3.11.

3.5.3 Optical Glasses

Optical glasses are a range of noncrystalline transparent solids used to fabricate various optical components, such as lenses, prisms, light pipes and windows. Most (but not all) optical glasses are either crown (K) types or flint (F) types. K-glasses are usually soda-lime-silica glasses, whereas flint glasses contain substantial lead oxide; hence F-glasses are denser and have higher refractive powers and dispersions. Barium glasses contain barium oxide instead of lead oxide and, like lead glasses, have high refractive indices, but lower dispersions. There are other high refractive index glasses, such as lanthanum- and rare earth-containing glasses. Optical glasses can also be made from various other glass formers, such as boron oxide, phosphorus oxide and germanium oxide. The Schott glass code or number is a special number designation (511 604.253 for Schott glass K7) in which the first three numbers (511) represent the three decimal places in the refractive index ($n_d = 1.511$), the next three numbers (604) represent the Abbe number ($\nu_d = 60.4$), and the three numbers after the decimal (253) represent the density ($\rho = 2.53 \text{ g/cm}^3$). A different numbering system is also used, where a colon is used to separate n_d and ν_d; for example, 517:645 for a particular borosilicate crown means $n_d = 1.517$, $\nu_d = 64.5$ (see also Sect. 3.5.1).

In the Schott glass coding system, optical glasses are represented by letters in which a last letter of K refers to crown, and F to flint. The first letters usually represent the most important component in the glass, such as P in the case of phosphate. The letters Kz ("Kurtz"), L ("leicht") and S ("schwer") before K or F represent short, light and dense (heavy) respectively (from German). S after K or F means "special". Examples include: BK, borosilicate crown; FK, fluor crown; PK, phosphate crown; PSK, dense phosphate crown; BaLK, light barium crown; BaK, barium crown; BaSK, dense barium crown; SSK, extra dense barium crown; ZnK, zinc crown; LaK, lanthanum crown, LaSK, dense lanthanum crown; KF, crown flint; SF, dense flint; SFS, special dense flint; BaF, barium flint; BaLF, barium light flint; BaSF, dense barium flint; LLF, extra light flint; LaF, lanthanum flint.

References

3.1 S. Adachi: *Properties of Group IV, III–V and II–VI Semiconductors* (Wiley, Chichester, UK 2005)
3.2 M. Fox: *Optical Properties of Solids* (Oxford Univ. Press, Oxford 2001)
3.3 J. H. Simmons, K. S. Potter: *Optical Materials* (Academic, San Diego 2000)
3.4 D. E. Aspnes, A. A. Studna: Phys. Rev **B27**, 985 (1983)
3.5 A. L. Cauchy: Bull. Sci. Math. **14**, 6 (1830)
3.6 A. L. Cauchy: *M'emoire sur la Dispersion de la Lumiere* (Calve, Prague 1836)
3.7 D. Y. Smith, M. Inokuti, W. Karstens: J. Phys.: Condens. Mat. **13**, 3883 (2001)
3.8 D. F. Edwards, E. Ochoa: Appl. Opt. **19**, 4130 (1980)
3.9 W. L. Wolfe: *The Handbook of Optics*, ed. by W. G. Driscoll, W. Vaughan (McGraw-Hill, New York 1978)
3.10 J. W. Fleming: Appl. Opt. **23**, 4486 (1984)
3.11 K. L. Wolf, K. F. Herzfeld, H. Geiger, K. Scheel (eds.): *Handbuch der Physik*, Vol. 20 (Springer, Berlin, Heidelberg 1928)
3.12 M. Herzberger: Opt. Acta **6**, 197 (1959)
3.13 P. J. Herve, L. K. J. Vandamme: J. Appl. Phys. **77**, 5476 (1995)
3.14 D. Dale, F. Gladstone: Philos. Trans. **153**, 317 (1863)
3.15 S. H. Wemple, M. DiDominico Jr.: Phys. Rev. **3**, 1338 (1971)
3.16 W. J. Turner, W. E. Reese: Phys. Rev. **127**, 126 (1962)
3.17 W. G. Spitzer, H. Y. Fan: Phys. Rev. **106**, 882 (1957)
3.18 J. D. Wiley, M. DiDominico: Phys. Rev. **B1**, 1655 (1970)
3.19 H. R. Riedl: Phys. Rev. **127**, 162 (1962)
3.20 R. L. Weihler: Phys. Rev. **152**, 735 (1966)
3.21 E. Hagen, H. Rubens: Ann. Phys. **14**, 986 (1904)

3.22 R. J. Elliott, A. F. Gibson: *An Introduction to Solid State Physics and Its Applications* (Macmillan, London 1974)
3.23 H. B. Briggs, R. C. Fletcher: Phys. Rev. **91**, 1342 (1953)
3.24 C. R. Pidgeon: Optical Properties of Solids. In: *Handbook on Semiconductors*, Vol. 2, ed. by M. Balkanski (North Holland, Amsterdam 1980) Chap. 5, pp. 223–328
3.25 H. E. Ruda: J. Appl. Phys. **72**, 1648 (1992)
3.26 H. E. Ruda: J. Appl. Phys. **61**, 3035 (1987)
3.27 W. Kaiser, R. J. Collins, H. Y. Fan: Phys. Rev. **91**, 1380 (1953)
3.28 I. Kudman, T. Seidel: J. Appl. Phys. **33**, 771 (1962)
3.29 A. E. Rakhshani: J. Appl. Phys. **81**, 7988 (1997)
3.30 R. H. Bube: *Electronic Properties of Crystalline Solids* (Academic, San Diego 1974)
3.31 F. Urbach: Phys. Rev. **92**, 1324 (1953)
3.32 J. Pankove: Phys. Rev. **140**, 2059 (1965)
3.33 R. A. Street: *Hydrogenated Amorphous Silicon* (Cambridge Univ. Press, Cambridge 1991)
3.34 D. A. Papaconstantopoulos, E. N. Economou: Phys. Rev. **24**, 7233 (1981)
3.35 M. H. Cohen, Fritzsche, S.R. Ovshinsky: Phys. Rev. Lett. **22**, 1065 (1969)
3.36 N. F. Mott, E. A. Davis: *Electronic Processes in Non-Crystalline Materials* (Clarendon, Oxford 1979)
3.37 J. Singh, K. Shimakawa: *Advances in Amorphous Semiconductors* (Taylor & Francis, London 2003)
3.38 S. Sherman, S. Wagner, R. A. Gottscho: Appl. Phys. Lett. **69**, 3242 (1996)
3.39 T. Tiedje, J. M. Cebulla, D. L. Morel, B. Abeles: Phys. Rev. Lett. **46**, 1425 (1981)
3.40 K. Winer, L. Ley: Phys. Rev. **36**, 6072 (1987)
3.41 D. P. Webb, X. C. Zou, Y. C. Chan, Y. W. Lam, S. H. Lin, X. Y. Lin, K. X. Lin, S. K. O'Leary, P. K. Lim: Sol. State Commun. **105**, 239 (1998)
3.42 W. B. Jackson, S. M. Kelso, C. C. Tsai, J. W. Allen, S.-H. Oh: Phys. Rev. **31**, 5187 (1985)
3.43 S. K. O'Leary, S. R. Johnson, P. K. Lim: J. Appl. Phys. **82**, 3334 (1997)
3.44 S. M. Malik, S. K. O'Leary: J. Non-Cryst. Sol. **336**, 64 (2004)
3.45 S. R. Elliott: *The Physics and Chemistry of Solids* (Wiley, Sussex 1998)
3.46 J. Singh: Nonlin. Opt. **29**, 119 (2002)
3.47 J. Tauc: Phys. Stat. Solidi **15**, 627 (1966)
3.48 F. Orapunt, S. K. O'Leary: Appl. Phys. Lett. **84**, 523 (2004)
3.49 A. P. Sokolov, A. P. Shebanin, O. A. Golikova, M. M. Mezdrogina: J. Phys. Conden. Mat. **3**, 9887 (1991)
3.50 G. D. Cody: Semicond. Semimet. **21**, 11 (1984)
3.51 K. Morigaki: *Physics of Amorphous Semiconductors* (World Scientific, London 1999)
3.52 L. Ley: *The Physics of Hydrogenated Amorphous Silicon II*, ed. by J. D. Joannopoulos, G. Lukovsky (Springer, Berlin, Heidelberg 1984) p. 61
3.53 T. Aoki, H. Shimada, N. Hirao, N. Yoshida, K. Shimakawa, S. R. Elliott: Phys. Rev. **59**, 1579 (1999)
3.54 K. Shimakawa, Y. Ikeda, S. Kugler: *Non-Crystalline Materials for Optoelectronics*, Optoelectronic Materials and Devices, Vol. 1 (INOE Publ., Bucharest 2004) Chap. 5, pp. 103–130
3.55 J. Singh: J. Mater. Sci. Mater. El. **14**, 171 (2003)
3.56 S. M. Malik, S. K. O'Leary: J. Mater. Sci. Mater. El. **16**, 177 (2005)
3.57 S. Abe, Y. Toyozawa: J. Phys. Soc. Jpn. **50**, 2185 (1981)
3.58 A. O. Kodolbas: Mater. Sci. Eng. **98**, 161 (2003)
3.59 J. Singh: *Excitation Energy Transfer Processes in Condensed Matter* (Plenum, New York 1994)
3.60 J. Singh: Nonlin. Opt. **18**, 171 (1997)
3.61 J. Singh, T. Aoki, K. Shimakawa: Philos. Mag. **82**, 855 (2002)
3.62 R. J. Elliott: *Polarons and Excitons*, ed. by K. G. Kuper, G. D. Whitfield (Oliver Boyd, London 1962) p. 269
3.63 T. Aoki, S. Komedoori, S. Kobayashi, T. Shimizu, A. Ganjoo, K. Shimakawa: Nonlin. Opt. **29**, 273 (2002)
3.64 P. C. Becker, N. A. Olsson, J. R. Simpson: *Erbium-Doped Fiber Amplifiers. Fundamentals and Technology* (Academic, New York 1999)
3.65 B. Henderson, G. F. Imbusch: *Optical Spectroscopy of Inorganic Solids* (Clarendon, Oxford 1989)
3.66 S. Hüfner: *Optical Spectra of Rare Earth Compounds* (Academic, New York 1978)
3.67 G. Blasse, B. C. Grabmaier: *Luminescent Materials* (Springer, Berlin, Heidelberg 1994)
3.68 B. DiBartolo: *Optical Interactions in Solids* (Wiley, New York 1968)
3.69 W. T. Carnall, G. L. Goodman, K. Rajnak, R. S. Rana: J. Chem. Phys. **90**, 3443 (1989)
3.70 M. J. Weber: Phys. Rev. **157**, 262 (1967)
3.71 E. Desurvire: *Erbium-Doped Fibre Amplifiers* (Wiley, New York 1994)
3.72 K. A. Gschneidner, Jr., LeRoy, Eyring (Eds.): *Handbook on the Physics and Chemistry of Rare Earths*, Vol. 25 (Elsevier, Amsterdam 1998)
3.73 R. Swanepoel: J. Phys. E **17**, 896 (1984)

4. Magnetic Properties of Electronic Materials

This work reviews basic concepts from both traditional macroscopic magnetism and unconventional magnetism, in order to understand current and future developments of submicronic spin-based electronics, where the interplay of electronic and magnetic properties is crucial. Traditional magnetism is based on macroscopic observation and physical quantities are deduced from classical electromagnetism. Physical interpretations are usually made with reference to atomic magnetism, where localized magnetic moments and atomic physics prevail, despite the fact that standard ferromagnetic materials such as Fe, Co and Ni are not localized-type magnets (they have extended s and localised d electronic states). While this picture might be enough to understand some aspects of traditional storage and electromechanics, it is not sufficient when describing condensed matter systems with smaller length scales (progressing toward the nanometer range). In this case, the precise nature of the magnetism (localized, free or itinerant as in Fe, Co and Ni transition metals) should be accounted for, along with the simultaneous presence of charge and spin on carriers. In addition, when we deal with the thin films or multilayers found in conventional electronics, or with objects of reduced dimensionality (such as wires, pillars, dots or grains), the magnetic properties are expected to be different from conventional three-dimensional bulk systems.

This chapter is organized as follows. We begin (in the Introduction) by highlighting the new era of submicronic spin-based electronics, and we present a table of papers on the topics we cover in the chapter, for the reader who wishes to learn more. The traditional elements of magnetism, such as the hysteresis loop, conventional types of magnetism and magnetic materials, are then presented (in Sect. 4.1). We then briefly describe (in Sect. 4.2) unconventional magnetism, which can be used to understand new high-tech materials that will be used in future devices based on spintronics and quantum information.

4.1	Traditional Magnetism	81
	4.1.1 Fundamental Magnetic Quantities	81
	4.1.2 The Hysteresis Loop	83
	4.1.3 Intrinsic Magnetic Properties	87
	4.1.4 Traditional Types of Magnetism and Classes of Magnetic Materials	90
4.2	Unconventional Magnetism	93
	4.2.1 Conventional and Unconventional Types of Exchange and Coupling in Magnetic Materials	93
	4.2.2 Engineering and Growth of Thin Magnetic Films	94
	4.2.3 Electronic Properties: Localized, Free and Itinerant Magnetism and Spin-Polarised Band Structure	95
	4.2.4 Prospects for Spintronics and Quantum Information Devices	98
References		99

Digital information technology involves three main activities:

- Processing of information (using transistors, logic gates, CPU, RAM, DSP...)
- Communication of information (using networks, switches, cables, fibers, antennae...)
- Storage of information (using tapes, hard disks, CD, DVD, flash memory...)

The application of magnetism to such technologies has traditionally been confined to information storage, originating from the development of bubble and ferrite core technologies, when RAM memory was based on

Table 4.1 Selected topics in magnetism, with corresponding applications and references for further reading

Topic	Applications/comments	Reference
Amorphous magnets	Shielding, sensing, transformers, transducers	*Boll, Warlimont* [4.1]
Coherent rotation model	Hysteresis loop determination	*Stoner, Wohlfarth* [4.2]
Coupling and exchange in multilayers	Biquadratic exchange, exchange bias, spring magnets	*Platt* et al. [4.3], *Slonczewski* [4.4], *Koon* [4.5]
Giant magnetoimpedance	Sensing, detectors	*Tannous, Gieraltowski* [4.6]
Giant magnetoresistance (GMR)	Recording heads, spin valves, spin filters	*White* [4.7]
Giant magnetostriction	Smart plane wings, MEMS, actuators, transducers, resonators	*Schatz* et al. [4.8], *Dapino* et al. [4.9]
Itinerant magnetism	Magnetism in transition metals	*Himpsel* et al. [4.10]
Localized magnetism	Atoms/molecules/ions/insulators	*Jansen* [4.11]
Losses in magnetic materials	Eddy currents, hysteresis loss	*Goodenough* [4.12]
Magnetic recording	Hard disk technology	*Richter* [4.13]
Magnetic thin films	Growth and characterization	*Himpsel* et al. [4.10]
Magnetoelastic effects	Cantilevers, MEMS	*Farber* et al. [4.14], *Dapino* et al. [4.9]
Microwave devices	Communications, bubble memory	*Coeure* [4.15], *Wigen* [4.16]
Permanent or hard magnets	Relays, motors, transformers	*Gutfleisch* [4.17]
Quantum computing/communications	Quantum devices, magnetic RAM	*Burkard, Loss* [4.18]
Sensors	Field detectors, probes	*Hauser* et al. [4.19]
Soft magnets	Shielding, sensing, transformers, transducers	*Boll, Warlimont* [4.1]
Spintronics	Spin diode, spin LED, spin transistor, magnetic RAM	*Prinz* [4.20], *Zutic* et al. [4.21]
Technology overview	Permanent and soft magnets	*Kronmueller* [4.22], *Simonds* [4.23]
Types of magnetic order	Ferromagnetism, antiferromagnetism, diamagnetism, paramagnetism	*Hurd* [4.24]

magnetism. RAM memory is a special type of nonpermanent (primary) information storage device, which can be distinguished from permanent (secondary) or mass information storage devices such as tapes, hard disks, floppy and zip disks, CDs, DVDs, and flash memory.

The field of applied magnetism is currently undergoing much transformation due to several recent developments, among which is the progress toward the nanometer scale in integrated circuits. At this length scale, quantum effects become extremely important and carrier spin becomes of interest since it may be conserved over this length scale and so could be used to carry and manipulate information. This would pave the way towards the fabrication of new devices based on charge and spin (spintronic devices) instead of just charge, as used in traditional microelectronics. This means that new types of junctions and transistors could be built that would use magnetism to tackle the processing of information.

Quantum effects have already been used in many products, since they are the basis of the *GMR* effect (Table 4.1) that is the basis for the recent surge towards extremely high densities in hard disks; however, the longer-term intention is to make use of these effects in basic components of *quantum computers* and *quantum communication devices*. The quantum computer is based on the *qubit* (quantum bit), which is the basic unit of information used in a quantum computer (equivalent to the classical bit used in conventional computing). If we consider a sphere, the classical bit can be viewed as an object with two possible states at the north and the south poles of the sphere, whereas a qubit is an object that can sit anywhere on the surface of the sphere (called the *Bloch sphere*; see *Nielsen* and *Chuang* [4.25]). A quantum computer can perform a massive number of computations in parallel, since one is allowed to access any superposition of states at any time in quantum mechanics due to its linearity, in contrast to a classical von Neumann type of computer, where one only has access to a single state at any time (Table 4.1). Quantum communications are extremely secure, since a caller may build a coherent state with the callee where any intrusion can perturb the coherence, providing very efficient detection.

From a fundamental point of view, one can start by representing a magnetic material by a single magnetic moment and then studying its behavior, before investigating the many interacting moments that are the building blocks of magnetic materials. From an applied point of view, the orientation of the moment defines the value of the bit. Once the orientation of the moment has been linked to a bit value, it becomes important to understand the physical processes, the energetics and the dynamics of the change in moment orientation (moment reversal from left to right or moment flip from up to down) in order to be able to control, alter and predict the bit value.

4.1 Traditional Magnetism

Classical magnetism relates to magnetic moments and their behavior when an external field, mechanical stress or some other perturbing effect is applied. The idea is to investigate the way that the energy of the magnetic moment changes with time. In order to describe the different magnetic energy terms that control the behavior of a moment, we can start from a single isolated moment in vacuum, at zero temperature ($T = 0$ K), zero applied magnetic field ($H = 0$) and zero frequency ($f = 0$). It is obvious that the energy is zero from a classical point of view (quantum mechanically, however, it is worth noting that, even at $T = 0$ K, quantum fluctuations exist that could disorient, flip or reverse the moment).

The next step is to apply a magnetic field H ($T = 0$ K, $f = 0$); we then obtain the Zeeman energy $E_Z = -\bm{M} \cdot \bm{H}$, with $\bm{M}$ representing the moment. If we place the moment in an unbounded (of infinite size) crystal, it is clear that the energy of the moment is anisotropic, since the direction of the moment (called an easy axis direction) is imposed by the internal symmetry of the crystal, in contrast to the case in vacuum, where $\bm{M}$ is free to point in any direction. The crystal may possess a set of easy axes (easy planes), in which case the moment will point in one of several directions (or any direction in one of several planes).

If the body containing the moment has a finite size, a new energy must be accounted for: the demagnetization energy. Magnetic surface charges (poles) induced on the surface bounding the body create a demagnetizing field inside the body (this is called the stray field outside of the body). The demagnetization energy is also called the shape anisotropy energy or the magnetostatic energy because it is (approximately) expressed (for ellipsoidal bodies) as $2\pi N_{\alpha\beta} M_\alpha M_\beta$, where $N_{\alpha\beta}$ is a set of factors (demagnetization coefficients) that depend on the shape of the body (the Einstein summation convention is used for repeated indices).

Now suppose we include several local moments in a material. If sites i and j carrying moments $\bm{M}_i$ and $\bm{M}_j$ are close enough, we then have an interaction energy between them called the exchange energy $-A_{ij} \bm{M}_i \bm{M}_j$, which will align moments $\bm{M}_i$ and $\bm{M}_j$ if A_{ij} is positive (ferromagnetic interaction). If A_{ij} is negative, the moments will align antiparallel to each other (antiferromagnetic interaction). In a ferromagnet we have a net moment, whereas in an antiferromagnet the net moment is zero.

The above description considers localized magnetism in distinct atoms (such as in a gas), ions, molecules or in special materials (like insulators or rare-earth solid state compounds possessing external-shell f electrons with highly atomic-like character). If we have a conducting material with free electrons interacting with localized atoms/ions/molecules (for transition metals with s and d electrons or magnetic semiconductors for example), a different type of magnetism called itinerant magnetism occurs. Nevertheless, it is possible to extend the notion of the magnetic moment to this case, accounting for the combined effects of free and localized charges modeled as an effective number of Bohr magnetons (see Sect. 4.2.1, Table 4.1 and Table 4.2).

The different physical mechanisms and types of magnetism briefly described above operate at different length scales. In order to gain some perspective and be able to ponder what lies ahead in terms of possible developments and hurdles, Fig. 4.1 gives a summary of different mechanisms, characterization methods and manufacturing processes along with their corresponding length scales. Note that, on the nanometer scale, the device size becomes comparable to most ranges of interactions encountered in magnetic materials, and this will trigger the development of novel effects and devices.

4.1.1 Fundamental Magnetic Quantities

Magnetization is the fundamental property exhibited by a magnetic substance. It originates from its electrons, as with the electric dipole moment. It can be

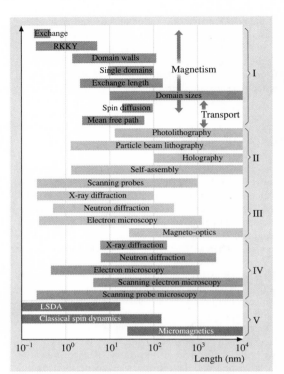

Fig. 4.1 Typical lengths of interest in magnetic materials and spintronics (in relation to basic phenomena, interactions and transport). Diagnostic techniques in III are depth profiling techniques and IV is for surface probing techniques) of interest in magnetic materials are also shown along with their probing lengths. II indicates growth methods and V theoretical characterization techniques. **LSDA** (local spin density approximation) is the spin counterpart of the density function theory (**DFT**) used to calculate band structure. Since most interaction lengths (shown in I) range between 1 and 100 nm, device size is comparable to interaction range in nanoelectronics (after [4.28])

intrinsic (it can exist without the application of any external field) or it can be induced by an external magnetic field. In atoms/molecules/ions and rare-earth solids, magnetization is created from individual atom-like localized magnetic moments. In the solid state, the moment is generally defined by the band structure and an effective moment can be defined in terms of a *Bohr magneton* $\mu_B = e\hbar/2m$, the magnetic moment carried by a single electron (e and m are the electron charge and mass). Magnetization is a thermodynamic quantity that changes with temperature, mechanical stress and chemical processes such as alloying (see *Chikazumi* [4.26]).

When a field is applied to a magnetic substance, the largest acquired magnetization M_s measured along the direction of the applied field is the saturation magnetization, meaning that all moments are aligned parallel to the field. Since increasing the temperature T causes more moments to misalign, one can define the saturation at $T = 0$ K with $M_s = N n_B \mu_B$, where N is the number of ions, atoms or molecules, and n_B the effective number of Bohr magnetons (see Table 4.2).

The number n_B is different for atoms/ions/molecules and solids and is determined by the electronic state; for example, for Fe^{3+} ions it is 5 (according to Hund rules, see *Jansen* [4.11]), since we have a $3d^5$ configuration (the orbital contribution is neglected), whereas for Fe atoms it is 2 ($3d^8$ configuration). Trivalent rare-earth ions with highly localized 4f orbitals possess magnetic moments that are determined by $g_J[J(J+1)]^{1/2}$, where g_J is the Landé factor and J the total angular momentum. Transition metal ions possess magnetic moments that are determined by $2[S(S+1)]^{1/2}$, where S is the total spin of the ion (there is no orbital contribution because of orbital moment quenching, see *Kittel* [4.27]). In the solid state, for example for ferromagnetic metals, n_B is determined from the band structure (Table 4.1). At lower dimensions (clusters, dots, thin films...), n_B tends to increase because of lower symmetry and coordination (high symmetry and coordination tend to decrease it). Among the elements, the rare-earth ions (Dy^{3+}, Ho^{3+} and Er^{3+}) possess large values of n_B – on the order of 10 – whereas solids such as Gd-based garnets (see *Wigen* [4.16]) have n_B values on the order of 15 (at $T = 0$ K).

Table 4.2 Selected ferromagnetic solids with their saturation magnetizations M_s, effective Bohr magnetons n_B and Curie temperatures [4.27]

Substance	M_s (G) at 300 K	M_s (G) at 0 K	n_B at 0 K	Curie T_c (K)
Fe	1707	1740	2.22	1043
Co	1400	1446	1.72	1388
Ni	485	510	0.62	627
Gd	–	2060	7.63	292
Dy	–	2920	10.2	88
MnAs	670	870	3.4	318
MnBi	620	680	3.52	630
CrO_2	515	–	2.03	386
EuO	–	1920	6.8	69

The moment due to angular momentum is given in the localized case (atoms/ions/molecules) by: $-g_L\mu_B L$ (for orbital angular momentum L), where $g_L = 1$ (the orbital Landé factor), or by $-g_S\mu_B S$ (for spin angular momentum S), where $g_S = 2$ (the spin Landé factor). In the case of an atom/ion with a total angular momentum $J = L + S$, the moment is given by $-g_J\mu_B J$, where:

$$g_J = 1 + [J(J+1) + S(S+1) - L(L+1)]/2J(J+1)$$

which is the total Landé factor. In the solid state, the Landé factor is determined from the band structure (using the spin-splitting of electron or hole bands in the presence of a magnetic field). It may become negative, anisotropic (a tensor) and depend on the wavevector k. Typical values of g for semiconductors at $k = 0$ include: InSb, $g = -44$; GaAs, $g = 0.32$; InAs, $g = -12$ [4.11, 21].

4.1.2 The Hysteresis Loop

When a magnetic field is applied to a ferromagnetic material [4.27], a change in magnetization takes place. The straightforward way to understand the underlying phenomena is to plot the magnetization M along the direction of the applied field H. The locus of the magnetization depicted in the M–H plane is the hysteresis loop (Fig. 4.2).

The term hysteresis means that when the material is field-cycled (in other words, when the field H is increased then decreased), two different nonoverlapping curves $M(H)$ are obtained. Another approach involves plotting the induction B versus H, since it is B that is actually measured. Several points in the M–H plane shown in the figure are worth mentioning. Point d is where the magnetization reaches its largest value M_s (the saturation magnetization). Points i and f denote the magnetic fields $\pm H_c$ for which the magnetization is zero. The magnetic field H_c is called the coercive field; it is generally large for hard materials and small for soft magnets. Points e and h denote the remanent magnetization M_r, which is the magnetization that remains after the applied field is switched off, which is the hallmark of permanent magnets.

The ratio M_r/M_s, called the squareness, is close to 1 when the applied magnetic field is closely aligned with the easy axis (and so the hysteresis loop is squarer). Once the easy axis (or set of easy axes) is determined, the angle between the easy axis and the magnetic field (ϕ, say) is varied and the hysteresis loop is plotted for different angles. When the angle ϕ is increased, the opening of the hysteresis loop is reduced; it is largest when the magnetic field is parallel to the easy axis and smallest when the magnetic field is parallel to the so-called hard

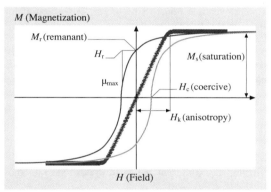

Fig. 4.3 Hysteresis loop obtained for an arbitrary angle ϕ between the magnetic field and the anisotropy axis (easy axis). Associated quantities such as coercive field H_c, remanent magnetization M_r and field H_r (given by the intersection of the tangent to the loop at $-H_c$ and the M_r horizontal line) are shown. The tangent to the loop at $-H_c$ is also called the maximum differential permeability. The *thick line* is the hysteresis loop when the field is along the hard axis and H_k (the anisotropy field) is the field value at the slope break where the magnetization reaches its saturation value M_s. Quantities such as H_c, M_r, H_r and μ_{max} depend on the angle ϕ. When the field is along the easy axis, the coercive field reaches its largest value (the hysteresis loop is broadest), at which point the magnetization jumps (at $\pm H_c$)

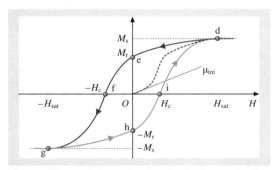

Fig. 4.2 Hysteresis loop showing the fundamental points in the M–H plane, such as the saturation magnetization M_s, the remanent M_r, the coercive field H_c (for which $M = 0$) and the saturation field H_{sat} (for which $M = M_s$), as well as the initial magnetization curve (*dashed*) with initial permeability μ_{ini} (after [4.29])

axis. All of these characteristics of the hysteresis loop are depicted in Fig. 4.3, and for a given temperature and frequency of the applied field H, quantities such as the remanent magnetization M_r, the remanent field H_r, the coercive field H_c and the maximum differential permeability μ_{max} vary with the angle ϕ;

In addition, the hysteresis loop changes and may even disappear altogether above a given temperature (Curie temperature), and it can be seriously altered by changing the frequency of the field (see below).

The Magnetization Process Viewpoint

When a magnetic field is applied to a material, a magnetization process takes place. This means that a change in magnetization can occur according to any of the following routes. At low applied magnetic field (and low frequency), domain boundaries bulge (this is a reversible regime, implying that if we decrease the field again, the magnetization will go back to the initial state by following the same path). This is called the domain nucleation and pinning process (see Fig. 4.4). At higher fields the walls are depinned and free to move. This is an irreversible regime (in contrast to the low-field case), meaning that, if we decrease the field, the change in magnetization upon field variation will not follow the same path. This regime is the free domain wall movement regime. At higher fields (or frequencies), the magnetization changes through the rotation of moments. In this case we have two possibilities: if the material is homogeneous and therefore behaves as if there is a single moment in the material, we get the coherent rotation regime (this is well described by the Stoner–Wohlfarth model, see Table 4.1). On the other hand, if the magnetization changes in a material in an inhomogeneous way (different points in the sample undergo different magnetization changes), we obtain a curling process [4.26]. Finally, if the magnetic field is applied along the easy axis and then suddenly reversed, the magnetization changes with a switching process, meaning the moment jumps from a positive value to a negative value of the same magnitude without any rotation process, since the magnetization is already along the easy axis.

The Energetics Viewpoint

Hysteretic behavior can be viewed as stemming from the motion of the energy minimum as the field is varied. In Fig. 4.5 the field is cycled in a clockwise fashion. The energy minimum is displayed for several values of the field. Hysteresis is shown to arise from the asymmetric behavior of the energy with the magnetization as the field is cycled. The loss of hysteresis is given by the area circumscribed by the hysteresis loop. Since soft materials have thinner hysteresis loops (equivalent to smaller coercivity) than hard materials their losses are smaller.

The Signal Processing Viewpoint

If we apply a time-dependent field $H(t)$ to the magnetic material and monitor the magnetization $M(t)$ as the output, then the relation $M(t) = F[H(t)]$ is that of a nonlinear filter. The magnetic material acts as a nonlinear filter because M is not proportional to H except at very low fields. This nonlinearity is easily observed if $M(t)$ is monitored as a square signal with a sinusoidal input excitation $H(t)$ (Fig. 4.6). In addition, the material imposes a delay in signal propagation that is proportional to the width of the hysteresis loop (twice the coercive field). Hysteretic behavior means that different values of output are obtained when the input excitation is in-

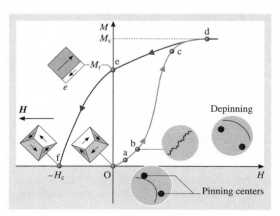

Fig. 4.4 Reversible and irreversible motion of the magnetization M with the field H. For small fields, $H < H_c$ and initial magnetization, Rayleigh's law is valid along a–b, but for large H, saturation is induced in M through irreversible domain boundary motion along b–c followed by rotation of magnetization along c–d. When the field is reversed along d–e–f towards $-H_c$, the variation in magnetization is not exactly the same as seen previously, and this is why a different $M(H)$ curve is obtained. Reversible movement at low field means that domain walls are pinned by impurities and they simply bulge under the action of $H(0-a)$, whereas jerky irreversible movement is associated with depinning of the domain walls (a–b), creating Barkhausen noise. Free irreversible movement beyond the pinning centers is depicted on the right (after [4.29])

creased or decreased, and this phenomenon is widely exploited (in control systems for instance).

The Information Storage Viewpoint

Obtaining a nonzero response (M) at zero excitation ($H = 0$) is a phenomenon known as remanence (we call this remanent value M_r). The major advantage of remanence (in addition to its usefulness in permanent magnets) is that information (the value of M_r) can be stored without any excitation source ($H=0$). Since we have M_r when the signal is decreased (we call this M_{r1}), we ought to get a different M_r when the signal is increased (we call this M_{r2}) due to hysteresis. In this case we get two bit values (0 for M_{r1} and 1 for M_{r2}). This contrasts sharply with electronic storage, where a voltage must be maintained in order to maintain the charge (representing the information); see Chapt. 51.

The Electromagnetic Compatibility and Frequency Synthesis Viewpoint

Hysteretic behavior can be viewed as the flux ϕ (corresponding to the magnetization M) induced in a circuit by an exciting current $i(t)$ at time t (corresponding to field H). The resulting relationship $\phi[i(t)]$ is a nonlinear characteristic that can be expanded as:

$$\phi[i(t)] = a_0 + a_1 i(t+\tau) + a_2 [i(t+\tau)]^2 + a_3 [i(t+\tau)]^3 + a_4 [i(t+\tau)]^4 + \ldots,$$

where τ accounts for the response delay of the flux (the larger the delay, the broader the hysteresis curve). Ordinary linear inductance of the circuit corresponds to the first derivative $[d\phi/di]$, which equals a_1 for a short delay τ. Higher order terms define nonlinear inductor behavior that results in (for a short delay)

$$L(i) = a_1 + 2a_2 i + 3a_3 i^2 + 4a_4 i^3 + \ldots.$$

This nonlinear inductor response is interesting from a frequency synthesis point of view, since it can generate harmonics and subharmonics from the exciting current $i(t)$ (see, for instance *Chua* [4.30]).

Time-Dependent Viewpoint

In spite of the variety of different viewpoints describing hysteresis given above, some experimentally observed phenomena cannot be explained using any of them, including one particularly important phenomenon known as the magnetic after-effect. Experimentally, one observes a change in magnetization with time $M(t)$, despite the fact that the applied field H is kept constant, implying the presence of a time-dependent hysteresis loop stemming solely from $M(t)$. Therefore, the material must contain units (single-domain grains) that respond differently over time, as if each unit had its own intrinsic delay; in this case the magnetization could be written in the form $M(t) = M_0 S(t) \ln(t)$, where $S(t)$ is called the magnetic viscosity. If the response time of each

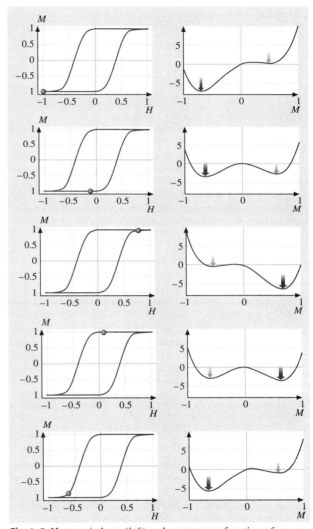

Fig. 4.5 Hysteresis loop (*left*) and energy as a function of magnetization (*right*). As the field H is varied in the hysteresis loop (as indicated by the *spot* ●) in a clockwise fashion, the energy changes its form and its absolute minimum (indicated by the *fat arrow*) moves accordingly. Hysteresis is due to asymmetric energetic behavior as the field is cycled

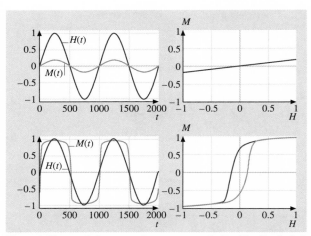

Fig. 4.6 Hysteresis loop as an input–output relationship. On the *left*, the magnetic field $H(t)$ and the magnetization $M(t)$ are displayed explicitly as functions of time t. On the *right*, the resulting input–output relationship (or M–H characteristic), the parametric curve $M(H)$, is shown. At the *top*, we have an output ($M(t)$) that is undelayed and proportional to the input $H(t)$, resulting in a paramagnetic (with no hysteresis) M–H characteristic. *Below*: an output $M(t)$ delayed and nonlinearly distorted version of the input $H(t)$, resulting in a hysteretic M–H characteristic. (After [4.31])

grain is considered to be of the thermal activation type, in other words it follows the Néel–Arrhenius formula (see Chapt. 51) $\tau = \tau_0 \exp(\Delta E / k_\mathrm{B} T)$, where τ_0 is an "attempt time" to cross some energy barrier ΔE (see Fig. 4.5) at a temperature T, then the after-effect implies that we have a distribution of τ_0 and ΔE values. Calling the probability density function $g(\tau)$, and using a scaling initial magnetization M_0, $M(t)$ is thus defined as:

$$M(t) = M_0 \int_0^\infty g(t) \exp(-t/\tau) \, \mathrm{d}\tau \, .$$

In the simple case where we have a single response time τ^*, in other words $g(\tau) = 2\delta(\tau - \tau^*)$, $M(t)$ behaves as $M(t) = M_0 \exp(-t/\tau^*)$.

After analyzing the different aspects of the hysteresis loop, we now move on to investigate the physical processes that affect the magnetization behavior. We start with small magnetic fields H; in other words $H < H_\mathrm{c}$ (the coercive field). The initial magnetization curve follows Rayleigh's law, $M(H) = \chi_0 H + \alpha_\mathrm{R} H^2$, where χ_0 is the low-field susceptibility and α_R is the Rayleigh coefficient.

Since this law is valid for small fields, it describes reversible changes in the magnetization.

We now consider the physical region inside a material where the magnetization is oriented along some direction. Typically, when the extent of the material is smaller than the exchange length, one expects a single domain structure (which contains about 10^{12}–10^{18} atoms, see also Fig. 4.1). For instance, a recording medium is considered to be made up of small grains that are made of single domains. As the recording density is increased, the grain size in the recording medium decreases, and if is small enough, its magnetization becomes sensitive to thermal energy – it can flip or reverse simply upon changes in temperature. This is called the superparamagnetic effect, and it traditionally limits longitudinal recording to $100\,\mathrm{Gbit/in}^2$. This issue has recently been partially circumvented by introducing antiferromagnetic coupling between the storage layer and a stabilization layer. A thin film of ruthenium (7–9 Å thick) is deposited between the storage layer and the stabilization layer (Chapt. 51). If the extent of the magnetic material is large, one expects to have a multidomain structure separated by domain walls. This multidomain structure emerges in order to minimize the magnetostatic long-range dipolar interaction energy between the different moments existing in the different domains [4.32].

A domain is a piece of a material that is magnetized along a given direction (see Fig. 4.7). In general, a magnetic material contains many regions where the magnetization is aligned in a particular direction in order to minimize the magnetostatic energy. Regions with different magnetization orientations can be positioned close to one another, albeit separated by a boundary called a domain wall (typically containing about 10^2–10^3 atoms, Fig. 4.1). Saturation occurs when all of these regions align along some direction imposed by the external applied field. The width of a domain wall is equal (in CGS units) to $\pi(A/K)^{1/2}$, where A is the typical nearest neighbor Heisenberg exchange and K the typical anisotropy, meaning that it is the result of exchange and anisotropy; it is thinner for higher anisotropy or smaller exchange (in Fe it is about 30 nm, whereas in the hard material $\mathrm{Nd}_2\mathrm{Fe}_{14}\mathrm{B}$ it is only 5 nm). Its energy is equal to $4(AK)^{1/2}$. For bulk materials it is of the Bloch type (see Fig. 4.8), whereas for thin films (with thicknesses on the order of the exchange length) it is of the Néel type, with a width proportional to $\left[A/(2\pi M_\mathrm{s}^2)\right]^{1/2}$. A single parameter $Q = 2K/M_\mathrm{s}^2$ allows us to discriminate between simple ($Q < 1$) and complex wall profiles ($Q > 1$) [4.33].

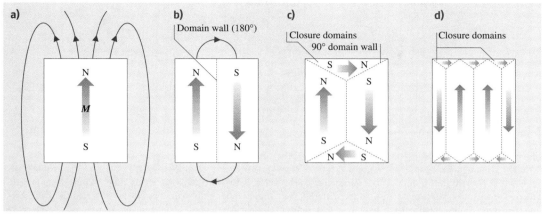

Fig. 4.7a–d Possible domain structures and wall structures in a magnetic material. (**a**) Initial magnetic configuration with a large stray field (outside the material). In order to reduce the magnetostatic energy (and the stray field), domains form, as in a highly anisotropic material (**b**). If n domains are formed, the magnetostatic energy is reduced by n [4.27]. In a material with small anisotropy, 45° closure domains are formed to minimize the magnetostatic energy (**c**). As more domains are formed, more walls are also formed, until a compromise is reached since it is necessary to minimize the sum of the magnetostatic energy and the wall energies (**d**). (After [4.29])

During a magnetization process, in a hysteresis loop cycle for instance, the irreversible jerky movements of domain walls due to local instabilities created by impurities, defects, inclusions or interactions between domain walls lead to Barkhausen noise. In devices and recording media, it is better to induce magnetization changes through rotation processes because they are less noisy.

4.1.3 Intrinsic Magnetic Properties

The induction B in a linear isotropic material is related to H through the relation $B = \mu_0(H + M)$, where μ_0 is the free space permeability and M is the magnetization of the material. Using the relation $M = \chi H$, which relates the magnetization to the applied field, one gets

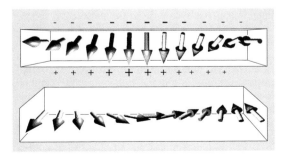

$B = \mu_0(1 + \chi)H$. This leads to the definition of the total permeability $\mu = \mu_0(1 + \chi)$, which yields the constitutive relation $B(H) = \mu H$. The quantity χ is called the susceptibility. The nonlinear constitutive equations $M(H)$ and $B(H)$ apply to a general nonlinear material. Such constitutive relations are similar to the $I(V)$ characteristics of electronic devices. For a linear material, μ is generally frequency-dependent. If losses occur during the propagation of an electromagnetic wave in a material due to the absorption of magnetic energy, it is possible to extend the definition of permeability to the complex plane (as in dielectrics). Losses are attributed to peaks in the imaginary part of the permeability at the absorption frequencies [4.12]. At low frequencies, losses are attributed to domains, whereas at high frequencies, losses are attributed to rotation processes. Finally, since the permeability relates two vector quantities B and H, if we consider a linear anisotropic material (such as a linear crystal), the relation $B(H) = \mu H$ becomes $B_\alpha = \mu_{\alpha\beta} H_\beta$ (Einstein summation convention used),

Fig. 4.8 Possible domain wall shapes in a magnetic material. *Above*: the Bloch type for bulk materials, with the magnetization rotating in a vertical plane with the associated poles. *Below*: the Néel type for thin films. For these walls, the magnetization rotates in a horizontal plane when the width of the film is smaller than the exchange length. (After [4.34])

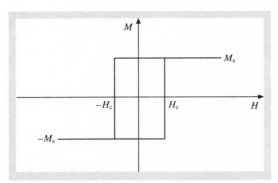
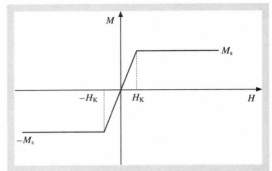

Fig. 4.9 Easy axis hysteresis loop. When the magnetic field is along the anisotropy axis (easy axis) of the material, the hysteresis loop is broadest and the magnetization jumps (switches) to $\pm M_s$ at the coercive field $\pm H_c$

Fig. 4.10 Hard axis hysteresis loop. When the magnetic field is along the hard axis of the material, the hysteresis loop is the thinnest and the magnetization rotates smoothly from one value to another between $\pm H_k$ (the anisotropy field)

and $\mu_{\alpha\beta}$ is a second-rank permeability tensor that has different components according to the relative directions of $\boldsymbol{B}$ and $\boldsymbol{H}$ (nine components in total, or 3^2 in three dimensions). For a nonlinear anisotropic material, we have $\mu_{\alpha\beta} = \partial B_\alpha / \partial H_\beta$.

When a material is run through a field cycle (a hysteresis loop) for the first time, the magnetization $\boldsymbol{M}$ follows a path different to the $\boldsymbol{M}(\boldsymbol{H})$ lines described above. This path represents the initial magnetization, and the associated permeability is the initial permeability (Fig. 4.2). If a magnetic material is subjected to a static magnetic field $\boldsymbol{H}$ and a small time-dependent field $\boldsymbol{h}(t)$ (alternating at frequency f) parallel to $\boldsymbol{H}$, the magnetic response is called the longitudinal permeability (at the frequency f). This kind of excitation is encountered in ferromagnetic resonance (FMR) problems (see *Wigen* [4.16]).

Alternatively, one can describe magnetic behavior in terms of susceptibility. This relates the magnetization to the applied field in a linear material through $\boldsymbol{M} = \chi \boldsymbol{H}$. For a nonlinear isotropic material, we have $\chi = \partial M / \partial H$, and for a nonlinear anisotropic material we have $\chi_{\alpha\beta} = \partial M_\alpha / \partial H_\beta$, with the $\boldsymbol{M}$ component along the direction α and the $\boldsymbol{H}$ component along β. Any jump in χ signals an important magnetic change in the material. When the temperature is decreased, the Curie temperature signals the change to ferromagnetic order from paramagnetic disorder, and the Néel temperature signals the change to antiferromagnetic order from paramagnetic disorder. Magnetic susceptibility spans several orders of magnitude (from about 10^{-5} to 10^6 cm^3/mol).

The magnetization occurs along the direction that minimizes the sum of the magnetic anisotropy energy and the Zeeman energy provided by the applied field. Since the hysteresis loop changes its shape as we vary the angle of the applied field, at some point it reaches its largest width. The field then indicates the easy axis direction (also called the anisotropy axis). The extremum value of the magnetic field is the coercive field $\pm H_c$, where the magnetization (switches) jumps (see Fig. 4.9) to its saturation value $\pm M_s$.

Alternatively, let us consider the direction along which it takes most effort to align the magnetization. This is easily determined as the field direction where the hysteresis loop when it is narrowest (see Fig. 4.10). The point at which the slope changes dramatically gives the value of the anisotropy field $\pm H_K$.

Crystals are generally anisotropic because their microscopic structures are not the same in all directions. Hence, the properties (including magnetic properties) will change with the orientation of the crystal. Indeed, there will be special directions along which the magnetization prefers to point. This means that some kind of energy will be minimized when the magnetization settles along these directions: the anisotropy energy. As well as being inherent to many crystal structures, anisotropy can also be induced by an external field, a change of symmetry (for example at the surface with respect to bulk) or mechanical deformation. Microscopically, anisotropy originates from spin-orbit coupling. Coercivity increases with anisotropy; in other words the hardness increases with anisotropy (Fig. 4.11). The anisotropy energy can be evaluated from the hysteresis loop by determining the loop for various angles between the applied field and the easy axis. In general, it is given by an expansion of the form $\sum_{\alpha,\beta} K_{\alpha\beta} M_\alpha M_\beta + \sum_{\alpha,\beta,\gamma,\delta} K_{\alpha\beta\gamma\delta} M_\alpha M_\beta M_\gamma M_\delta \ldots$, where $K_{\alpha\beta}$, $K_{\alpha\beta\gamma\delta}$ are the anisotropy coefficients (second- and fourth-order respectively), and M_α, M_β are the components (direction cosines) of the normalized magnetization (by the saturation magnetization M_s). α, β, γ are the indices along the crystal x, y, z axes. This means that the norm of $\boldsymbol{M}$, $|\boldsymbol{M}| = \left(M_x^2 + M_y^2 + M_z^2\right)^{1/2} = 1$ in this case.

The anisotropy energy of a crystal with a single dominant axis (such as hexagonal – like cobalt, tetragonal and rhombohedral crystals) is called uniaxial. To the lowest approximation, it is given by $K_1 \sin^2 \theta$, where K_1 is the anisotropy strength, and the angle θ is the angle between the magnetization and the easy axis. The next lowest approximation allows fourth-order terms to be added to the $K_1 \sin^2 \theta$ term. In hexagonal crystals, the fourth-order term is of the form $K_2 \sin^4 \theta$. Rhombohedral symmetry allows two fourth-order terms of the form $K_2 \sin^4 \theta + K_4 \cos \theta \sin^3 \theta \cos 3\phi$. In tetragonal crystals, the fourth-order terms are of the form $K_2 \sin^4 \theta + K_4 \cos \theta \sin^4 \theta \cos 4\phi$. The easy axis is taken to be along the z-axis, the angle ϕ is the magnetization polar angle in the xy-plane, and θ is the azimuthal angle.

The anisotropy energy for a crystal with two dominant axes (including monoclinic, triclinic and orthorombic crystals) is called biaxial. This means that two anisotropy directions exist and are competing. Some other cases are given below (see also Table 4.1 for other definitions).

Cubic Anisotropy

The anisotropy energy for the most symmetric crystals, cubic crystals, is given by $K_1\left(M_1^2 M_2^2 + M_1^2 M_3^2 + M_2^2 M_3^2\right) + K_2 M_1^2 M_2^2 M_3^2$ (the energy is to the sixth order). Here the coefficients M_1, M_2, M_3 are the components of the normalized magnetization with respect to the crystal axes. That is, $M_1 = \sin \theta \cos \phi$, $M_2 = \sin \theta \sin \phi$, $M_3 = \cos \theta$. The easy axis, as above, is taken to be along the z-axis, and the angle ϕ is the polar angle in the xy-plane.

Helical Anisotropy

In a magnetic wire, the magnetization may prefer to lie radially in a plane perpendicular to the wire axis, resulting in a radial easy axis (radial anisotropy). If it lies along the tangents to circles that lie on planes perpendicular to the wire axis, then we have circular anisotropy. However, if it lies in a plane tangent to the lateral surface of the wire (parallel to the wire axis) and it makes an angle that is different to 0° or 90° degrees with any plane perpendicular to the wire axis, we then have helical anisotropy.

Shape Anisotropy

A finitely sized magnetic body (an ellipsoidal shape is assumed here for simplicity) possessing uniform magnetization $\boldsymbol{M}$ (represented by its components M_α) contains a magnetic energy (also called the magnetostatic energy) given by $2\pi N_{\alpha\beta} M_\alpha M_\beta$ (Einstein summation). The $N_{\alpha\beta}$ coefficients are the demagnetization coeffi-

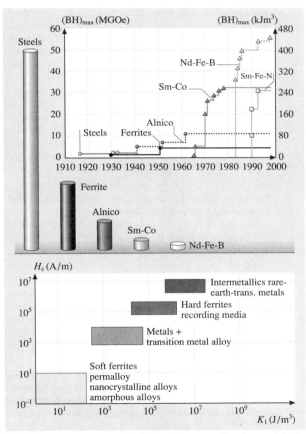

Fig. 4.11 In the upper graph, the development of hard materials for permanent magnets is associated with the value of the energy density $(BH)_{\max}$. In the middle, the progress is indicated by equivalent volume change for a fixed energy density. In the lower graph, the correlations between the coercivity H_c and the anisotropy coefficient K_1 for various materials, their magnetic hardnesses and corresponding applications. (After [4.17, 22], see Table 4.1)

cients of the body, which are determined by its shape. The origin of this terminology is its resemblance to the familiar anisotropy energy of the form $K_{\alpha\beta}M_\alpha M_\beta$ (Einstein summation). If a magnetic material contains N moments (atoms, ions or molecules, each carrying a moment $\boldsymbol{\mu}_i$), the energy originating from the dipolar coupling energy between the different moments is written as:

$$W_{\text{dip}} = \frac{1}{2}\sum_{i=1}^{N}\sum_{j=1}^{N}\frac{\boldsymbol{\mu}_i \cdot \boldsymbol{\mu}_j}{r_{ij}^3} - \frac{3(\boldsymbol{\mu}_i \cdot \boldsymbol{r}_{ij})(\boldsymbol{\mu}_j \cdot \boldsymbol{r}_{ij})}{r_{ij}^5}$$

$\boldsymbol{\mu}_i$ and $\boldsymbol{\mu}_j$ are two moments ($i \neq j$) in the material separated by a distance $\boldsymbol{r}_{ij}$. This energy can only be represented by constant coefficients ($N_{\alpha\beta}$) if the body has an ellipsoidal shape. Hence one can write $W_{\text{dip}} \approx 2\pi N_{\alpha\beta}M_\alpha M_\beta$ (Einstein convention), where the magnetization M is the sum of all individual moments $\boldsymbol{\mu}_i$.

Surface Anisotropy

A finitely sized magnetic body with a bulk anisotropy (which is not too large) will realign the magnetization close to its surface in order to minimize the magnetostatic energy. In other words, the body has a surface anisotropy that is different from the bulk one. This originates from an abrupt change of symmetry at the interface between the bulk and free surface.

Anisotropy From Demagnetization

The demagnetization energy is expressed using coefficients that describe the demagnetization field $\boldsymbol{H}_d$ inside a finitely sized material, created by a bulk magnetization acting against an applied external magnetic field. The components of the demagnetization field (in the ellipsoidal case) are given by (with Einstein summation) $[\boldsymbol{H}_d]_\alpha = -2\pi N_{\alpha\beta}M_\beta$ (much like the depolarization field in the electrical case). Constant coefficients ($N_{\alpha\beta}$) are only valid when the body has an ellipsoidal shape. The coefficients depend on the geometry of the material. There are usually three positive coefficients along three directions N_{xx}, N_{yy} and N_{zz} (assuming the off-diagonal terms are all 0) in simple geometries such as wires, disks, thin films and spheres. All three coefficients are positive and smaller than 1, and their sum is 1. For a sphere, all three of the coefficients are equal to $1/3$. For a thin film (or a disk) they are given by 0,0,1 when the z-axis is perpendicular to a film (or the disk) lying in the xy-plane. For a cylindrical wire of infinite length that has its axis aligned with the z-direction, the values are $1/2, 1/2, 0$.

4.1.4 Traditional Types of Magnetism and Classes of Magnetic Materials

The main traditional types of magnetism are ferromagnetism, antiferromagnetism, ferrimagnetism, paramagnetism and diamagnetism. However, other types are also described in the review by *Hurd* [4.10], and with the expected advances in materials science we may expect to encounter other new classes in the future, as described in Sect. 4.2 of this work, dedicated to unconventional magnetic types (Table 4.1).

A ferromagnet is an assembly of magnetic moments interacting with a positive exchange integral that minimize their energies by adopting a common parallel configuration resulting in a net large value of total magnetization. Such a definition is valid for localized magnetism but not for itinerant ferromagnets (such as the transition metals Fe, Ni and Co), since one does not have distinct localized moments that can define an exchange integral in these materials. A ferromagnetic material (itinerant or localized) displays a characteristic hysteresis curve and remanence ($M \neq 0$ for $H = 0$) when one varies the applied magnetic field. When heated, the material generally loses this ordered alignment and becomes paramagnetic at the Curie temperature. Ferromagnets are usually metallic, but there are ferromagnetic insulators, such as $CrBr_3$, EuO, EuS and garnets [4.27, 35].

An antiferromagnet is made up of an assembly of magnetic moments interacting with a negative exchange integral that minimize their energies by adopting an antiparallel configuration. Again, such a definition is not valid for itinerant antiferromagnets (such as Cr and Mn) since one does not have distinct localized moments and so an exchange integral cannot be defined. The net total magnetization is zero, so we do not get hysteresis. In the localized magnetism case, it is possible to consider the entire crystal as made of two interpenetrating sublattices containing moments that are all parallel inside each sublattice but where the magnetizations from all sublattices cancel each other out. When heated, the material generally loses this alternately ordered alignment and becomes paramagnetic at the Néel temperature. Oxides are generally antiferromagnetic insulators (an exception is EuO) [4.29, 35].

It is possible to have intermediate order between ferromagnets and antiferromagnets; this occurs in the ferrimagnets used in microwave devices [4.16]. In the localized case, one considers the crystal as being made of two sublattices (as in the case of an antiferromagnet), with total magnetizations that oppose one another. How-

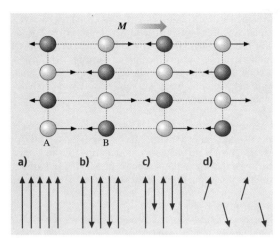

Fig. 4.12a–d Shown *above* is the magnetic moment arrangement in an ferrimagnet. The total net magnetization is nonzero. *Below*: a comparison of the magnetic arrangements between a ferromagnet (**a**), an antiferromagnet (**b**), a ferrimagnet (**c**) and a weak ferromagnet (**d**). (After [4.29])

ever, the magnitude of each magnetization is different, resulting into a net total magnetization, in contrast to an antiferromagnet.

Another intermediate case is a canted antiferromagnet, also called a weak ferromagnet. This is an assembly of magnetic moments that alternate in direction making a small inclination in the direction of the moment (canting). The small angle (Fig. 4.12d) between two neighboring moments results in a small ferromagnetic moment. Canting results from a Dzyaloshinski–Moriya exchange interaction (see Sect. 4.2.1 and [4.24, 35]). In the absence of that interaction, the moments are aligned antiparallel, like in a perfect antiferromagnet.

If we have an assembly of magnetic moments oriented at random with a zero total magnetic moment, we call it a paramagnet. Such materials do not display hysteresis or remanence when one varies the applied magnetic field. The magnetic field simply imposes some alignment onto the randomly oriented moments. For a paramagnet, the susceptibility χ is positive and many orders of magnitude smaller than for ferromagnets, and it behaves as $1/(T - T_c)$ where T_c is the Curie temperature. Some materials never order magnetically, not even at low temperatures (which implies that $T_c = 0$ K), where order prevails. On the other hand, some materials, known as superparamagnetic, show no hysteresis or remanence but have $T_c = \infty$. This does not mean that they are always ordered, but their individual moment is so large (a classical moment corresponds to a quantum angular momentum $J = \infty$, resulting in $T_c = \infty$) that it takes an infinite temperature to destroy it completely [4.27, 35].

A material that combats the influence of a magnetic field by trying to maintain zero induction ($\boldsymbol{B} = 0$) while the field is applied is known as diamagnetic (copper is diamagnetic and superconductors are perfect diamagnets). The term diamagnetic is actually a misnomer, since "dia" means across, implying that the field propogates across the material, whereas the opposite is true. When a diamagnetic substance such as a silicon crystal is placed in a magnetic field, the magnetization vector $\boldsymbol{M}$ in the material points in the opposite direction to the applied field. A negative susceptibility can be interpreted as the diamagnetic substance trying to expel the applied field from itself. A substance exhibits diamagnetism whenever the constituent atoms in the material have closed subshells and shells. This means that each constituent atom has no permanent magnetic moment in the absence of an applied field. Typical diamagnetic materials include covalent crystals and many ionic crystals, because the constituent atoms in this substances

Table 4.3 Examples of soft magnetic materials and their hierarchy according to the saturation magnetostriction coefficient λ_s. The main composition is successively based on Fe, NiFe and finally Co. μ_{max} is the maximum differential permeability and H_c is the coercive field. $Fe_{80}B_{20}$ is also called Metglass 2605. $Fe_{40}Ni_{40}P_{14}B_6$ is also called Metglass 2826. The highest μ_{max} is attained by $(Fe_{0.8}Ni_{0.2})_{78}Si_8B_{14}$, reaching 2×10^6 after annealing [4.35]

Alloy	H_c (mOe)	μ_{max} at 50 Hz	Saturation magnetostriction coefficient
$Fe_{80} B_{20}$	40	320 000	$\lambda_s \approx 30 \times 10^{-6}$
$Fe_{81} Si_{3.5} B_{13.5} C_2$	43.7	260 000	
$Fe_{40} Ni_{40} P_{14} B_6$	7.5	400 000	$\lambda_s \approx 10 \times 10^{-6}$
$Fe_{40} Ni_{38} Mo_4 B_{18}$	12.5–50	200 000	
$Fe_{39} Ni_{39} Mo_4 Si_6 B_{12}$	12.5–50	200 000	
$Co_{58} Ni_{10} Fe_5 (Si, B)_{27}$	10–12.5	200 000	$\lambda_s \approx 0.1 \times 10^{-6}$
$Co_{66} Fe_4 (Mo, Si, B)_{30}$	2.5–5	300 000	

have no unfilled sub-shells. Since the diamagnetic material tries to minimize the effect of H, it expels field lines – a phenomenon that can be exploited in magnetic levitation. A superconductor is a perfect diamagnet, and a metal exposed to high frequencies is partially diamagnetic, since the applied field can only penetrate it to skin depth. The susceptibility χ is constant for a diamagnet (it does not vary with temperature) and is slightly negative. Superconductors have $\chi = -1$ (below critical temperature), whereas semiconductors have the following values of susceptibility (in cm^3/mol) at room temperature: Si, -0.26×10^{-6}; Ge, -0.58×10^{-6}; GaAs, -1.22×10^{-6}; as given by *Harrison* [4.36]).

Materials with a relatively small coercive field (typically smaller than 1000 A/m), as preferred in transformer cores and magnetic read heads, are called soft magnetic materials. Most (but not all) simple metals, transition metals and their compounds are soft. Permalloys, amorphous and nanocrystalline alloys and some ferrites are soft. Amorphous materials are soft because their disordered structure does not favor any direction (no anisotropy energy), whereas nanocrystals possess anisotropy over a short length scale (although it can be larger than its bulk counterpart). Softness is also measured by the maximum permeability attainable (see Table 4.3 of soft elements).

On the other hand, a material with a relatively large coercive field (typically larger than 10 000 A/m), as preferred in permanent magnets, motors and magnetic recording media (disks and tapes), is called a hard magnetic material. This means that stored data is not easily lost since a large field is required to alter the magnitude of magnetization. Most (but not all) rare-earth metals, their compounds and intermetallics are hard. There are also hard ferrites. Permanent magnets are used in power systems (in power relays), motors and audio/video equipment (such as headphones, videotapes); see Table 4.1 and Fig. 4.11.

Ferrites are ferrimagnetic ceramic-like alloys. During the early development of ferrites, the compositions of all ferrites could be described as $FeOFe_2O_3$. However, more modern ferrites are better described as $MOFe_2O_3$, where M is a divalent metal (note that the Fe in Fe_2O_3 is trivalent). Ferrites are used in microwave engineering and recording media due to their very low eddy current losses [4.12] and the fact that they operate over a large frequency interval (kHz to GHz). The ratio of the resistivity of a ferrite to that of a typical metal can reach as high as 10^{14}. They are made by sintering a mixture of metallic oxides $MOFe_2O_3$ where M=Mn, Mg, Fe, Zn, Ni, Cd... Ferrite read heads are however limited to frequencies below 10 MHz as far as switching is concerned, and this is why several new types of read head (thin films, AMR, GMR, spin valves, magnetic tunnel junctions) have been or are being developed in order to cope with faster switching (Chapt. 51). Other conventional magnetic materials similar to ferrites include spinels and garnets, as described below.

Spinels
Spinels are alloys with the composition $(MO)_x$ $(MO)_{1-x}Fe_2O_3$ (a generalization of the ferrite composition), that have the structure of $MgAl_2O_4$ (which provides the origin of the word spinel) [4.16].

Garnets
Garnets are oxides that have compositions related to the spinel family, of the form ($3\ M_2O_3, 5\ Fe_2O_3$), that crystallize into the garnet cubic structure [$Ca_3Fe_2\ (SiO_4)_3$]. They are ferromagnetic insulators of general formula $M_3Fe_5O_{12}$, where M is a metallic trivalent ion (M=Fe^{3+} for example). Garnets have been used in memory bubble technology, lasers and microwave devices (because their ferromagnetic resonance linewidth with respect to the field is small, on the order of a fraction of an Oersted, when the resonance frequency is several tens of GHz [4.16].), especially those of general formula $M_3Ga_5O_{12}$. For instance, when produced as a thin film a few microns thick, $Gd_3Ga_5O_{12}$, called GGG (gadolinium gallium garnet), exhibits perpendicular anisotropy with domains (bubbles) that have up or down magnetization (perpendicular with respect to the film plane). Thus, a bit can be stored in a bubble and can be controlled using a small magnetic field. GGG is considered to be one of the most perfect artificially made crystals, since it can be produced with extremely few defects (less than 1 defect per cm^2). Another nomenclature, called the [cad] notation, is used with rare-earth iron garnets of general formula $X_3Y_2Z_3O_{12}$. This notation means: dodecahedral (c site is surrounded by 12 neighbours and represented by element X); octahedral (a site is surrounded by 8 neighbours and represented by element Y); and tetrahedral (d site is surrounded by 4 neighbours and represented by element Z). The most important characteristic of these garnets is the ability to adjust their compositions and therefore their magnetic properties according to selected substitutions on the c, a or d sites. The element X is a rare earth, whereas Y and Z are Fe^{3+}. The magnetization is changed by placing nonmagnetic ions on the tetrahedral d site: increasing the amount of Ga^{3+}, Al^{3+}, Ge^{4+}, or Si^{4+} will decrease the magnetization. On the other hand, increas-

ing the amount of Sc^{3+} or In^{3+} at the octahedral a site will increase the magnetization. Ion substitution can also be used to tailor other magnetic properties (including anisotropy, coercivity and magnetostriction). Garnets are typically grown using liquid phase epitaxy at a growth speed that easily reaches a micron in thickness in one minute, and a very high yield is achieved. These materials have not only been used in bubble materials but also in magneto-optical displays, printers, optical storage, microwave filters and integrated optics components. Despite all of these attractive properties, their easy tunabilities and very high yields, problems soon arose with the limited access times of bubble memories, since the switching frequency was found to be limited to less than about 10 MHz [4.16] (Chapt. 51).

4.2 Unconventional Magnetism

4.2.1 Conventional and Unconventional Types of Exchange and Coupling in Magnetic Materials

In conventional magnetic materials, the strength of the magnetic interactions between two neighboring localized moments i and j (as in atoms/ions/molecules and rare-earth solids) is described by a direct exchange interaction. The latter is essentially a Coulomb (electrostatic) interaction between the electrons at the i and j sites. The word exchange is used because the corresponding overlap integral describing this interaction involves wavefunctions with permuted (exchanged) electron coordinates (in order to respect the Pauli exclusion principle). The exchange energy between sites i and j is given by $-A_{ij}M_i \cdot M_j$ where M_i and M_j are, respectively, the magnetization at the i and j sites [4.29]. When $A_{ij} > 0$, the energy is minimized when the moments are parallel, and when $A_{ij} < 0$ an antiparallel configuration of the moments is favored. The energy arising from exchange over a distance r in the continuum limit is approximately AM^2/r^2.

The RKKY (Ruderman–Kittel–Kasuya–Yoshida, [4.27]) oscillatory interaction occurs between two localized moments mediated by a surrounding electron gas. It varies in 3-D systems as $\cos(2k_F r)/r^3$, where r is the distance between the moments and k_F is the Fermi wavevector of the electron gas. It was recently discovered that a counterpart of the RKKY interaction exists in 2-D between two magnetic thin films separated by a metallic spacer [4.37]. The RKKY-like interaction between the two magnetic films across a metallic spacer is oscillatory with respect to the spacer thickness z. Thus it becomes possible to decide to couple the magnetic films positively (ferro) or negatively by changing the thickness z of the metallic spacer (see Fig. 4.1 for typical lengths). This is extremely useful for thin film devices (see Chapt. 51 and Table 4.1).

The Dzyaloshinski–Moriya exchange interaction is a vectorial exchange interaction between two neighboring localized moments (M_i and M_j) of the form $D_{ij} M_i \times M_j$, which contrasts with the scalar ordinary exchange interaction of the form $A_{ij}M_iM_j$. This cants (producing a small inclination between) two neighboring antiferromagnetic moments that are usually antiparallel, resulting in weak ferromagnetism. This is due to asymmetric spin-orbit effects [4.24].

Present interest is focused on magnetic thin films and their interactions. Information storage, sensing, spintronics, quantum computing and other applications of magnetic thin film devices are the main spur to understand the nature and extent of magnetic exchange interactions and coupling effects as well as those that arise between magnetic 2-D layers in order to tailor appropriate devices. Novelty is expected since the device size is comparable to the interaction length (Fig. 4.1).

The coupling strength of the interaction between two neighboring magnetic films i and j can be modeled by a factor J_{ij}. This is similar to the exchange integral A_{ij} between two neighboring localized moments, but it involves entire layers generally made from itinerant magnets and not the single moments used with A_{ij}. The exchange interaction is of the form $J_{ij}M_i \cdot M_j$, where M_i and M_j are the magnetizations per unit surface of films i and j. The main interest in J_{ij} lies in the fact that its range is longer in reduced dimensions (1-D and 2-D) than in 3-D (for instance, an RKKY-like interaction between two magnetic films across a metallic spacer is oscillatory with a longer range than it is in 3-D, since it varies like $1/r^2$ instead of $1/r^3$), and its physics is entirely different from the standard RKKY interaction between localized moments [4.37]. The sign of the interaction depends on the thickness of the metallic spacer. Other types of exchange exist between films, such as biquadratic or higher order with a generalized Heisenberg form $I_{ij}[M_iM_j]^n$ where $n \geq 2$ and I_{ij} is a coupling constant

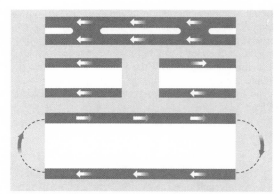

Fig. 4.13 Coupling and exchange between magnetic layers separated by a nonmagnetic (metallic or insulating) spacer. The *top* drawing shows ferromagnetic coupling arising from pinholes that exist across the spacer, whereas the *middle* drawing shows exchange-coupled layers that might be ferro- (*left*) or antiferromagnetic (*right*), depending on the (generally metallic) spacer thickness. Finally, the *bottom* drawing schematizes magnetostatic dipolar coupling between the top and bottom layers. (After [4.10].)

that is a function of n and the separation between the films (Table 4.1).

Interactions between two ferromagnetic (or other types of magnetic) films are not limited to metallic spacers; they can occur across an insulating or semiconducting spacer (Fig. 4.13). Magnetic pinholes generally mediate direct ferromagnetic coupling between neighboring multilayers. They are viewed as shorts across the insulating or semiconducting spacer. Néel proposed that conformal roughness (the orange-peel model, see *Schulthess* and *Butler* [4.38] for instance) at interfaces can result in ferromagnetic coupling for a moderate thickness of spacer material. Magnetostatic dipolar coupling occurs between the roughness features or between domain walls in the two magnetic layers. In this case, stray flux fields from walls in one film can influence the magnetization reversal process in the other. The Slonczewski loose-spin model (see Table 4.1, related to exchange and coupling in multilayers) is based on the change in angular momentum experienced by spin-polarized electrons tunneling across an insulating spacer, resulting in a magnetic exchange coupling. The magnitude and sign of the coupling oscillates depending on the interfacial barrier properties ([4.10] and Table 4.1).

In the absence of a spacer, the exchange energy per unit surface area of ferromagnetic and antiferromagnetic films gives rise to a new phenomenon called exchange bias (the shift of the hysteresis curve of the ferromagnetic film, see entry in Table 4.1). An important consequence of this bias is a pinning-down of the direction of the ferromagnetic moment, which is used in low-noise read heads because it hinders noisy (jerky) domain wall motion (Chapt. 51). Exchange bias is also called anisotropic exchange, and is still poorly understood from on a fundamental level despite the fact it was discovered more than 50 years ago.

4.2.2 Engineering and Growth of Thin Magnetic Films

A magnetic film is considered to be thin if its thickness is smaller than the (bulk-defined) exchange length; in other words its thickness corresponds to the spatial extent of a single domain [4.10]. Magnetic thin films possess very attractive and distinct physical properties with respect to their bulk counterparts. For instance, reduced dimensionality, coordination and symmetry lead to magnetocrystalline anisotropy energies that are two or three orders of magnitude larger than for the bulk (the phenomenon of quenching of the orbital angular momentum – as seen in transition metal ions – is absent, see *Kittel* [4.27]). Saturation magnetization M_s is also enhanced with respect to bulk, since the effective Bohr magneton number n_B per atom/molecule is larger at lower dimensions (for instance in nickel clusters n_B can reach 1.8, whereas in the bulk it is only 0.6). Couplings are also enhanced in thin films, as in the case of the RKKY-like interaction across a metallic spacer. Coupling between two magnetic layers across a semiconducting spacer or an insulator leads to entirely new interactions such as those described by the Slonczewski loose-spin model or the Bruno quantum interference model. New couplings also arise, such as anisotropic exchange or exchange bias between a ferro and an antiferromagnetic layer, and new types of multilayers can be made (spring magnets) by alternating hard and soft thin films (Table 4.1).

When we consider the growth of magnetic materials as thin films, we should recall that these films possess distinct physical properties with respect to the materials used in conventional microelectronics, such as surface energies. The free energies γ of the surface and the interface play dominant roles. They determine the growth modes in thermal equilibrium; the morphology of material B grown on material A depends on the balance between the free surface energies of the substrate, the overlayer and the interface [4.10]. Transition metal-based magnetic materials exhibit a relatively

Table 4.4 Surface energies γ (in J/m^2) for magnetic and nonmagnetic materials, listed with respect to their atomic number for the low-energy cleavage surface. These are approximate values, which are difficult to measure in general and depend on surface orientation and reconstruction [4.10]

Magnetic metal γ(J/m^2)	Cr 2.1	Mn 1.4	Fe 2.9	Co 2.7	Ni 2.5	Gd 0.9		
Transition metal γ(J/m^2)	Ti 2.6	V 2.9	Nb 3.0	Mo 2.9	Ru 3.4	Rh 2.8	Pd 2.0	Ta 3.0	W 3.5	Pt 2.7
Simple or noble metal γ(J/m^2)	Al 1.1		Cu 1.9		Ag 1.3		Au 1.6		
Semiconductor γ(J/m^2)	Diamond 1.7		Si 1.2		Ge 1.1		GaP 1.9		GaAs 0.9
Insulator γ(J/m^2)	LiF 0.34		NaCl 0.3		CaF$_2$ 0.45		MgO 1.2		Al$_2$O$_3$ 1.4

high surface energy, owing to their partially filled d shells. Noble metal substrates have smaller surface energies, and insulating substrates even less. Additionally, when one performs epitaxial growth, another concern is lattice-matching the different underlayers, as displayed in Table 4.4 and Table 4.5, along with typical quantities of interest in representative magnetic materials.

4.2.3 Electronic Properties: Localized, Free and Itinerant Magnetism and Spin-Polarised Band Structure

Building a working device requires an understanding of not only magnetic properties but also electronic properties and their interplay. We expect that new devices will be constructed from a variety of magnetic (conventional and unconventional) materials as well as others already known in microelectronics. Insulating oxides (except EuO) and rare-earth compounds with well-localized external-shell f electrons are solid state materials with atom-like magnetism. Magnetic atoms, ions and molecules or associated with well-defined localized orbitals and individual moments arising from orbital, spin or total angular momentum. When these moments get close to one another, as in the solid state, they interact as defined by Heisenberg: $A_{ij}\mathbf{M}_i.\mathbf{M}_j$ [4.29]. The latter is altered by the presence of the surrounding free-electron gas. Therefore we must understand magnetism in a free-electron gas, its counterpart arising from localized moments, and finally its nature when we have the hybrid case (itinerant magnetism), which occurs in a transition metal (free s and localized d electrons). This problem is very complicated, and so we will rely upon a "one-electron approximation" of band structure, and more precisely spin-polarized band structure [4.11].

In a free-electron gas, one can assume independent noninteracting electrons, so many-electron and nonlocal effects (arising from exchange) do not need to be taken into account. Magnetism in this case is due to individual electron spins and it is straightforward to establish that so-called Pauli paramagnetism holds [4.27]. In addition,

Table 4.5 Lattice-matched combinations of magnetic materials, substrates and spacer layers. There are two main groups of lattice-matched systems with lattice constants close to 4.0 Å or 3.6 Å respectively, after making 45° rotations of the lattice or doubling the lattice constant (After *Himpsel* et al.[4.10] with minor editing)

First group					
Magnetic metal $2^{1/2}$ a (Å) (a (Å))	Cr (bcc) 4.07 (2.88)	Fe (bcc) 4.05 (2.87)	Co (bcc) 3.99 (2.82)		
Simple or noble metal a (Å)	Al 4.05	Ag 4.09	Au 4.07		
Semiconductor a /$2^{1/2}$ (Å) (a (Å))	Ge 3.99 (5.65)	GaAs 4.00 (5.65)	ZnSe 4.01 (5.67)		
Insulator a (Å) (a /$2^{1/2}$ (Å))	LiF 4.02 (2.84)	NaCl 5.65 (3.99)	MgO 4.20 (2.97)		
Second group					
Material a (Å)	Fe (fcc) 3.59	Co (fcc) 3.55	Ni (fcc) 3.52	Cu 3.61	Diamond 3.57

orbital effects give rise to Landau diamagnetism. The spin-polarized density of states is shifted by the field contribution $\pm\mu_B B$ depending on the spin orientation with respect to the field B. In the free-electron Pauli case, the magnetization can be evaluated (at $T = 0\,\text{K}$) from

$$\begin{aligned} M &= \mu_B(N_\uparrow - N_\downarrow) \\ &= \frac{\mu_B}{2} \int_{-\mu_B B}^{E_F} dE\, N(E + \mu_B B) \\ &\quad - \frac{\mu_B}{2} \int_{\mu_B B}^{E_F} dE\, N(E - \mu_B B) \end{aligned}$$

which yields $M \approx \mu_B^2 N(E_F) B$, where μ_B is the Bohr magneton, $N(E)$ is the free electron and $N_{\uparrow\downarrow}(E)$ the spin-polarized density of states. The contribution from Landau (orbital) diamagnetism is a lot more complicated to evaluate, but it is exactly minus one third of the paramagnetic expression [4.27].

States derived from the 3d and 4s atomic levels are responsible for the physical properties of transition metals (itinerant magnets). The 4s electrons are more spatially extended (have a higher principal quantum number) and determine (for instance) the compressibility, whereas the 3d states determine the magnetic properties. The 3d electrons still propagate throughout the material; hence the term itinerant magnetism. The spin-polarized band structure of a transition metal is different to that of a conventional electronic material (see Fig. 4.14 and Fig. 4.15 for the electronic band structures of iron, nickel and cobalt and Table 4.6 for their physical properties).

Many metals have an odd number of electrons (in the atomic state), and so one would expect them to be magnetic due to their unpaired spins; however, only five of them are actually magnetic (Cr, Mn, Fe, Ni and Co, which are all transition metals with s–d hybridization). Note that copper is diamagnetic, with $\chi = -0.77 \times 10^{-6}\,\text{cm}^3/\text{mol}$ at room temperature; Fe, Ni and Co are ferromagnets; whereas Cr and Mn are antiferromagnets.

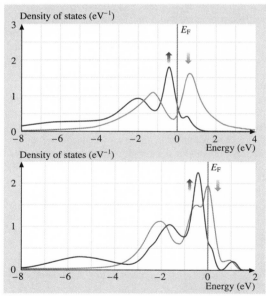

Fig. 4.14 Spin-polarized density of states for bcc Fe and fcc Ni. The bands for the spin up electrons are separated from the spin down band by the exchange splitting energy. The method used is superior to the **LSDA** method which yields a noisy density of states. The authors call it **LDA+DMFT** (local density approximation + dynamical mean field theory). (After [4.39])

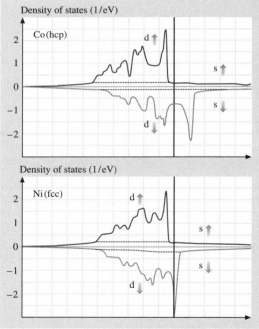

Fig. 4.15 Symmetric double-sided representation of the spin-polarized density of states for Co and Ni (obtained with the LMTO method or linearized muffin-tin orbitals), with the Fermi level indicated. (After [4.40])

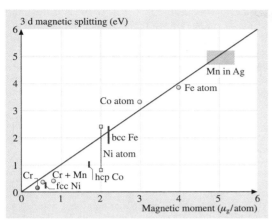

Fig. 4.16 Exchange splitting versus the magnetic moment (in Bohr magnetons μ_B per atom) in 3d transition metals. The *diagonal line* is $1\,\text{eV}/\mu_B$, and the magnetic moment values are given for the element (in the atomic state) and the corresponding crystal. (After [4.10])

It is not straightforward to model magnetism and related properties in transition metal alloys due to the itinerant nature of the electrons. In solids, magnetism arises mainly from electrostatic (Coulomb) electron–electron interactions, namely the exchange interactions, and in magnetic insulators these can be described rather simply by associating electrons with particular atomic sites so that Heisenberg exchange $A_{ij}\mathbf{M}_i\cdot\mathbf{M}_j$ can be used to describe the behavior of these systems. In metallic systems it is not possible to distribute the itinerant electrons in this way, and such simple pairwise interactions between sites cannot be defined. Metallic magnetism is a complicated many-electron effect that has required significant effort over a long period to understand and describe it [4.10]. A widespread approach is to map this problem onto one involving independent electrons moving in the fields set up by all of the other electrons. It is this aspect that gives rise to the spin-polarized band structure that is often used to explain the properties of metallic magnets (such as the non-integer values of multiples of the Bohr magneton μ_B). In addition, compositional structure plays a major role in itinerant magnetism. Consider for instance the ordered Ni–Pt alloy. This is an antiferromagnet, whereas its disordered counterpart is ferromagnetic. It can be said that itinerant magnetism occurs between two extreme limits: the localized and the completely free electron cases (more appropriate to alkali metals), as described below. As far as transport is concerned, one can use the Stoner two-band model, which is a two-fluid model (as used for electrons and holes in semiconductors) with one population of electrons with spin up and another with spin down, along with the corresponding spin-polarized densities of states (Fig. 4.14, Fig. 4.15 and *Himpsel* et al. [4.10].)

In ordinary crystalline nonmagnetic materials, the electronic band structure is built from the lattice (periodic) potential seen by a single electron (with any spin state) in the structure. Mathematically, this amounts to solving the Schrödinger equation corresponding to that potential and calculating its eigenfunctions (spin-independent wavefunctions). The eigenvalues are the ordinary bands. In a magnetic material, one must include exchange effects (which derive from Pauli exclusion principle-dependent electrostatic interactions that yield non-local interactions), and one must account explicitly for spin in the wavefunctions. Several methods exist to do this, such as the LSDA (local spin density approximation, see Fig. 4.1) which is a spin extension of the LDA (local density approximation) where one builds a local approximation to the exchange interaction from the local density of electrons [4.39]. The local spin density approximation (see Fig. 4.1) formalism provides a reliable description of the magnetic properties of transition metal systems at low temperatures. It also provides a mechanism for generating non-integer effective numbers n_B of Bohr magnetons, together with a plausible account of the many-electron nature of magnetic moment forma-

Table 4.6 Fundamental magnetic properties of the transition metals Fe, Co and Ni [4.35]. Note that the domain wall thicknesses and energies are approximate

Magnetic property	Fe	Co	Ni
n_B at $T = 0\,\text{K}$ (in Bohr magnetons)	2.22	1.72	0.62
M_s at 300 K (in 10^6 A/m)	1.71	1.40	0.48
Exchange energy A (meV)	0.015	0.03	0.020
Curie temperature (K)	1043	1404	631
Anisotropy K_1 (J/m^3) at 300 K	4.8×10^4	45×10^4	-0.5×10^4
Lattice spacing (nm) a / c	0.29	0.25 / 0.41	0.35
Domain wall thickness • in nm • in lattice parameters	40 / 138	15 / 36	100 / 285
Domain wall energy (J/m^2)	3×10^{-3}	8×10^{-3}	1×10^{-3}

tion at $T = 0$ K. The bands obtained depend on spin, as depicted in the figures cited earlier. One can obtain the spin-dependent band structure from the spin-polarized density of states for each spin polarization (up ↑ or down ↓, also called the majority and minority, like in ordinary semiconductors). Tables 4.7 and 4.8 give the polarizations and some spin-dependent band-structure quantities for representative transition metals and their alloys.

This semiconductor-like nomenclature (majority–minority) will eventually become confusing when we start dealing with metals and semiconductors simultaneously. For the time being, however, this nomenclature is acceptable so long as we are dealing solely with magnetic metals, and one can define a new type of gap (originating from the exchange interaction) called the exchange splitting or (spin) gap between two spin-dependent bands (Fig. 4.14 and Fig. 4.15). This explains the existence of novel materials such as half-metals, which can be contrasted with semi-metals (graphite) where we have a negative electronic gap because of valence and conduction band overlap.

Half-metals (such as CrO_2 and NiMnSb) possess one full spin-polarized band (up for instance) while the other (down) is empty. These materials are very important for spintronics and (especially) when injecting spin-polarized carriers.

4.2.4 Prospects for Spintronics and Quantum Information Devices

Presently, we are witnessing the extension of electronics to deal with spin and charge instead of charge only, the realm of traditional electronics. The reason spin becomes interesting and useful stems from the following ideas. As device integration increases and feature length decreases towards the nanometer scale, the spins of individual carriers (electrons or holes) become good quantum numbers. This means that spin value is conserved over the nanoscale (the spin diffusion length is typically 5–50 nm), whereas it was not previously (in the micron regime), so it can be used in the nanometer regime to carry useful information. This means that carriers transport energy, momentum, charge and additionally spin. In addition, there is the potential that quantum computers could be constructed from spintronic components [4.25]. In perfect analogy with ordinary electronics, spintronics is based on four pillars:

1. Spin injection: How do we create a non-equilibrium density of spin-polarized carriers – electrons with spin up $n_↑$ (or down $n_↓$) or holes with spin up $p_↑$ (or down $p_↓$)? This can be viewed as the spin extension of the Haynes–Shockley experiment, and it can be done optically or electrically using thin magnetic layers serving as spin filters or analyzers/polarizers as in spin valves (Table 4.1). Spin injection can also be achieved with carbon nanotubes since they do not alter the spin state over large distances (Chapt. 51).
2. Spin detection: How do we detect the spin and charge of a non-equilibrium density of spin-polarized carriers?
3. Spin manipulation: How do we alter and control the spin and charge states of a non-equilibrium density of spin-polarized carriers?
4. Spin coherence: How do we maintain the spin and charge states of a non-equilibrium density of spin-polarized carriers over a given propagation length?

Spintronics is based, like microelectronics, on particular materials, their growth techniques (epitaxial or other) and their physical properties (electrical, mechanical, magnetic, thermal), as well as the ability to fabricate thin films and objects of reduced dimensionality (quantum dots, quantum wires, quantum pillars, clusters) and a knowledge of different processing steps (oxidation, diffusion, doping, implantation, etching, passivation, thermal insulation, annealing, texturing, sputtering, patterning) that can be used to build useful devices. All of the techniques established in the

Table 4.7 Spin polarization expressed in % for several ferromagnetic materials according to several authors. The discrepancies between the different results stem from the various approaches used to estimate the density of states at the Fermi level, and points to how difficult it is to obtain a unanimous figure. Note that for a half-metal like CrO_2 or NiMnSb, the polarization is 100%

	Meservey, Tedrow [4.41]	Moodera, Mathon [4.42]	Monsma, Parkin [4.43]
Fe	40	44	45
Co	35	45	42
Ni	23	33	31
$Ni_{80}Fe_{20}$	32	48	45
$Co_{50}Fe_{50}$	47	51	50

Table 4.8 Magnetic splittings δk_{ex}, full width half maxima $\delta k_\uparrow$ and $\delta k_\downarrow$, and spin-dependent mean free paths $\lambda_\uparrow$ and $\lambda_\downarrow$ for NiFe and NiCr alloys (± 0.01 Å^{-1}) (After [4.44])

	δk_{ex} (Å^{-1})	$\delta k_\uparrow$ (Å^{-1})	$\delta k_\downarrow$ (Å^{-1})	$\lambda_\uparrow$ (Å)	$\lambda_\downarrow$ (Å)
Ni	0.14	0.046	0.046	> 22	> 22
$Ni_{0.9}Fe_{0.1}$	0.14	0.04	0.10	> 25	10
$Ni_{0.8}Fe_{0.2}$	0.14	0.03	0.22	> 33	5
$Ni_{0.93}Cr_{0.07}$	0.09	0.096	0.086	11	10
$Ni_{0.88}Cr_{0.12}$	≤ 0.05	0.12	0.11	8	9

field of microelectronics must be enlarged and extended to magnetism-based electronics (magnetoelectronics), which highlights the challenge of controlling magnetic interactions that have an anisotropic vectorial character in contrast to the scalar electrical interactions (based solely on charge) present in conventional microelectronic devices. Magnetoelectronics introduces the notions of anisotropy (magnitude, nature and direction), coercivity, saturation magnetization, and so on, which need to be controlled during growth (magnetic field-assisted growth or epitaxial growth must further developed in order to favor magnetic anisotropy along desired directions). For instance, there is, in some devices, the need to grow amorphous metallic magnetic layers (in order to get a very small anisotropy, resulting in a very magnetically soft layer), and these may be harder to grow than amorphous semiconductors or insulators (in metals, a cooling speed of one million degrees per second is typically needed to get an amorphous material). Temperature is a very important parameter, since ferromagnetism (antiferromagnetism) is lost above the Curie (Néel) temperature. New types of materials emerge when spin-polarized carriers are used: for instance, half-metallic materials (such as CrO_2 and NiMnSb) that possess carriers that are completely polarized in terms of spin (all up or all down). Additionally, magnetic interactions can be based on localized, free, itinerant, para-, dia-, ferro-, antiferro- or ferrimagnetic materials, which can be electrically metallic, insulating or semiconducting. For instance, the possibility of controlling ferromagnetic interactions between localized spins in a material using transport carriers (electrons or holes), as well as the demonstration of efficient spin injection into a normal semiconductor, have both recently renewed the interest in diluted magnetic semiconductors. If made functional at a reasonably high temperature, ferromagnetic semiconductors would allow one to incorporate spintronics into usual electronics, which would pave the way to integrated quantum computers (Table 4.1).

References

4.1 R. Boll, H. Warlimont: IEEE Trans. Magn. **17**, 3053 (1981)
4.2 E. C. Stoner, E. P. Wohlfarth: Phil. Trans. R. Soc. London A **240**, 599 (1948)
4.3 C. L. Platt, M. R. Mc Cartney, F. T. Parker, A. E. Berkowitz: Phys. Rev. B **61**, 9633 (2000)
4.4 J. C. Slonczewski: Phys. Rev. B **39**, 6995 (1989)
4.5 N. C. Koon: Phys. Rev. Lett. **78**, 4865 (1997)
4.6 C. Tannous, J. Gieraltowski: J. Mater. Sci. Mater. El. **15**, 125 (2004)
4.7 R. White: IEEE Trans. Magn. **28**, 2482 (1992)
4.8 F. Schatz, M. Hirscher, M. Schnell, G. Flik, H. Kronmueller: J. Appl. Phys. **76**, 5380 (1994)
4.9 M. J. Dapino, R. C. Smith, F. T. Calkins, A. B. Flatau: J. Intel. Mat. Syst. Str. **13**, 737 (2002)
4.10 F. J. Himpsel, J. E. Ortega, G. J. Mankey, R. F. Willis: Adv. Phys. **47**, 511 (1998)
4.11 H. F. Jansen: Physics Today (Special Issue on Magnetoelectronics) **April**, 50 (1995)
4.12 J. B. Goodenough: IEEE Trans. Magn. **38**, 3398 (2002)
4.13 H. J. Richter: J. Phys. D **32**, 147 (1999)
4.14 P. Farber, M. Hörmann, M. Bischoff, H. Kronmueller: J. Appl. Phys. **85**, 7828 (1999)
4.15 P. Coeure: J. Phys. (Paris) Coll. C-6 **46**, 61 (1985)
4.16 P. E. Wigen: Thin Solid Films **114**, 135 (1984)
4.17 O. Gutfleisch: J. Phys. D **33**, 157 (2000)
4.18 G. Burkard, D. Loss: Europhys. News **Sept–Oct**, 166 (2002)
4.19 H. Hauser, L. Kraus, P. Ripka: IEEE Instru. Meas. Mag. **June**, 28 (2001)
4.20 G. A. Prinz: Science **282**, 1660 (1998)
4.21 I. Zutic, J. Fabian, S. Das Sarma: Rev. Mod. Phys. **76**, 323 (2004)

4.22 H. Kronmueller: J. Magn. Magn. Mater. **140-144**, 25 (1995)
4.23 J. L. Simonds: Physics Today (Special Issue on Magnetoelectronics) **April**, 26 (1995)
4.24 C. M. Hurd: Contemp. Phys. **23**, 469 (1982)
4.25 M. A. Nielsen, I. L. Chuang: *Quantum Computation and Quantum Information* (Cambridge Univ. Press, New York 2000)
4.26 S. Chikazumi: *Physics of Ferromagnetism*, Int. Ser. Monogr. Phys., 2nd edn. (Oxford Univ. Press, Clarendon 1997)
4.27 C. Kittel: *Introduction to Solid State Physics*, 6th edn. (Wiley, New York 1986)
4.28 M. R. Fitzsimmons, S. D. Bader, J. A. Borchers, G. P. Felcher, J. K. Furdyna, A. Hoffmann, J. B. Kortright, I. K. Schuller, T. C. Schulthess, S. K. Sinha, M. F. Toney, D. Weller, S. Wolf: J. Magn. Magn. Mater. **271**, 103 (2004)
4.29 S. O. Kasap: *Principles of Electronic Materials and Devices*, 3rd edn. (McGraw-Hill, New York 2001)
4.30 L. O. Chua: *Introduction to Non-Linear Network Theory* (McGraw-Hill, New York 1969)
4.31 B. K. Chakrabarti, M. Acharyya: Rev. Mod. Phys. **71**, 847 (1999)
4.32 A. Hubert, R. Schäfer: *Magnetic Domains* (Springer, Berlin, Heidelberg 1998)
4.33 A. P. Malozemoff, J. C. Slonczewski: *Magnetic Domains in Bubble-Like Materials* (Academic, New York 1979)
4.34 D. Buntinx: . Ph.D. Thesis (Université Catholique de Louvain, Louvain 2003)
4.35 D. Jiles: *Introduction to Magnetism and Magnetic Materials*, 2nd edn. (Chapman and Hall, New York 1991)
4.36 W. A. Harrison: *Electronic Structure and the Properties of Solids* (Freeman, New York 1980)
4.37 P. Bruno: Phys. Rev. B **52**, 411 (1995)
4.38 T. C. Schulthess, W. H. Butler: J. Appl. Phys. **87**, 5759 (2000)
4.39 A. I. Lichtenstein, M. I. Katsnelson, G. Kotliar: Phys. Rev. Lett. **87**, 067205 (2001)
4.40 A. Barthelemy: GDR Pommes Proceedings **CNRS** publication (June, Aspet, France, 2001)
4.41 R. Meservey, P. M. Tedrow: Phys. Rep. **238**, 173 (1994)
4.42 J. Moodera, G. Mathon: J. Magn. Magn. Mater. **200**, 248 (1999)
4.43 D. J. Monsma, S. S. P. Parkin: Appl. Phys. Lett. **77**, 720 (2000)
4.44 K. N. Altmann, N. Gilman, J. Hayoz, R. F. Willis, F. J. Himpsel: Phys. Rev. Lett. **87**, 137201 (2001)

5. Defects in Monocrystalline Silicon

The aggregation of instrinsic point defects (vacancies and Si interstitials) in monocrystalline silicon has a major impact on the functioning of electronic devices. While agglomeration of vacancies results in the formation of tiny holes (so-called "voids", around 100 nm in size, which have almost no stress field), the aggregation of Si interstitials exerts considerable stress on the Si matrix, which, beyond a critical size, generates a network of dislocation loops around the original defect. These dislocation loops are typically microns in size. Consequently, they are much more harmful to device functioning than vacancy clusters. However, the feature size in electronic devices has now shrunk down to the 100 nm scale, meaning that vacancy aggregates are also no longer acceptable to many device manufacturers.

This chapter is intended to give an introduction to the properties of intrinsic point defects in silicon and the nucleation and growth of their aggregates. Knowledge in this field has grown immensely over the last decade. It is now possible to accurately simulate the aggregation process so that the defect behavior of semiconductor silicon can be precisely tailored to the needs of the device manufacturer. Additionally, the impact of various impurities on the aggregation process is elucidated.

5.1	**Technological Impact of Intrinsic Point Defects Aggregates**	102
5.2	**Thermophysical Properties of Intrinsic Point Defects**	103
5.3	**Aggregates of Intrinsic Point Defects**	104
	5.3.1 Experimental Observations	104
	5.3.2 Theoretical Model: Incorporation of Intrinsic Point Defects	107
	5.3.3 Theoretical Model: Aggregation of Intrinsic Point Defects	109
	5.3.4 Effect of Impurities on Intrinsic Point Defect Aggregation	112
5.4	**Formation of OSF Ring**	115
References		117

As the feature size continues to shrink in device industry, a thorough understanding of defect behavior in bulk silicon becomes more and more important. Three major defect types relevant to device performance have been identified: vacancy aggregates (known as "voids", which usually have a size of less than 150 nm); Si interstitial clusters embedded in a network of dislocation loops, each of which extend over several microns (L-pits); and large grown-in oxygen precipitates. The latter generate stacking faults (OSF) during wafer oxidation. The voids form in the center of the crystal, while L-pits are observed in the outer region. The two concentric defect regions are usually separated by a small OSF ring. The type of defect that develops in the growing crystal is determined by a simple parameter: the ratio of the pull rate to the temperature gradient at the growth interface. In industry, crystals with only one type of defect – voids – are produced almost exclusively. The formation and behavior of voids has been studied intensively and is accurately described by current theoretical models. As the feature size is now approaching the void size, the growth of so-called "perfect silicon" with almost no detectable defects may be adopted. Furthermore, the doping of crystals with impurities like nitrogen or carbon is being widely investigated. These impurities can significantly reduce the defect size, but they may also have harmful effects, such as enhancing the generation of OSFs. Some models have recently been proposed which may allow us to predict some of the effects of impurities.

5.1 Technological Impact of Intrinsic Point Defects Aggregates

Single intrinsic point defects in silicon – vacancies and interstitials – have not been found to have any negative impact on device performance so far. However, if they aggregate into clusters they can be even detrimental to device functionality. This is also true of extended defects like dislocations. When silicon wafer technology was in its formative years, much of the work devoted to improve wafer quality focused on controlling dislocation density in the silicon crystals, as it was not possible to grow dislocation-free crystals. However, with the introduction of dislocation-free crystal growth processes into mass production, the issue of extended dislocations in relation to bulk silicon quality vanished. As the feature size decreased and the demand for higher device performance increased, it soon became apparent that intrinsic point defects and their aggregation during the cool-down phase of the crystal growth process were having an increasingly negative impact on device performance and yield. Historically, one of the first serious challenges in this regard was the aggregation of Si interstitials in floating zone (FZ)-grown crystals, which results in a local network of dislocation loops (secondary defects), so-called "A-swirls" [5.1] or L-pits [5.2]. Although the diameters of these dislocation loops are only a few microns, they are large enough to generate hot spots in the space charge regions of high-power devices [5.3, 4]. In the second half of the 1980s, the industry began to encounter problems with the early breakdown of the gate oxide in memory devices based on Czochralski (CZ)-grown silicon [5.5]. After intensive gate oxide integrity (GOI) investigations, it was found that the root cause of the gate oxide degradation was tiny micro holes – voids – which were formed by vacancy aggregation during crystal growth [5.6, 7]. Each void is thermally stabilized by an oxide layer present on its inner surface. After wafer polishing, the voids show up as dimples or laser light scattering (LLS) defects on the wafer surface, causing a local thinning of the gate oxide [5.8, 9]. Voids are considerably smaller (less than 150 nm) than A-swirl defects, and so their impact on device performance is only apparent if the location of a void coincides with that of an active element, such as a transistor. In addition, most of these defective transistors can be repaired due to the built-in redundancy of memory chips. Consequently, vacancy aggregates are tolerable for many devices, so long as their density is insignificant compared to those of device process-induced defects. This should be contrasted with A-swirls or L-pits, which always result in permanent device damage due to their large sizes.

Empirically, it has been found that gate oxides 40–50 nm thick are most susceptible to void defects [5.9]. Thinner oxides show higher GOI yields, and when the thickness drops below 5 nm the influence of voids on the GOI yield disappears [5.10, 11]. However, as the feature size continues to shrink, additional adverse effects have been identified, such as shorts between trench capacitors and lack of device reliability [5.12, 13]. As the design rule becomes equal to or less than the void size, these problems are expected to aggravate and device manufacturers may have to switch to materials with extremely small defect sizes or those that contain virtually no defects.

There are three main ways to achieve silicon with no harmful intrinsic point defect aggregates. The first is to grow silicon crystals in a regime where Si interstitials and vacancies are incorporated in equal concentrations (see Sect. 5.3.2), resulting in almost complete mutual annihilation of point defects (so-called "perfect" or "pure" silicon) [5.14–17]. The method inevitably involves lower pull rates and very tight control over crystal growth parameters, which yields considerably lower throughput and higher costs, in particular for 300 mm crystal growth.

The second approach is the growth of nitrogen-doped crystals with very fast pull rates (high cooling rates) and subsequent high-temperature ($\approx 1200\,°\text{C}$) wafer annealing [5.18]. Nitrogen doping in conjunction with a fast pull rate decreases the vacancy aggregate size (Sect. 5.3.3 and Sect. 5.3.4) [5.19, 20], meaning that they are easy to dissolve using a high-temperature wafer treatment. Void annihilation first requires the dissolution of the inner oxide layer, which in turn, necessitates the outdiffusion of oxygen. Thus, annealed wafers only exhibit a near-surface defect-free region $\approx 10\,\mu\text{m}$ in depth, which is, however, sufficient for device manufacturing. Annealed wafers also take advantage of the notable mechanical strengthening effects of nitrogen doping [5.21–23], which helps to suppress slippage generation during high-temperature treatment. One very recent development is rapid thermal wafer annealing at a temperature of around $1300\,°\text{C}$. At this temperature, outdiffusion of oxygen is not necessary because the oxygen concentration is usually below that required for oxygen solubility, so the inner oxide layer dissolves throughout the bulk and the voids collapse. This process yields silicon of a similar quality to that resulting from perfect silicon crystal growth.

The third method is the well-known epi wafer approach.

All of these methods require rather precise defect engineering in order to obtain the properties demanded by the device industry, except in the case of pp+ epi wafers. Here, the high boron concentration of the substrate suppresses intrinsic point defect aggregation (Sect. 5.3.4) and enhances oxygen precipitation in the bulk. Therefore, this wafer type not only provides a defect-free epi layer, but also gives metallic contaminants superior internal gettering (impurity removal) abilities and high slip resistances.

5.2 Thermophysical Properties of Intrinsic Point Defects

Understanding intrinsic point defect aggregation undoubtedly requires rather exact knowledge of their respective thermophysical properties.

The intrinsic point defects – vacancies and Si intersititials – can exist in different configurations. Generally, six localized point defect configurations of high symmetry are considered: the vacancy and the split-vacancy on the one hand, and the tetrahedral, the hexagonal, the bond-centered and the [100] split or dumbbell Si interstitial on the other [5.24]. While the localized configuration works rather well for vacancies, theoretical calculations strongly favor an extended configuration of lower symmetry for Si interstitials [5.25, 26]. The extended self-interstitial model was originally proposed to explain the high pre-exponential factor in the coefficient of self-diffusion, and this model now has support from theoretical calculations [5.24, 27, 28]. According to theory, the high value of the pre-exponential factor results from the multitude of self-interstitial configurations with similar energies and the significant lattice relaxations that accompany some of these configurations.

Vacancies and Si interstitials can also exist in various charged states (such as V^{2+}, V^+, V^0, V^-, V^{2-}), and at the high temperatures (> 1000 °C) where point defects start to aggregate all states should be present in dynamic equilibrium [5.29]. Due to this equilibrium, it is not meaningful to assign a specific charge to vacancies and Si-interstitials, respectively. However, atomistic calculations show that the charged states are much higher in energy and so their populations should be negligible. So far, there is no indication that charged states have any impact on defect aggregation and so they are not considered in current defect nucleation models.

Unfortunately, it is generally not possible to observe intrinsic point defects directly and so their thermophysical properties cannot be measured directly either. Thus, indirect approaches must be used that involve fitting defect concentrations along with many other parameters. Various experimental systems have been used to infer the thermophysical properties of point defects. The most common of these are metal diffusion experiments where a metallic contaminant such as zinc, gold or platinum is introduced into the bulk via high-temperature drive-in diffusion [5.30–32]. The diffusion rate of the metallic impurity, which is easily detectable using standard methods, is related to the mobility and the concentration of intrinsic point defects (kick-out and Frank–Turnbull mechanism), which provides a way to indirectly probe the behavior of the point defects. These experiments yield good estimates for the product $C_{eq}D$ of the equilibrium concentration C_{eq} and the diffusivity D for self-interstitials I and vacancies V, respectively. The following values are derived from zinc diffusion results [5.33]:

$$C_I^{eq} D_I = 1.5 \times 10^{26} \exp(-4.95\,\text{eV}/k_B T)\,\text{cm}^{-1}\text{s}^{-1}$$

and

$$C_V^{eq} D_V = 1.3 \times 10^{23} \exp(-4.24\,\text{eV}/k_B T)\,\text{cm}^{-1}\text{s}^{-1}.$$

Another frequently used experimental method is the defect analysis of CZ crystals grown with varying pull rates. In this case, the observables are the dynamic responses of the oxidation-induced stacking fault (OSF) ring and the interstitial–vacancy boundary as a function of changes in crystal growth process conditions [5.2, 34–36] (Sect. 5.4). These observables have been quantitatively correlated to intrinsic point defect distributions in crystals and, can therefore be used to derive thermophysical properties [5.37]. Of particular importance is the complementary nature of crystal growth and metal diffusion experiments with regard to parametric sensitivity. The high temperature dependence of the IV–boundary and its sensitivity to self-interstitial and vacancy competition implies that these experiments are suitable for determining some pre-exponential coefficients. The metal diffusion experiments, which can be carried out over a wide range of temperatures, are particularly useful for probing activation energies.

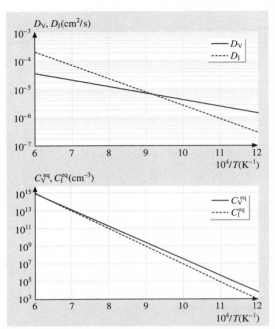

Fig. 5.1 Diffusivities and equilibrium concentrations of vacancies and Si interstitials, respectively. (After [5.41])

In recent years, RTA experiments with fast cooling rates have been carried out with silicon wafers to investigate the influence of the free vacancy concentration on oxygen precipitation [5.17, 38]. By heating the wafers to above 1150 °C, detectable concentrations of Frenkel pairs are created in the wafer bulk. The wafer is then held at this temperature for around 10 s to allow the self-interstitials to diffuse out, resulting in a near-equilibration of both interstitials [$C_I^{eq}(T)$] and vacancies [$C_V^{eq}(T)$]. Following this rapid equilibration, the wafer is rapidly cooled (at rates of $\approx 100\,°\mathrm{C/s}$) to freeze in the vacancy excess which can subsequently be probed via its impact on oxygen precipitation. This technique is especially useful for gaining information on [$C_V^{eq}(T) - C_I^{eq}(T)$] at various temperatures [5.39]. Quenching experiments at different holding temperatures are also reported for hydrogen-doped samples, which allow us to quantitatively measure the VH$_4$ complex [5.40]. From these results, the vacancy formation energy was deduced as $\approx 4\,\mathrm{eV}$ with relatively good accuracy.

Experimental data that relate to the recombination coefficient of vacancies and Si interstitials are scarce [5.42, 43]. Nevertheless, there is general agreement that recombination between vacancies and Si interstitials is so fast that the product of the concentrations of both species is always in thermal equilibrium. The exact value is therefore not relevant to subsequent discussion here.

There have been numerous attempts to compute the thermophysical properties of point defects directly by atomistic simulations. The most important aspect of these atomistic calculations is the accuracy of the assumed interatomic interaction. As the ab initio approach which explicitly considers electronic interactions is extremely computationally expensive and therefore limited to small systems and zero temperature, most simulations are carried out using empirical potentials, such as the Stillinger–Weber or Tersoff potential functions, discarding explicit representations of electronic interactions. An excellent overview of the current status of this field is given by Sinno [5.44]. A parameter set that yields rather good results for defect aggregation is shown in Fig. 5.1. It is clear that the difference between the vacancy and Si interstitial concentrations at the melting point is only around 30% in favor of vacancies. On the other hand, vacancies diffuse more slowly than Si interstitials. These two properties are of great relevance to the peculiar defect behavior of silicon.

In the past, fitted values for equilibrium concentrations and diffusivities inferred from various experimental data often spread over many orders of magnitude, meaning that they were of little help. However, in recent years, theoretical methods and the accuracy of experimental data have significantly improved and the predicted values are now in much better agreement with those gleaned from experiments. Today, the uncertainties in thermophysical data are usually less than 10%.

5.3 Aggregates of Intrinsic Point Defects

5.3.1 Experimental Observations

As pointed out in Sect. 5.1, the dominant intrinsic point defect aggregates in silicon single crystals are voids and L-pits/A-swirls. Figure 5.2 shows a TEM image of a void with a 5 nm oxide layer on the inner surface. Oxidation of the inner surface is not observed in FZ crystals due to their very low oxygen contents. The

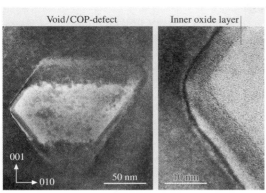

Fig. 5.2 TEM picture of a void/COP (*left*). The inner surface of the void is cover by an oxide layer (*right*). (After [5.52])

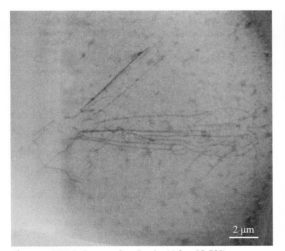

Fig. 5.3 TEM picture of an L-pit. (After [5.53])

depending on their delineation or the detection technique used: crystal originated particle (COP) [5.49], laser light scattering tomography defect (LSTD) [5.2], flow pattern defect (FPD) [5.50], and D-defects [5.51]. Today, it is generally accepted that they all denote the same defect type: vacancy aggregates.

The TEM image of an L-pit is depicted in Fig. 5.3. The network of perfect dislocation loops characteristic of this defect type is clearly visible. The core of the defect is a self-interstitial aggregate which forms an extrinsic stacking fault [5.53, 54]. When the stacking fault reaches a critical size, the strain exerted on the crystal lattice is relaxed by the generation of dislocation loops. It is believed that the B-swirl observed in FZ

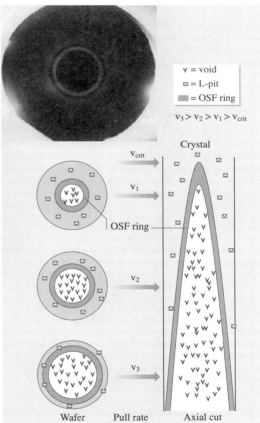

Fig. 5.4 Variation in the radial defect behavior of a CZ crystal as a function of pull rate. The *insert on the left* shows a wafer from the crystal part grown with medium pull rate. The defects were delineated by Secco etching

thickness of the oxide layer decreases with increasing cooling rate of the growing crystal. A strain field is generally not observed around the void. Usually, voids exhibit octahedral morphologies, but they can change to platelet or even rodlike shapes if the crystal is doped with nitrogen (Sect. 5.3.4). Under standard growth conditions, twin voids consisting of two partial octahedral voids are predominantly observed [5.45, 46]. At lower oxygen contents and higher cooling rates of the growing crystals, single octahedral voids are preferentially formed [5.47, 48]. Typical sizes range from 70–200 nm depending on the crystal growth conditions. Historically, different notations were introduced for this type of defect crystals [5.55, 56] is most likely related to self-interstitial aggregates with sizes below the critical limit [5.57, 58]. While the B-swirl can be annihilated by appropriate high-temperature treatment, the L-pit/A-swirl is extremely stable.

In the case of epi wafers, voids at the substrate surface are covered by the epilayer and no defects are generated in the epilayer. However, if the layer thickness is very thin ($< 1\,\mu m$), a dimple remains at the epilayer surface which is detectable by laser light scattering [5.60]. In contrast to voids, L-pits always generate clearly visible and harmful defects in the epilayer [5.61] and are therefore unacceptable to device manufacturers. It is well-established that the pull rate V and the axial temperature gradient G at the growth interface of the growing crystal have the biggest influence on the defect types that develop in the growing crystal [5.34, 35, 62] and their spatial [5.36] as well as their density/size distributions [5.59]. At high pull rates, vacancy-related defects (voids) are observed over the entire crystal volume (Fig. 5.4). When the pull rate is reduced, oxidation-induced stacking faults (OSFs) develop in a small ring-like region near the crystal rim [5.63]. If the pull rate is decreased further, the ring diameter shrinks and L-pits are detected in the area outside the OSF ring [5.2]. No voids are found in this outer region. Thus, the OSF ring obviously represents the spatial boundary between vacancy- and self-interstitial-type defects. At a critical pull rate, the void region and the OSF ring disappear completely and only L-pits are detected. In FZ crystals with their inherently low oxygen contents, no OSF ring is observed as a boundary; instead a defect free-zone is observed as the boundary [5.64].

The radial position of the OSF ring can be approximately described by the equation [5.36]:

$$V/G(r) = \xi_{tr} = 1.34 \times 10^{-3}\,\mathrm{cm^2 K^{-1} min^{-1}}\,,$$

Here r is the radial position. Thus, the parameter V/G controls the type of grown-in defect: if $V/G > \xi_{tr}$ vacancy aggregates develop, while for $V/G < \xi_{tr}$ Si interstitial-related defects are observed. This was first recognized by *Voronkov* in 1982 [5.62].

A further important parameter for the control of grown-in defects is the thermal history of the crystal. While its influence on the type of defect that develops in the growing crystal is negligible, detailed investigations have revealed a strong effect of cooling rate on the defect density/size distribution. For a crystal cooling in the temperature range between roughly $1000\,°C$ and $1100\,°C$, there seems to be an exponential relation between the void density N_{void} and the dwell time t

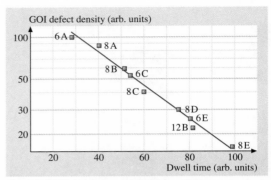

Fig. 5.5 Void density as derived from GOI measurements as a function of the dwell time of the growing crystal at $\approx 1100\,°C$. The *straight line* denotes simulation results. The various data points relate to different growth processes. (After [5.59])

(Fig. 5.5) [5.65]. The defect density data in Fig. 5.5 were derived from GOI measurements that can detect much smaller defect sizes than delineation techniques based on etching and/or laser light scattering. The correlation between the various detection methods down to LSE sizes of $0.12\,\mu m$ was found to be rather good [5.19].

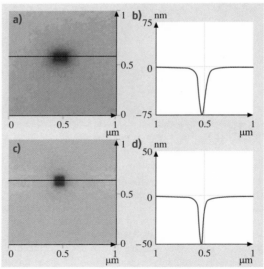

Fig. 5.6 Atomic force microscopy measurements of voids at the wafer surface. The depth of the surface void is rather close to its original size in the crystal. Samples are from a slowly cooled (*left*) and a rapidly cooled crystal (*right*). (After [5.60])

The exact upper value of the above temperature range depends on the crystal growth conditions and will be shown (Sect. 5.3.3) to coincide with the void nucleation temperature. The width of the selected temperature interval, on the other hand, does not have much impact on the above empirical relation. As will be discussed below, the relevant temperature interval is rather small [5.66]. Hence, the dwell time might just as well be replaced by the inverse of the cooling rate at the nucleation temperature.

It was further observed that the average void size increases as dwell time increases (with slower cooling rates) and vice versa (Fig. 5.6) [5.59, 67]. Another striking feature is the characteristic change in the defect density/size distribution across the crystal diameter (Fig. 5.7) [5.68]. While large voids of low density prevail in the crystal center, the distribution gradually shifts to small sizes and high density towards the boundary of the void region. As the cooling rates of the crystal center and rim are almost the same, this remarkable variation in the density/size distribution must be related to the radial inhomogeneity of $V/G(r)$. Unfortunately, similar data are not yet available for L-pits. The behavior of this defect type is more complex, as small interstitial aggregates which do not generate dislocation loops are not detected as L-pits.

Valuable information about the defect nucleation temperature is obtained from transient growth experiments where the pull rate is reduced to zero and the growth process is halted when a certain crystal length has been reached [5.69–71]. After a certain amount of time has elapsed, growth is resumed and the crystal is grown to full length. During the halting period, the various crystal positions are at different temperatures according to the axial temperature gradient of the crystal. By analyzing the axial defect behavior of the as-grown crystal, one can identify the temperature ranges where specific defect types nucleate and grow in size. The results of those experiments clearly demonstrated that the nucleation temperature of voids lies between $1000\,°C$ and $1100\,°C$ and varies with the growth conditions. It was also demonstrated that the nucleation and growth period of the voids occurs over a temperature interval of less than 50 K [5.72]. Void growth is then obviously stopped by the formation of the inner oxide layer. The oxide layer continues to grow down to relatively low temperatures. The scarce data available for L-pits indicate a similar temperature range for nucleation [5.73].

5.3.2 Theoretical Model: Incorporation of Intrinsic Point Defects

The incorporation of intrinsic point defects is the result of two competing fluxes at the growth interface: a vacancy and an interstitial flux which are both directed into the crystal (Fig. 5.8). Each of the two fluxes consists of two components. The first flux component is driven by the advancing growth interface, which generates a convective flux proportional to the pull rate V, and the second is driven by the vacancy/interstitial recombination behind the interface, which creates a concentration gradient and, in turn, a Fickian diffusion flux. The latter is proportional to G. If the pull rate is low,

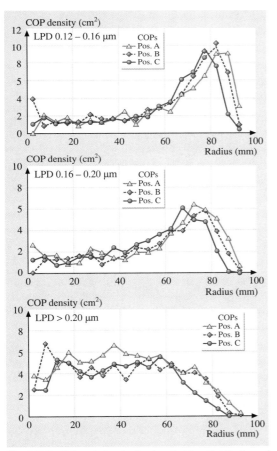

Fig. 5.7 Radial variation in the density/size distribution of voids/COPs across a wafer as measured by laser light scattering. The light point defect size (LPD) correlates well to the actual void size. The positions *A, B* and *C* relate to different axial crystal positions

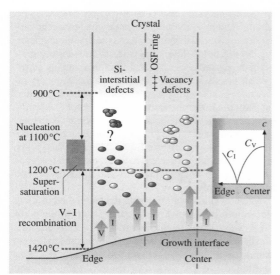

Fig. 5.8 Schematic picture of the incorporation of vacancies (V) and Si interstitials (I), respectively, into the growing crystal. The *insert on the right* shows the radial variation of the remaining species after V–I recombination has ceased and supersaturation starts

then the Fickian diffusion dominates over the convective flux. As self-interstitials diffuse significantly faster than vacancies, the self-interstitial flux wins over the vacancy flux. During crystal cooling, V–I recombination virtually eliminates the vacancies and the surviving self-interstitials are driven into supersaturation and finally aggregate. On the other hand, at fast pull rates the convective flux dominates and, due to the larger equilibrium concentration of vacancies compared to that of

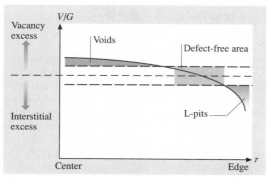

Fig. 5.9 Schematic radial variation in defect types as a function of the radial variation of V/G

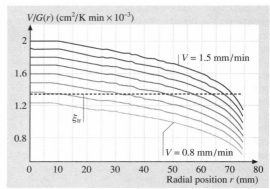

Fig. 5.10 Radial variation in the crossover between the transition value ξ_{tr} and $V/G(r)$ as function of decreasing pull rate V

self-interstitials, more vacancies flow into the crystal, resulting in vacancy aggregates. A similar change in the prevailing defect type is obtained when G is varied. The above model, originally proposed by *Voronkov* [5.62], yields the previously mentioned parameter V/G which determines the defect type.

In standard growth processes, G always exhibits a notable increase from the center towards the crystal rim, which (at sufficiently low pull rates) results in a vacancy excess in the center, while Si interstitials dominate at the crystal rim, as indicated in Fig. 5.9. Figure 5.10 illustrates how the radial position of the crossover between ξ_{tr} and $V/G(r)$ and hence the boundary between the void and L-pit regions is shifted when V is varied.

The exact value of ξ_{tr} is still under discussion, because G has to be taken directly at the interface where the thermal field in the crystal changes very rapidly [5.36, 74]. Therefore, as yet it has not been possible to accurately measure G at the growth interface [5.75]. Fortunately, computer simulations with sufficiently refined meshes permit a rather accurate calculation of G. Based on these calculations, a value of $0.13 \text{ mm}^2\text{min}^{-1}\text{K}^{-1}$ has been determined from growth experiments, with hot zones of different Gs [5.34], which is widely accepted for now.

It should be noted that the above behavior is probably unique to silicon as it stems from the peculiar phenomenon that the vacancy and Si interstitial equilibrium concentrations in crystalline silicon are very similar, but C_v^{eq} is slightly higher than C_I^{eq}, although the diffusivity of Si interstitials exceeds that of vacancies at the melting point. In addition, the quantitative relationship between the relevant parameters which allows the changeover in defect behavior to occur in accessible growth rate regimes is quite astonishing. The time-dependent behaviors of the point defect concentrations are determined by the species conservation equations [5.76]

$$\frac{DC_I}{Dt} = \nabla \left[D_I(T) \nabla C_I - \frac{D_I(T) C_I Q_I^*}{k_b T^2} \nabla T \right]$$
$$- k_{IV} \left(C_I C_V - C_I^{eq} C_V^{eq} \right)$$
$$- \frac{\partial}{\partial t} \int_2^\infty n f_I(n,r) \, \mathrm{d}n \,, \qquad (5.1)$$

where Q_I^* is the reduced heat of transport for Si interstitials, which describes the rate of material flow due to a temperature gradient. A similar expression applies for vacancies (just exchange the I and V notations). The substantial derivative $D/Dt = [\partial/\partial t + V(\partial/\partial z)]$ is defined here as the rate of change of species concentration at a point moving with the velocity (V) of the advancing growth interface. This term relates to the incorporation of species due to convection. The first term on the left-hand side accounts for Fickian diffusion and the second for thermodiffusion. The last term represents the loss of point defects into their aggregates of all possible sizes. This term can be neglected as long as the supersaturation of the surviving species has not increased to the point where cluster nucleation starts. A detailed analysis also reveals that thermodiffusion has no detectable impact. Hence, the simple picture drawn above – which only takes into account convection and Fickian diffusion – is quite reasonable. Based on these simplifications, the incorporated vacancy concentration can be calculated from a simple expression [5.74, 77].

$$C_V(r) \approx \left[C_V^{eq}(T_m) - C_I^{eq}(T_m) \right] \left(1 - \frac{\xi_{tr} G(r)}{V} \right) . \qquad (5.2)$$

This expression is also applicable to Si interstitials (at $V/G < \xi_{tr}$). It then defines the incorporated Si interstitial concentration C_I with a minus sign. If V/G approaches ξ_{tr}, then the vacancy and Si interstitial fluxes become equal and both species are annihilated by V–I recombination. Thus, the incorporated V and I concentrations are negligible and no aggregates can form. This particular condition must be met for the growth of so-called perfect silicon.

5.3.3 Theoretical Model: Aggregation of Intrinsic Point Defects

Due to the technological dominance of vacancy-rich silicon, previous theoretical investigations of defect aggregation have mainly focused on the formation of voids. In addition, it was also easier to verify the theoretical results for vacancy aggregation because voids are not obscured by the secondary defect generation that occurs for L-pits.

From Fig. 5.11, it can be seen that, for a standard growth process, the V-supersaturation already begins to build up at around 1200 °C [5.78], which is well above the experimentally determined nucleation temperature of voids (≈ 1100 °C). Thus, the supersaturation C_V/C_V^{eq} necessary to nucleate voids is appreciable, ≈ 10 [5.76], as estimated from Fig. 5.11. This fact also allows us to decouple the phase where V–I recombination is the major vacancy loss mechanism from the phase where nucleation dominates the vacancy loss, so that one can assume that the cluster formation starts with a fixed concentration which can be calculated from (5.2). For special growth processes that are designed to minimize or even eliminate voids, the starting vacancy concentration after V–I recombination is so small that the beginning of supersaturation (and hence void nucleation) occurs at notably lower temperatures [5.77].

The first vacancy aggregation calculations were performed with a model that was originally developed to simulate oxygen precipitation [5.59, 79]. For the formation of small clusters (less than 20 vacancies), the kinetics are described by chemical rate equations; for

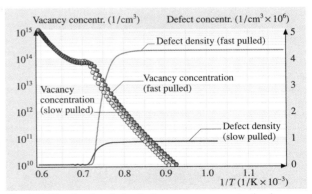

Fig. 5.11 Simulated variation in the vacancy concentration in growing crystals as a function of temperature for rapidly and slowly cooling growth processes. The density of voids nucleated at around 1100 °C and below is shown on the *right scale*

larger clusters, the rate equations are expanded into a continuum formulation that yields a single Fokker–Planck equation. The void shape is assumed to be spherical. Input data are the computed temperature field, which can be quite accurately calculated using commercially available standard codes, and the surviving vacancy concentration after V–I recombination has ceased. Besides the physical properties of the vacancies and Si interstitials, the surface energy must also be known, which is estimated to be around $950 \, \text{erg/cm}^2$.

As can be seen from Fig. 5.11, the computer simulations yield a nucleation temperature of around $1100\,°\text{C}$, which agrees fairly well with the experimental results. The prediction of the correct aggregation temperature has been a matter of discussion. With the above input data, one ends up with a far lower aggregation temperature than those observed [5.80, 81]. The main reason for this is that the nucleation process does not start as a small octahedral void as previously assumed, but rather as an "amorphous cloud" of vacancies which, after some time, relaxes into an octahedron with (111) facets [5.82]. Proper modeling of this initial nucleation process with atomistic calculations does indeed shift the calculated aggregation temperature into the experimentally observed range.

The results in Fig. 5.11 also demonstrate that a higher cooling rate obviously decreases the nucleation temperature. The origin of this effect is the increased supersaturation, which builds up because there is less time for vacancy consumption. Due to the shift in the nucleation temperature, the residual vacancy concentration after void formation remains higher for fast cooling rates than for slow cooling. This has an important effect on the nucleation of oxygen precipitates, as the latter is strongly enhanced by a higher free vacancy concentration. Figure 5.12 illustrates the evolution of the void density/size distribution during the nucleation period. It is apparent that the maximum of the distribution is shifted to larger void sizes not only due to the growth of the previously nucleated voids, but also due to the fact that the newly nucleated clusters have a larger size. The latter effect results from the decreased supersaturation with higher void density, which, in turn, increases the critical radius for stable nuclei. Thus, only increasingly large clusters can nucleate towards the end of the nucleation period.

As depicted in Fig. 5.13, a fast cooling rate decreases the cluster size but increases their density and vice versa, which correlates well with the experimental data. The reason for this is the above-mentioned higher supersaturation at fast cooling rates, which entails a larger nucleation rate. The higher cluster density then effectively lowers the residual vacancy concentration. The clusters, however, have no time to grow and so remain small.

A similar effect is obtained when the initial vacancy concentration before the onset of nucleation is reduced [5.83]. A lower vacancy concentration causes a downward shift in the nucleation temperature, where the vacancy diffusivity is lower. As a consequence, they need more time to diffuse to sinks, which in turn drives up supersaturation and so more clusters of a smaller size are formed again. The behavior of the size distribution as a function of the initial vacancy concentration gives a simple explanation for the experimentally found radial variation in the void size distribution in Fig. 5.7. As

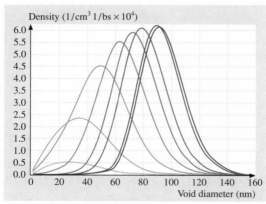

Fig. 5.12 Simulated evolution of the void density–size distribution at various temperatures of the growing crystal. The temperature difference between each curve is 4 K

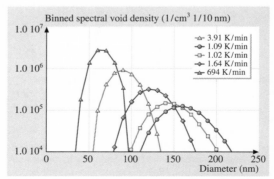

Fig. 5.13 Simulated density/size distributions of voids for growth processes with different cooling rates

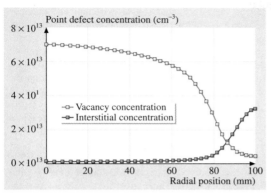

Fig. 5.14 Simulated radial variations in the vacancy and Si interstitial concentrations, respectively. (After [5.68])

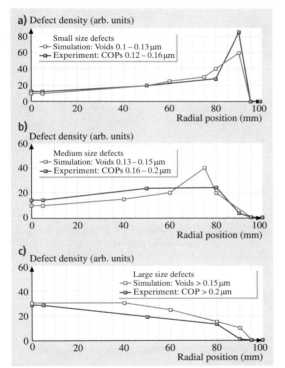

Fig. 5.15 Simulated radial variation in the density/size distribution of voids and comparison with experimental data. (After [5.68])

is low in the center, but increases towards the crystal rim until the vacancy concentration drops below a critical value, where the average void size is below the detection limit. On the other hand, the number of large voids shrinks with decreasing vacancy concentration. Therefore, the density of large COPs is highest in the center. This behavior is also fairly well reproduced by simulation (Fig. 5.15).

A simple expression that predicts a void density proportional to $q^{3/2} C_V^{-1/2}$ ($q =$ cooling rate, $C_V =$ initial vacancy concentration before nucleation starts) has been derived by *Voronkov* et al. [5.66]. This formula is obviously in excellent agreement with the experimental results (Fig. 5.16). A deviation from the above expression was reported for detached crystals, where significantly higher cooling rates can be achieved (2–70 K/min) [5.84]. In the latter case, this discrepancy was tentatively explained by a suppression of voids in favor of enhanced formation of oxygen precipitates at high cooling rates (> 40 K/min) [5.85]; in other words, so-called "oxide particles" nucleate at around $1100\,°\mathrm{C}$ rather than voids.

The $C^{-1/2}$ dependence is more difficult to verify experimentally because a change in the initial vacancy concentration entails a change in the nucleation temperature and so temperature-dependent parameters like diffusivities and equilibrium concentrations must be adjusted. Most of the current aggregation models do not consider the oxidation of the inner void surface, which stops void growth before the vacancies are essentially depleted. This has less impact on the void size distribution but considerably affects the residual vacancy concentration and therefore oxygen precipitation (Sect. 5.4).

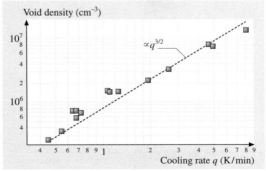

Fig. 5.16 Void density as derived from GOI measurements as a function of cooling rate of the growing crystal. The *straight line* relates to predictions of the $q^{3/2}$ law

the initial vacancy concentration decreases towards the crystal rim (Fig. 5.14), the average size of a void shrinks but void density grows. Thus, the density of small voids

The occurrence of double voids is still under discussion. One attempt to explain this phenomenon is based on a partial oxidation of the (111) facets during void growth, which allows the incorporation of further vacancies at the unoxidized residual facet surface [5.86].

5.3.4 Effect of Impurities on Intrinsic Point Defect Aggregation

Nitrogen

Nitrogen has long been known to simultaneously suppress interstitial- and vacancy-related defects – A-swirls and D-defects/voids, respectively [5.88] – in floating zone (FZ)-grown crystals. The nitrogen concentrations required are very low ($< 2 \times 10^{14}$ at/cm^3) and were found to be in the same range as the surviving residual intrinsic point defect concentrations.

In the case of Czochralski (CZ)-grown silicon, nitrogen has been an undesirable dopant in the past, because it enhances the formation of oxidation-induced stacking faults (OSF) [5.89]. As the nitrogen concentration increases, the OSF ring widens [5.90] until it extends over the entire crystal diameter. In addition, due to the comparatively high oxygen concentration ($3-8 \times 10^{17}$ at/cm^3), the desirable defect suppressing effects of nitrogen are largely offset by the interaction of nitrogen and oxygen [5.91]. However, nitrogen still has a notable effect on the defect size: the higher the nitrogen concentration, the lower the defect size [5.19, 92]. This effect has recently become important, because a small defect size is highly favorable for defect annealing [5.16]. It was further found that nitrogen not only

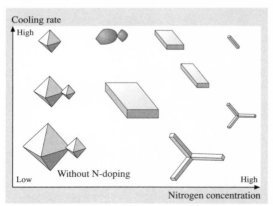

Fig. 5.17 Variation in void morphology as a function of nitrogen concentration and cooling rate of the growing crystal. (After [5.87])

reduces the defect size but also systematically changes the void morphology from octahedron to platelet and, finally, to rods or even rod clusters with increasing concentration (Fig. 5.17) [5.87]. Nitrogen doping has gained additional interest as it also strongly enhances oxygen precipitation [5.93–95] which can be used to improve the efficiency of gettering of metallic contaminants in low thermal budget device processes [5.96, 97].

Unfortunately, quantitative data on the chemical and physical properties of nitrogen in silicon are rare. Experimentally, it was found that nitrogen exists as N–N dimers in as-grown silicon [5.98], which has a diffusivity three orders of magnitude higher than oxygen at 1100 °C [5.99]. These diffusion experiments indicate that the N–N dimer is stable up to at least 1270 °C [5.100], which is in line with first-principles calculations of the N–N dimer configuration [5.101–103]. In contrast with the nitrogen molecule, the N–N dimer in silicon does not have any direct bonds between the two neighboring nitrogen atoms. Experiments with nitrogen-implanted silicon samples heated to melt temperature by a laser pulse and subsequently quenched to room temperature proved that roughly 10% of the nitrogen can be frozen on substitutional sites as single nitrogen atoms [5.104, 105]. The fact that substitutional nitrogen is generally not observed in regularly cooled samples suggests that substitutional nitrogen is not stable in silicon at lower temperatures [5.106].

As the nitrogen concentration needed to suppress A-swirl and D-defects is in the same range as the intrinsic point defect concentration before aggregation, it was suggested that nitrogen directly interacts with vacancies (V) and Si interstitials (I) and thus prevents their aggregation [5.107, 108]. One model proposes that the storage of vacancies in impurity complexes would gradually shift the transition value ξ_{tr} of V/G to higher values which, in turn, would explain the shrinkage of the void region with increasing nitrogen content [5.109]. On the other hand, the model also seems to imply that the A-swirl region simultaneously expands toward the crystal center, which is in conflict with experimental results. In particular, the Li-drift experiments of *Knowlton* et al. [5.110] proved that, despite the suppression of detectable COPs, the extent of the vacancy-rich region does not change due to nitrogen doping. Other attempts are based on the reduction of vacancy diffusivity due to the vacancy–nitrogen interaction [5.111] or on increasing the reaction barrier to vacancy absorption at the void interface [5.112]. However, the fact that nitrogen doping has no impact on the inner oxide layer growth of the voids [5.87] does not support the notion that nitrogen modifies the properties of the void interface. None of these proposals gives any explanation for influence of nitrogen on Si interstitial aggregation.

First-principles calculations by *Sawada* et al. and *Kageshima* et al. [5.101, 102] have shown that the N–N dimer can form stable N_2V and N_2V_2 complexes, the latter being more stable due to a larger enthalpy of formation. Based on these results, it has been proposed that the vacancy aggregation is suppressed by the reaction

I) $N_2 + V \leftrightarrow N_2V$,

whereas the formation of Si interstitial agglomerates is prevented by the reaction

II) $N_2V + I \leftrightarrow N_2$.

Nitrogen is not likely to be incorporated as a N–N dimer into the crystal. This is concluded from the experimental fact that nitrogen does not evaporate from the silicon melt and so it must be strongly bound to silicon atoms, which is not possible for a nitrogen molecule. It was therefore proposed that single nitrogen atoms are incorporated on interstitial (N_I) as well as substitutional sites (N_s) at the growth interface [5.113]. Behind the growth interface, the highly mobile interstitial nitrogen atoms then quickly form N–N dimers via $N_I + N_I \leftrightarrow N_2$ and nitrogen-vacancy complexes via $N_I + N_s \leftrightarrow N_2V$ inside the crystal. A ratio of $[N_I]:[N_s] = 7:1$, which corresponds to 12% of the nitrogen at substitutional sites, has been determined from FZ crystals grown with various nitrogen contents. This is in good agreement with the above-mentioned experimental findings for substitutional nitrogen. As vacancies and Si interstitials have very similar concentrations at the melting point, reactions (I) and (II) mainly act as an additional V–I recombination path during crystal cooling, with only marginal impact on the levels of N_2 and N_2V. However, when recombination is complete and the vacancies or the Si interstitials have been wiped out, reaction (I) or (II) prevents the aggregation of the surviving species. Using this model, it has been possible to determine the radial variations of vacancies and Si interstitials in FZ crystals before the onset of defect aggregation [5.113]. This was achieved by measuring the radial extension of the COP and A-swirl region as a function of the nitrogen concentration. The result is in excellent agreement with theoretical predictions (Fig. 5.18). Surprisingly, the experimental data could not be fitted by assuming that the formation of N_2V_2 complexes prevails, as predicted by theoretical calculations. It is not clear whether the reaction to N_2V_2 is hampered by a high-energy barrier.

While this model works well for FZ crystals, it does not explain why the effect of nitrogen on defect aggregation is strongly diminished in CZ crystals. It is clear that the well-known interaction between nitrogen and oxygen, which results in the formation of N_xO_y complexes [5.114], is probably the origin of the vanishing nitrogen effect. However, it is also known that, despite oxygen levels of $5-6 \times 10^{17}$ at/cm^3, ample unreacted nitrogen is left in oxygen- and nitrogen-doped crystals to suppress defect aggregation [5.115]. As the experimentally confirmed N_2O complex is only stable up to about 700 °C [5.114, 116], it has been proposed [5.113] that nitrogen reacts with oxygen at high temperatures according to

III) $2NO \leftrightarrow N_2 + 2O_i$.

The existence of the NO complex has been theoretically predicted by *Gali* et al. [5.117]. It has a similar structure to the N_2 complex but with a smaller binding energy. The NO formation can occur directly close to the growth interface via $N_i + O_i \leftrightarrow NO$ or $N_s + O_i \leftrightarrow NO + V$. The latter reaction obviously generates additional vacancies, which should, however, only have a small impact on the total vacancy concentration because the close proximity of the growth interface and the high diffusivity of the vacancies will not allow large deviations from the equilibrium value. Due to the large excess of oxygen over nitrogen, it has been assumed that the equilibrium of reaction (III) is on the left-hand side near the melting point and so very little N_2 is available. As the temperature decreases, the

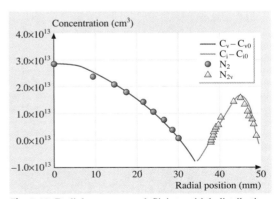

Fig. 5.18 Radial vacancy and Si interstitial distributions as derived from a FZ crystal grown with an axially varying nitrogen concentration. $C_{V0} = 8.8 \times 10^{12}$ at/cm^3 and $C_{I0} = 7.7 \times 10^{12}$ at/cm^3. The *solid* and *dashed lines* refer to simulation results. (After [5.113])

equilibrium gradually shifts to the right-hand side and more and more N_2 is generated. If the temperature at which N_2 starts to develop in sufficient quantities is below the vacancy aggregation temperature, the formation of voids is not suppressed. If the N_2 formation temperature is above the vacancy aggregation temperature, void formation depends on how much N_2 is available in relation to the vacancy concentration at the void nucleation temperature. The higher the N_2 formation temperature, the more N_2 is available to reduce the number of vacancies by reaction (I). According to reaction (III), the formation temperature of N_2 should increase with lower oxygen and higher nitrogen concentrations. Lower free vacancy concentrations and therefore lower aggregation temperatures result in voids of smaller size but higher density, in agreement with the experimental findings.

After void nucleation, the free vacancy concentration drops rapidly and the above reaction (I) reverses, giving $N_2V \rightarrow N_2 + V$, which slows down the decay of free vacancies. Thus, the N_2V-complex acts as an interim reservoir for vacancies – its formation before void formation removes vacancies and its dissociation after void formation releases them again [5.90]. In conjunction with the lower vacancy aggregation temperature, this results in a notable increase in free vacancy concentration during this cool-down phase compared to nitrogen-free material (Fig. 5.19). As free vacancies are known to strongly favor the nucleation of oxygen precipitates, the latter effect gives a simple explanation for the observed strong enhancement of oxygen precipitation in nitrogen-doped CZ crystals (Sect. 5.4).

The remarkable variation in the void morphology as a function of the nitrogen content is still under discussion. A similar change in morphology is also known for oxygen precipitates. In this case, the effect can be interpreted in terms of the balance between the relaxation energy of the lattice strain and the interfacial energy. However, TEM investigations do not reveal any lattice strain around voids, and so another mechanism is required to explain the varying void morphology.

Boron

Boron doping affects the intrinsic point defect aggregation in a similar way to nitrogen doping, but much higher concentrations are required [5.119]. It was found that the COP region as well as the diameter of the OSF ring starts to shrink when the boron content exceeds roughly 5×10^{18} at/cm^3 (Fig. 5.20). In contrast with nitrogen, OSF formation is not enhanced and the width of the OSF ring remains unchanged with increasing boron concentration. At around 10^{19} at/cm^3, the COPs and the OSF ring disappear in the center of the crystal. Although detailed investigations into the behavior of L-pits in highly boron-doped silicon are still required, we can infer from the defect-free quality of pp+ epi wafers that L-pits are simultaneously suppressed by the high boron level (Sect. 5.3.1).

Dornberger et al. [5.67] suspected that changes in the equilibrium concentrations and diffusivities of intrinsic point defects in the presence of high boron concentrations would modify the transition value ξ_{tr} of the parameter V/G as a function of the boron content. *Voronkov* and *Falster* [5.81] also proposed a modification of equilibrium concentrations as a result of the shift

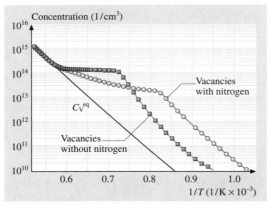

Fig. 5.19 Simulated variations in the vacancy concentration as a function of temperature for a nitrogen-doped and an undoped crystal

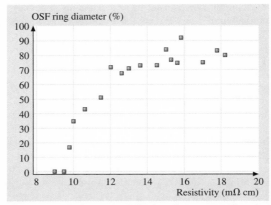

Fig. 5.20 Variation in the OSF ring diameter as a function of the boron concentration. (After [5.118])

in the Fermi level due to the high boron concentration. Based on this approach, the latter authors calculated a shift in the transition value ξ_{tr} which would account for the observed shrinkage of the void region. As with nitrogen doping, the ξ_{tr} shift entails the simultaneous appearance of L-pits in the outer crystal region and a shrinking void region, which is in conflict with experimental results. Another attempt by *Sinno* et al. [5.120] considers reversible reactions between boron and intrinsic point defects, in particular self-interstitials. The formation of BI and B_2I complexes consumes self-interstitials which would otherwise be annihilated by recombination. As recombination consumes the remaining self-interstitials, BI and B_2I start to dissociate again according to le Chatelier's principle. As a result, the self-interstitial concentration increases, which shifts the point defect balance towards the intersitial-rich side, causing the void region to shrink. No comment is made about whether or not the model is also able to account for the suppression of L-pit formation by storing supersaturated self-interstitials in boron complexes.

Carbon

Carbon doping is also known to impact intrinsic point defect aggregation. It has been reported that the void region in the crystal center disappears upon carbon doping [5.57, 121], while the region of Si interstitial aggregates is widened; in other words, carbon does not inhibit the formation of L-pits/A-swirl. This behavior appears to be in line with the model put forward in [5.109], which predicts an upward shift of ξ_{tr} for carbon doping and, in turn, an enlargement of the interstitial-rich region.

It was also found that carbon doping reduces the grown-in defect size [5.87, 122]. Although the size reduction is appreciable, the morphology of the voids is not changed, in contrast to the consequences of nitrogen doping. Only a tendency towards multiple void formation was observed. The thickness of the inner oxide layer was found to be similar to that of undoped crystals, which indicates that carbon doping has no influence on the growth of the inner oxide layer. As with boron, the effect on defect aggregation is only seen at significantly higher concentrations ($\approx 1 \times 10^{17}$ at/cm^3) than for nitrogen.

At higher concentrations, carbon is known to enhance oxygen precipitation too [5.123, 124]. As carbon predominantly resides on substitutional sites, it is very unlikely that, as in the case of nitrogen doping, a higher residual vacancy concentration is responsible for the stronger oxygen precipitation. On the other hand, the small carbon atom exerts a local tensile strain on the surrounding lattice and attracts Si interstitials to form C_xI_y complexes. Thus, Si interstitials ejected into the lattice by growing oxygen precipitates are effectively removed and, in turn, further precipitate growth is not retarded by a build-up of Si interstitial supersaturation. The enhanced oxygen precipitation may also be related to heterogeneous nucleation at small carbon aggregates [5.125].

5.4 Formation of OSF Ring

The ring-like distributed OSFs are oxygen precipitates with platelet shapes that grow particularly large at the edge of the void region, and exceed a critical size necessary to create stacking faults during subsequent wafer oxidation there [5.126]. The critical radius of the grown-in platelets is ≈ 70 nm. The formation of OSFs can be suppressed if the cooling rate of the growing crystal is increased (this means, for example, that OSFs are not found in oxygen-doped FZ crystals [5.118] which have very high cooling rates) or if the oxygen content is decreased.

The peculiar ring-like distribution is a consequence of the well-known strong enhancement of oxygen precipitation by vacancies [5.39, 127]. The reason for this is that the absorption of vacancies allows the oxygen precipitate, which occupies twice as much volume as the corresponding silicon lattice, to nucleate and grow without building up notable strain energy. As will be described below, the radially inhomogeneous oxygen precipitation is a consequence of a substantial radial inhomgeneity of the free vacancy concentration in contrast to the rather flat radial profile of oxygen.

The vacancy concentration, as pointed out previously, has its maximum at the center of the growing crystal after V–I recombination has ceased (Fig. 5.21). Therefore, the critical supersaturation for void formation is first reached at the crystal center at relatively high temperatures. Hence, the free vacancies are quickly consumed in this area. As the crystal cools, voids are also nucleated in the regions of lower initial vacancy concentration, meaning that vacancy consumption then also occurs further away from the crystal center. As the removal of vacancies is enhanced at higher temperatures, where the diffusivity is large, the ra-

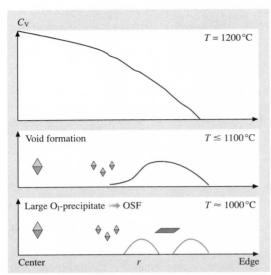

Fig. 5.21 Schematic evolution of the radial vacancy profile and accompanying void and large O_I-precipitate formation at decreasing temperatures

dial vacancy distribution finally drops to a minimum in the crystal center, and the maximum of the residual vacancies gradually moves toward the V/I boundary upon further cooling. Because of this shifted vacancy maximum, the first oxygen aggregates nucleate close to the V/I boundary on the vacancy-rich side. The relatively high nucleation temperature at this position results in larger sizes but lower densities for the corresponding grown-in oxygen precipitates compared to those nucleated at lower temperatures away from the V/I boundary. Consequently, they preferentially reach the critical size for OSF formation during subsequent

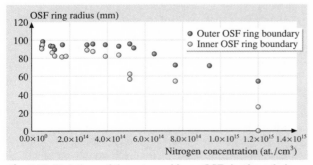

Fig. 5.22 Variations of the outer and inner OSF ring boundaries as a function of the nitrogen concentration. (After [5.90])

wafer oxidation. At sufficiently fast crystal cooling rates, the oxygen precipitates are nucleated at higher densities but smaller sizes and cannot grow to the critical size. Thus, OSF formation is prevented, in agreement with experimental findings. The same density/size effect is obtained if the oxygen content is lowered – similar to the behavior of voids – and so OSFs are also suppressed.

The nucleations of the first oxygen precipitates quickly results in the depletion of residual vacancies in their neighborhood. Thus, the previous vacancy maximum at the V–I boundary is then converted into a pronounced minimum which suppresses further nucleation in this area. As a result, two vacancy peaks with a region of large oxygen precipitates in-between is frozen in as the crystal cools down to room temperature (Fig. 5.21). Later on, when the wafers are subjected to further heat treatments, such as a nucleation step at 780 °C followed by a growth step at 1000 °C, new oxygen precipitates with high density are preferentially nucleated at the radial positions of the two remaining vacancy maxima. This results in a distinct profile for the interstitial oxygen concentration after these heat treatments. In the radial band where the large precipitates and the OSF ring are located, the interstitial oxygen concentration is only slightly reduced because further nucleation of precipitates during wafer heat treatment is suppressed there and the comparatively low density of precipitates does not consume much of the interstitial oxygen. On the other hand, the high density bands at the locations adjacent to the former vacancy peaks coincide with pronounced minima in the interstitial oxygen content.

Considerable changes in the widths and the radial positions of OSF rings are observed if the crystals are nitrogen-doped. With increasing nitrogen content, the outer and inner OSF ring boundaries are shifted towards the crystal center [5.90]. However, the shift is much larger for the inner OSF ring boundary, resulting in a significant widening of the OSF ring width (Fig. 5.22) until it extends over the entire crystal volume. This behavior was tentatively explained by a schematic model that takes into account the slower build-up of vacancy supersaturation due to the formation of N_2V complexes and the resulting lower void nucleation temperature. The model also predicts that the void and the OSF region overlap, roughly in the area between the inner boundary and the middle of the OSF ring, which is in agreement with experimental data (Fig. 5.23).

In contrast to nitrogen, carbon tends to suppress OSF formation. This is again an indication that the

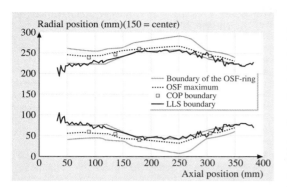

Fig. 5.23 Variations in the OSF ring width and its radial location in comparison to the variations of the COP and LLS boundaries along the crystal axis. The LLS defects were detected on the as-polished wafer surface, which detects only the larger voids, while the COP boundary was determined by LLS measurements after repeated SC1 cleaning, which enlarges the voids to above the LLS detection limit

mechanism of the effect of carbon is different. The above-mentioned idea of heterogeneous nucleation due to small carbon clusters may also explain this behavior. As heterogeneous nucleation needs less supersaturation, it prevails over homogeneous nucleation. Furthermore, the density of carbon clusters is much higher than that of homogeneously nucleated oxygen precipitates. Thus, the heterogeneously nucleated oxygen precipitates cannot grow to the critical size due to their large densities.

References

5.1 P. M. Petroff, A. J. R. deKock: J. Cryst. Growth **30**, 117 (1975)
5.2 S. Sadamitsu, S. Umeno, Y. Koike, M. Hourai, S. Sumita, T. Shigematsu: Jpn. J. Appl. Phys. **32**, 3675 (1993)
5.3 L. I. Bernewitz, B. O. Kolbesen, K. R. Mayer, G. E. Schuh: Appl. Phys. Lett. **25**, 277 (1975)
5.4 B. O. Kolbesen, A. Mühlbauer: Solid State Electron. **25**, 759 (1982)
5.5 W. Bergholz, W. Mohr, W. Drewes: Mater. Sci. Eng. **B4**, 359 (1989)
5.6 M. Itsumi, H. Akiya, T. Ueki, M. Tomita, M. Yamawaki: J. Appl. Phys. **78**(10), 5984 (1995)
5.7 M. Miyazaki, S. Miyazaki, Y. Yanase, T. Ochiai, T. Shigematsu: Jpn. J. Appl. Phys. **34**, 6303 (1995)
5.8 J. G. Park, J. M. Park, K. C. Cho, G. S. Lee, H. K. Chung: Effect of crystal defects in device characteristics, Proc 2nd Int. Symp. on Advanced Science and Technology of Silicon Materials, Kona-Hawaii, USA 1996, ed. by M. Umeno (145th Committee of the Japan Society for the Promotion of Science, 1996) 2519
5.9 M. Itsumi: Mater. Sci. Eng. **B73**, 184 (2000)
5.10 T. Bearda, M. Houssa, P. Mertens, J. Vanhellemont, M. Heyns: Appl. Phys. Lett. **75**(9), 1255 (1999)
5.11 U. Lambert, A. Huber, J. Grabmeier, J. Vanhellemont, R. Wahlich, G. Kissinger: Microelectron. Eng. **48**, 127 (1999)
5.12 E. Dornberger, D. Temmler, W. v. Ammon: J. Electrochem. Soc. **149**(4), 1 (2002)
5.13 C. Kupfer, H. Roth, H. Dietrich: Mat. Sci. Semicon. Proc. **5**, 381 (2003)
5.14 A. M. Eidenzon, N. I. Puzanov: Inorg. Mater. **33**(3), 272 (1997)
5.15 J. G. Park, H. K. Chung: Wafer requirements: Memory devices, Silicon Wafer Symp., Portland 1999 (SEMICON West 99, Portland 1999) D–1
5.16 J. G. Park: J. Jpn. Assoc. Crystal Growth **27**(2), 14 (2000)
5.17 R. Falster: Advances of the defect engineering of polished silicon wafers: Perfect silicon and magic denuded zones, Silicon Wafer Symp., Portland 1999 (SEMICON West 99, Portland 1999) E–13
5.18 D. Gräf, M. Suhren, U. Lambert, R. Schmolke, A. Ehlert, W. v. Ammon, P. Wagner: J. Electrochem. Soc. **145**(1), 275 (1998)
5.19 D. Gräf, M. Suhren, U. Lambert, R. Schmolke, A. Ehlert, W. v. Ammon, P. Wagner: J. Electrochem. Soc. **145**(1), 275 (1998)
5.20 X. Yu, D. Yang, X. Ma, L. Li, D. Que: Semicond. Sci. Technol. **18**, 399 (2003)
5.21 K. Sumino, I. Yonenaga, M. Imai, T. Abe: J. Appl. Phys. **54**(9), 5016 (1983)
5.22 L. Jastrzebski, G. W. Cullen, R. Soydan, G. Harbeke, J. Lagowski, S. Vecrumba, W. N. Henry: J. Electrochem. Soc. **134**(2), 466 (1987)
5.23 G. Wang, D. Yang, D. Li, Q. Shui, J. Yang, D. Que: Physica B **308–310**, 450 (2001)
5.24 D. Maroudas, R. Brown: Appl. Phys. Lett. **62**(2), 172 (1993)
5.25 A. Seeger, K. P. Chik: Phys. Status Solidi A **29**, 455 (1968)
5.26 H. R. Schober: Phys. Rev. B **39**, 13013 (1989)
5.27 R. Car, P. J. Kelly, A. Oshiyama, S. T. Pantelides: Phys. Rev. Lett. **52**, 1814 (1984)
5.28 D. Maroudas, R. Brown: Phys. Rev. B **47**(23), 15562 (1993)
5.29 G. D. Watkins: Mater. Sci. Semicon. Process. **3**, 227 (2000)

5.30 N. A. Stolwijk, J. Holzl, W. Frank, E. R. Weber, H. Mehrer: Appl. Phys. A **39**, 37 (1986)
5.31 H. Bracht, N. A. Stolwijk, H. Mehrer: Phys. Rev. B **52**, 16542 (1995)
5.32 H. Zimmermann, H. Ryssel: Appl.Phys. A **55**, 121 (1992)
5.33 H. Bracht: Native point defects in silicon, Proc. 3rd Int. Symp. on Defects in Silicon III, Seattle, Washington 1999, ed. by T. Abe, W. M. Bullis, S. Kobayashi, W. Lin, P. Wagner (The Electrochemical Society, Pennington 1999) 357
5.34 W. v. Ammon, E. Dornberger, H. Oelkrug, H. Weidner: J. Cryst. Growth **151**, 273 (1995)
5.35 M. Hourai, E. Kajita, T. Nagashima, H. Fujiwara, S. Umeno, S. Sadamitsu, S. Miki, T. Shigematsu: Mater. Sci. Forum **196–201**, 1713 (1995)
5.36 E. Dornberger, W. v. Ammon: J. Electrochem. Soc **143**(5), 1648 (1996)
5.37 T. Sinno, R. A. Brown, W. v. Ammon, E. Dornberger: Appl. Phys. Lett. **70**(17), 2250 (1997)
5.38 M. Akatsuka, M. Okui, N. Morimoto, K. Sueoka: Jpn. J. Appl. Phys. **40**, 3055 (2001)
5.39 R. Falster, V. V. Voronkov, F. Quast: Phys. Status Solidi B **222**, 219 (2000)
5.40 N. Fukata, A. Kasuya, M. Suezawa: Jpn. J. Appl. Phys. **40**, L854 (2001)
5.41 T. Frewen, T. Sinno, E. Dornberger, R. Hoelzl, W. v. Ammon, H. Bracht: J. Electrochem. Soc. **150**(11), G673 (2003)
5.42 D. A. Antoniadis, I. Moskowitz: J. Appl. Phys. **53**(10), 6780 (1982)
5.43 H. J. Gossmann, C. S. Rafferty, A. M. Vredenberg, H. S. Luftman, F. C. Unterwald, D. J. Eaglesham, D. C. Jacobson, T. Boone, J. M. Poate: Appl. Phys. Lett. **64**(3), 312 (1994)
5.44 T. Sinno: Thermophysical properties of intrinsic point defects in crystalline silicon, Proc 9th Int. Symp. on Silicon Materials Science and Technology, Semiconductor Silicon, Philadelphia 2002, ed. by H. R. Huff, L. Fabry, S. Kishino (The Electrochemical. Society, Pennington 2002) 212
5.45 T. Ueki, M. Itsumi, T. Takeda: Jpn. J. Appl. Phys. **37**, 1669 (1998)
5.46 M. Itsumi: J. Cryst. Growth **237–239**, 1773 (2002)
5.47 S. Umeno, Y. Yanase, M. Hourai, M. Sano, Y. Shida, H. Tsuya: Jpn. J. Appl. Phys. **38**, 5725 (1999)
5.48 M. Nishimura, Y. Yamaguchi, K. Nakamura, J. Jablonski, M. Watanabe: Electrochem. Soc. Symp. Proc. **98-13**, 188 (1998)
5.49 J. Ryuta, E. Morita, T. Tanaka, Y. Shimanuki: Jpn. Appl. Phys. **29**, L1947 (1990)
5.50 H. Yamagishi, I. Fusegawa, N. Fujimaki, M. Katayama: Semicond. Sci. Techn. **7**, A135 (1992)
5.51 P. J. Roksnoer, M. M. B. Van de Boom: J. Cryst. Growth **53**, 563 (1981)
5.52 H. Bender, J. Vanhellemont, R. Schmolke: Jpn. Appl. Phys. **36**, L1217 (1997)
5.53 R. Schmolke, W. Angelberger, W. v. Ammon, H. Bender: Solid State Phenom. **82–84**, 231 (2002)
5.54 K. Nakai, M. Hasebe, K. Ohta, W. Ohashi: J. Cryst. Growth **210**, 20 (2000)
5.55 H. Föll, B. O. Kolbesen: Appl. Phys. **8**, 319 (1975)
5.56 P. M. Petroff, A. J. R. de Kock: J. Cryst. Growth **36**, 1822 (1976)
5.57 J. Chikawa, T. Abe, H. Harada: Impurity effect on the formation of microdefects during silicon crystal growth. In: *Semiconductor Silicon*, ed. by H. R. Huff, T. Abe, B. Kolbesen (The Electrochemical Society, Pennington 1986) p. 61
5.58 H. Föll, U. Gösele, B. O. Kolbesen: J. Cryst. Growth **40**, 90 (1977)
5.59 E. Dornberger, J. Esfandyari, D. Gräf, J. Vanhellemont, U. Lambert, F. Dupret, W. v. Ammon: Simulation of grown-in voids in Czochralski silicon crystals, Crystalline Defects and Contamination Control: Their Impact and Control in Device Manufacturing II, Nürnberg 1997, ed. by B. O. Kolbesen, P. Stallhofer, C. Claeys, F. Tardiff (The Electrochemical Society, Pennington 1997) 40
5.60 R. Schmolke, M. Blietz, R. Schauer, D. Zemke, H. Oelkrug, W. v. Ammon, U. Lambert, D. Gräf: Advanced silicon wafers for 0.18 μm design rule and beyond: Epi and fLASH!, High Purity Silicon VI, Phoenix 2000, ed. by C. L. Claeys, P. Rai-Choudhury, M. Watanabe, P. Stallhofer, H. J. Dawson (The Electrochemical. Society, Pennington 2000) 1
5.61 W. v. Ammon, E. Dornberger, P. O. Hansson: J. Cryst. Growth **198/199**, 390 (1999)
5.62 V. V. Voronkov: J. Cryst. Growth **59**, 625 (1982)
5.63 M. Hasebe, Y. Takeoka, S. Shinoyama, S. Naito: Ring-like distributed stacking faults in CZ-Si wafers. In: *Defect Control in Semiconductors*, ed. by K. Sumino (Elsevier, Amsterdam 1990) p. 157
5.64 H. Yamagishi, I. Fusegawa, K. Takano, E. Iino, N. Fujimaki, T. Ohta, M. Sakurada: Evaluation of FDPs and COPs in silicon single-crystals, Semiconductor Silicon, San Francisco 1994, ed. by H. R. Huff, W. Bergholz, K. Sumino (The Electrochemical Society, Pennington 1994) 124
5.65 W. v. Ammon, E. Dornberger: *Properties of Crystalline Silicon*, EMIS Datareviews Series 20, ed. by R. Hull (INSPEC, London 1999) p. 37
5.66 V. V. Voronkov, R. Falster: J. Cryst. Growth **194**, 76 (1998)
5.67 E. Dornberger, D. Gräf, M. Suhren, U. Lambert, P. Wagner, F. Dupret, W. v. Ammon: J. Cryst. Growth **180**, 343 (1997)
5.68 E. Dornberger, J. Esfandyari, J. Vanhellemont, D. Gräf, U. Lambert, F. Dupret, W. v. Ammon: Simulation of non-uniform grown-in void distributions in Czochralski crystal growth, Semiconductor Silicon, San Francisco 1998, ed. by H. R. Huff, U. Gösele, H. Tsuya (The Electrochemical Society, Pennington 1998) 490
5.69 M. Hourai, T. Nagashima, E. Kajita, S. Miki: Oxygen precipitation behavior in silicon during Czochralski crystal growth, Semiconductor Silicon, San Francisco

5.70 T. Iwasaki, A. Tomiura, K. Nakai, H. Haga, K. Kojima, T. Nakashizu: Influence of coolingcondition during crystal growth of CZ-Si on oxide breakdown property, Semiconductor Silicon, San Francisco 1994, ed. by H. R. Huff, W. Bergholz, K. Sumino (The Electrochemical Society, Pennington 1994) 744
5.71 K. Takano, K. Kitagawa, E. Iino, M. Kimura, H. Yamagishi: Mater. Sci. Forum **196–201**, 1707 (1995)
5.72 M. Akatsuka, M. Okui, S. Umeno, K. Sueoka: J. Electrochem. Soc. **150**(9), G587 (2003)
5.73 J. Furukawa, H. Tanaka, Y. Nakada, N. Ono, H. Shiraki: J. Cryst. Growth **210**, 26 (2000)
5.74 V. V. Voronkov, R. Falster: J. Appl. Phys. **86**(11), 5975 (1999)
5.75 A. Natsume, N. Inoue, K. Tanahashi, A. Mori: J. Cryst. Growth **225**, 221 (2001)
5.76 T. Sinno, E. Dornberger, W. v. Ammon, R. A. Brown, F. Dupret: Mater. Sci. Eng. **28**, 149 (2000)
5.77 Z. Wang, R. Brown: J.Crystal Growth **231**, 442 (2001)
5.78 E. Dornberger, W. v. Ammon, D. Gräf, U. Lambert, A. Miller, H. Oelkrug, A. Ehlert: The impact of dwell time above 900 °C during crystal growth on the gate oxide integrity of silicon wafers, Proc. 4th Int. Symp. on High Purity Silicon, San Antonio 1996, ed. by C. L. Claeys, P. Rai-Choudhury, M. Watanabe, P. Stallhofer, H. J. Dawson (The Electrochemical Society, Pennington 1996) 140
5.79 J. Esfandyari, G. Hobler, S. Senkader, H. Pötzl, B. Murphy: J. Electrochem. Soc. **143**, 995 (1996)
5.80 V. V. Voronkov, R. Falster: J. Cryst. Growth **198/199**, 399 (1999)
5.81 V. V. Voronkov, R. Falster: J. Appl. Phys. **87**(9), 4126 (2000)
5.82 T. A. Frewen, S. S. Kapur, W. Haeckl, W. v. Ammon, T. Sinno: J. Crystal Growth **279**, 258 (2005)
5.83 N. I. Puzanov, A. M. Eidenzon: Semicond. Sci. Technol. **7**, 406 (1992)
5.84 K. Nakamura, T. Saishoji, J. Tomioka: J. Cryst. Growth **237–239**, 1678 (2002)
5.85 V. V. Voronkov: Mater. Sci. Eng. **B73**, 69 (2000)
5.86 V. V. Voronkov, R. Falster: J. Crystal Growth **226**, 192 (2001)
5.87 J. Takahashi, K. Nakai, K. Kawakami, Y. Inoue, H. Yokota, A. Tachikawa, A. Ikari, W. Ohashi: Jpn. J. Appl. Phys. **42**, 363 (2003)
5.88 T. Abe, M. Kimura: In: *Semiconductor Silicon, 1990*, ed. by H. R. Huff, K. Barraclough, J. Chikawa (The Electrochemical Society, Pennington 1990) p. 105
5.89 D.-R. Yang, Y.-W. Wang, H.-N. Yao, D.-L. Que: Progress in Natural Science **3**(2), 176 (1993)
5.90 W. v. Ammon, R. Hoelzl, T. Wetzel, D. Zemke, G. Raming, M. Blietz: Microelectron. Eng. **66**, 234 (2003)
5.91 W. v. Ammon, A. Ehlert, U. Lambert, D. Gräf, M. Brohl, P. Wagner: Gate oxide related bulk properties of oxygen doped floating zone and Czochralski silicon, Semiconductor Silicon, San Francisco 1994, ed. by H. R. Huff, W. Bergholz, K. Sumino (The Electrochemical Society, Pennington 1994) 136
5.92 K. Nakai, Y. Inoue, H. Yokota, A. Ikari, J. Takahashi, A. Tachikawa, K. Kitahara, Y. Ohta, W. Ohashi: J. Appl. Phys. **85**(8), 4301 (2001)
5.93 F. Shimura, R. S. Hockett: Appl. Phys. Lett. **48**, 224 (1986)
5.94 Q. Sun, K. H. Yao, H. C. Gatos, J. Lagowski: J. Appl. Phys. **71**(8), 3760 (1992)
5.95 K. Aihara, H. Takeno, Y. Hayamizu, M. Tamatsuka, T. Masui: J. Appl. Phys. **88**(6), 3705 (2000)
5.96 K. Nakai, Y. Inoue, H. Yokota, A. Ikari, J. Takahashi, W. Ohashi: Formation of grown-in defects in nitrogen doped CZ-Si crystals, Proc 3rd Int. Symp. on Advanced Science and Technology of Silicon Materials, Kona, Hawaii 2000, ed. by M. Umeno (145th Committee of the Japan Society for the Promotion of Science, Kona 2000) 88
5.97 D. Gräf, U. Lambert, R. Schmolke, R. Wahlich, W. Siebert, E. Daub, W. v. Ammon: 300 mm Epi pp-wafer: Is there sufficient gettering?, Proc. 6th Int. Symp. on High Purity Silicon, Seattle, Washington 2000, ed. by C. L. Claeys, P. Rai-Choudhury, M. Watanabe, P. Stallhofer, H. J. Dawson (The Electrochemical Society, Pennington 2000) 319
5.98 H. J. Stein: Nitrogen in crystalline silicon, Proc. Int. Symp. on Oxygen, Carbon, Hydrogen and Nitrogen in Crystalline Silicon, Boston 1986, ed. by J. C. Mikkelsen Jr., S. J. Pearton, J. W. Corbett, S. J. Pennycook (Materials Research Society, Pittsburg 1986) 523
5.99 Y. Itoh, T. Abe: Appl. Phys. Lett. **53**(1), 39 (1988)
5.100 A. Hara, A. Ohsawa: Interaction of oxygen and other point defects in silicon crystals, Proc. Int. Symp. on Advanced Science and Technology of Silicon Materials, Kona, Hawaii 1991, ed. by K. Kohra (145th Committee of the Japan Society for the Promotion of Science, Kona 1991) 47
5.101 H. Sawada, K. Kawakami: Phys. Rev. B **62**(3), 1851 (2000)
5.102 H. Kageshima, A. Taguchi, K. Wada: Appl. Phys. Lett. **76**(25), 3718 (2000)
5.103 R. Jones, S. Öberg, F. B. Rasmussen, B. B. Nielson: Phys. Rev. Lett. **72**, 1882 (1994)
5.104 K. L. Brower: Phys. Rev. B **26**, 6040 (1982)
5.105 H. J. Stein: Appl. Phys. Lett. **47**(12), 1339 (1985)
5.106 K. Murakami, H. Itoh, K. Takita., K. Masuda: Appl. Phys. Lett. **45**(2), 176 (1984)
5.107 W. v. Ammon, D. Gräf, W. Zulehner, R. Schmolke, E. Dornberger, U. Lambert, J. Vanhellemont, W. Hensel: Suppression of point defect aggregation in FZ silicon single crystals by nitrogen doping; Extended Abstracts, Semiconductor Silicon, San Diego 1998, ed. by H. R. Huff, U. Gösele, H. Tsuya (The Electrochemical Society, Pennington 1998) Abstract no. 512

5.108 K. Nakamura, T. Saishoji, S. Togawa, J. Tomioka: The effect of nitrogen on the grown-in defect formation in CZ silicon crystals. In: *Proceedings of the Kazusa Akademia Park Forum on the Science and Technology of Silicon Materials*, ed. by K. Sumino (Kazusa Akademia Park, Chiba 1999) p. 116

5.109 V. V. Voronkov, R. Falster: J. Electrochem. Soc. **149**(3), G167 (2002)

5.110 W. B. Knowlton, J. T. Walton, J. S. Lee, Y. K. Wong, E. E. Haller, W. v. Ammon, W. Zulehner: Mater. Sci. Forum **196–201**, 1761 (1995)

5.111 T. Ono, S. Umeno, T. Tanaka, E. Asayama, M. Hourai: Behavior of defects in nitrogen doped CZ-Si crystals, Proc. Int. Symp. of the Forum on the Science and Technology of Silicon Materials, Shonan Village Center, Kanagawa 2001, ed. by H. Yamata-Kaneta, K. Sumino (Japan Technical Information Service, Tokyo 2001) 95

5.112 K. Nakamura, T. Saishoji, S. Togawa, J. Tomioka: Influence of nitrogen on the pont defect reaction in silicon, Proc. Int. Symp. of the Forum on the Science and Technology of Silicon Materials, Shonan Village Center 2001, ed. by H. Yamata-Kaneta, K. Sumino (Japan Technical Information Service, Tokyo 2001) 109

5.113 W. v. Ammon, R. Hölzl, J. Virbulis, E. Dornberger, R. Schmolke, D. Gräf: J. Cryst. Growth **226**(1), 19 (2001)

5.114 P. Wagner, R. Oeder, W. Zulehner: Appl. Phys. A **46**, 73 (1988)

5.115 W. v. Ammon, P. Dreier, W. Hensel, U. Lambert, L. Köster: Mater. Sci. Eng. **B36**, 33 (1996)

5.116 M. W. Qi, S. S. Tan, B. Zhu, P. X. Cai, W. F. Gu, M. Xu, T. S. Shi, D. L. Que, L. B. Li: J. Appl. Phys. **69**, 3775 (1991)

5.117 A. Gali, J. Miro, P. Deak, C. Ewels, R. Jones: J. Phys. Condens. Mat. **8**, 7711 (1996)

5.118 W. v. Ammon: Crystal growth of large diameter CZ Si crystals, Proc 2nd Int. Symp. on Advanced Science and Technology of Silicon Materials, Kona, Hawaii 1996, ed. by M. Umeno (145th Committee of the Japan Society for the Promotion of Science, Kona 1996) 233

5.119 M. Suhren, D. Gräf, U. Lambert, P. Wagner: Crystal defects in highly boron doped silicon, Proc. 4th Int. Symp. on High Purity Silicon, San Antonio 1996, ed. by C. L. Claeys, P. Rai-Choudhury, M. Watanabe, P. Stallhofer, H. J. Dawson (The Electrochemical Society, Pennington 1996) 132

5.120 T. Sinno, H. Susanto, R. Brown, W. v. Ammon, E. Dornberger: Appl. Phys. Lett. **75**, 1544 (1999)

5.121 T. Abe, T. Masui, H. Harada, J. Chikawa: In: *VLSI Science and Technology, 1985*, ed. by W. M. Bullis, S. Broyda (The Electrochemical Society, Pennington 1985) p. 543

5.122 R. Takeda, T. Minami, H. Saito, Y. Hirano, H. Fujimori, K. Kashima, Y. Matsushita: Influence of LSTD size on the formation of denuded zone in hydrogen-annealed CZ silicon wafers, Proc. 6th Int. Symp. on High Purity Silicon, Phoenix 2000, ed. by C. L. Claeys, P. Rai-Choudhury, M. Watanabe, P. Stallhofer, H. J. Dawson (The Electrochemical Society, Pennington 2000) 331

5.123 S. Kishino, M. Kanamori, N. Yoshihizo, M. Tajima, T. Iizuka: J. Appl. Phys. **50**, 8240 (1978)

5.124 T. Fukuda: Appl. Phys. Lett. **65**(11), 1376 (1994)

5.125 F. Shimura: J. Appl. Phys. **59**, 3251 (1986)

5.126 K. Sueoka, M. Akatsuka, K. Nishihara, T. Yamamoto, S. Kobayashi: Mater. Sci. Forum **196–201**, 1737 (1995)

5.127 J. Vanhellemont, C. Claeys: J. Appl. Phys. **62(9)**, 3960 (1987)

6. Diffusion in Semiconductors

Atomic diffusion in semiconductors refers to the migration of atoms, including host, dopant and impurities. Diffusion occurs in all thermodynamic phases, but the solid phase is the most important in semiconductors. There are two types of semiconductor solid phase: amorphous (including organic) and crystalline. In this chapter we consider crystalline semiconductors and describe the processes by which atoms and defects move between lattice sites. The emphasis is on describing the various conditions under which diffusion can occur, as well as the atomic mechanisms that are involved, rather than on tabulating data. For brevity's sake, we also focus on the general features found in the principal semiconductors from Groups IV, III–V and II–VI; IV–VI and oxide semiconductors are excluded from consideration. It is not surprising that most of the data available in this field relate to the semiconductors that are technologically important – they are used to fabricate electronic and optoelectronic devices. One unavoidable consequence of this technological need is that diffusion data tend to be acquired in a piecemeal fashion.

6.1 Basic Concepts 122

6.2 Diffusion Mechanisms 122
    6.2.1 Vacancy and Interstitial Diffusion Mechanisms 122
    6.2.2 The Interstitial–Substitutional Mechanism: Dissociative and Kick-Out Mechanisms 122
    6.2.3 The Percolation Mechanism 123

6.3 Diffusion Regimes 123
    6.3.1 Chemical Equilibrium: Self- and Isoconcentration Diffusion ... 123
    6.3.2 Chemical Diffusion (or Diffusion Under Nonequilibrium Conditions)........ 123
    6.3.3 Recombination-Enhanced Diffusion 125
    6.3.4 Surface Effects 125
    6.3.5 Short Circuit Paths 125

6.4 Internal Electric Fields 126

6.5 Measurement of Diffusion Coefficients... 126
    6.5.1 Anneal Conditions 126
    6.5.2 Diffusion Sources 126
    6.5.3 Profiling Techniques 126
    6.5.4 Calculating the Diffusion Coefficient................................. 127

6.6 Hydrogen in Semiconductors 127

6.7 Diffusion in Group IV Semiconductors 128
    6.7.1 Germanium 128
    6.7.2 Silicon 128
    6.7.3 $Si_{1-x}Ge_x$ Alloys 129
    6.7.4 Silicon Carbide 129

6.8 Diffusion in III–V Compounds................ 130
    6.8.1 Self-Diffusion 130
    6.8.2 Dopant Diffusion 130
    6.8.3 Compositional Interdiffusion 131

6.9 Diffusion in II–VI Compounds................ 131
    6.9.1 Self-Diffusion 132
    6.9.2 Chemical Self-Diffusion.............. 132
    6.9.3 Dopant Diffusion 132
    6.9.4 Compositional Interdiffusion 132

6.10 Conclusions... 133

6.11 General Reading and References........... 133

References 133

Diffusion describes the movement of atoms in space, primarily due to thermal motion, and it occurs in all forms of matter. This chapter focuses on atom diffusion in crystalline semiconductors, where diffusing atoms migrate from one lattice site to another in the semiconductor crystal. The diffusion of atoms and defects is at the heart of material processing, whether at the growth or post-growth stage, and control over diffusion is the basis of process simulation and defect engineering. Such control calls for an understanding of the diffusion processes involved in a given situation. The needs of device technology have provided the main impetus for investigations into the diffusion of atoms in semiconductors. As the physical dimensions of devices have shrunk, the barriers to understanding diffusion mechanisms and processes in complex structures have greatly multiplied.

6.1 Basic Concepts

Consider a particle in a three-dimensional isotropic lattice which migrates by making jumps from one lattice site to a nearest neighbor site. If the distance between nearest neighbor sites is a and the particle makes n jumps in time t, then, assuming each jump is random (so the directions of successive jumps are independent of each other), the mean square displacement $\langle R^2 \rangle$ is equal to na^2 [6.1]. Fick's first law defines the associated diffusivity D to be $na^2/6t = \langle R^2 \rangle/6t = va^2/6$, where $v = n/t$ is the average jump rate of the particle. Taking the diffusion length as $2\sqrt{(Dt)}$, it follows that this is also equal to $2\sqrt{(\langle R^2 \rangle/6)}$. For $D = 10^{-12}\,\text{cm}^2/\text{s}$, $t = 10^4\,\text{s}$ and $a = 2 \times 10^{-8}\,\text{cm}$, the diffusion length is $2\,\mu\text{m}$, $n = 1.5 \times 10^8$ jumps and the total distance na traveled by the particle is $3\,\text{cm}$. However, it turns out that in most diffusion mechanisms successive jumps are correlated, not random. The effect of nonrandom jumps is to decrease the diffusivity of the particle relative to what it would be if the jumps were random. Taking this correlation into account leads to $D = fva^2/6$, where $f (\leq 1)$ is the correlation factor [6.1], v is temperature-dependent and f may or may not be, depending on the particular situation. Overall, the temperature dependence of D is found to obey the Arrhenius relation $D = D_0 \exp(-Q/kT)$.

6.2 Diffusion Mechanisms

Two categories of diffusion mechanism are recognized: defect and nondefect. A simple example of the latter class is the simultaneous jumps of two adjacent atoms in order to exchange sites. There is a general consensus, however, that nondefect mechanisms do not play any significant role in semiconductor diffusion, although recently nondefect contributions have been proposed for self- and dopant diffusions in Si [6.2]. In the elemental semiconductors Si and Ge, vacancies and interstitials are the primary defects. In binary compound semiconductors (such as GaAs and ZnSe) there are two sublattices, the anion and cation, so there are vacancies and interstitials for each sublattice together with antisite defects on each sublattice. Further complexity arises due to the various states of ionization of the defects: the mobility of a defect depends on its charge state. The relative concentrations of the different charge states will be determined by the position of the Fermi level.

6.2.1 Vacancy and Interstitial Diffusion Mechanisms

In the Si lattice, a vacancy V_{Si} can migrate by a nearest neighbor Si atom jumping into the vacancy (in other words, the Si atom and V_{Si} have exchanged sites so that the Si atom has also migrated). Equally, a substitutional dopant atom can migrate by jumping into a V_{Si} at a nearest neighbor site. Similarly, in a binary semiconductor such as GaAs, Ga atoms can migrate over the Ga sublattice via jumps into nearest neighbor Ga vacancies, as can dopant atoms substituting into the Ga sublattice. Anti-site defects can diffuse by jumping into vacancies in the same sublattice, such as the As anti-site defect in GaAs, As_{Ga}, diffusing by jumps into Ga vacancies. For self-interstitials, such as Si_i or Ga_i, their concentrations are sufficiently small for neighboring interstitial sites to always be empty, which means that the occupancy of nearest neighbor sites is not a factor when determining jump rates. If a self-interstitial, such as Si_i, pushes a Si atom on a normal lattice site into an interstitial site instead of jumping into a neighboring interstitial site, and therefore replaces the displaced Si atom, the process is known as the "interstitialcy mechanism". This concept extends to a substitutional dopant atom forming a pair with a self-interstitial, which then migrates with the dopant atom, alternating between substitutional and interstitial sites. It is also possible for point defects to form complexes which can diffuse as a single entity. Examples are the Frenkel pair $V_{Si}Si_i$, di-vacancies such as $V_{Ga}V_{Ga}$ or $V_{Ga}V_{As}$, and the defect pair formed between a substitutional dopant atom and an adjacent vacancy.

6.2.2 The Interstitial–Substitutional Mechanism: Dissociative and Kick-Out Mechanisms

The interstitial–substitutional diffusion mechanism arises when a dopant species Z occupies both interstitial and substitutional sites, represented by Z_i and Z_s respectively, and diffusion is restricted to jumps of Z_i. In this case, we may ask how the Z_s concentration $[Z_s]$ is linked to the Z_i migration. Consider the diffusion of Z in Si. The dissociative mechanism (also known as the Frank–Turnbull mechanism) is based on the defect interaction

$$Z_i + V_{Si} \rightleftharpoons Z_s$$

and application of the law of mass action (LMA) leads to $[Z_i][V_{Si}] \propto [Z_s]$. In the kick-out mechanism, the defect interaction is

$$Z_i + Si_s \rightleftharpoons Z_s + Si_i$$

and therefore $[Z_i] \propto [Z_s][Si_i]$ ($[Si_s]$ is omitted because it is effectively constant). In order to sustain growth in $[Z_s]$ by either mechanism, it is clearly necessary to have either a supply of V_{Si} or a means of removing Si_i. For simplicity, neutral charge states have been assigned to all of the defects in these two interactions. A detailed treatment of the kick-out mechanism has been given by *Frank* et al. [6.3]. For in-diffusion of Z, the Frank–Turnbull mechanism consumes vacancies and will therefore tend to reduce the local vacancy concentration, whereas the local self-interstitial concentration will be enhanced by the kick-out mechanism. Out-diffusion of Z reverses the effects on the native defect concentrations.

6.2.3 The Percolation Mechanism

The percolation mechanism [6.4] was proposed to explain group V dopant diffusion in Si at high dopant concentrations (in excess of $\approx 1\%$). At low concentrations diffusion is via dopant-V_{Si} pairs. As the dopant concentration increases, regions occur in the Si lattice where the proximity of the dopant atoms enhances the mobility and concentration of the V_{Si}. Within this network the diffusivity of dopant-V_{Si} pairs is thereby also enhanced so that the dopant diffusivity increases overall. The percolation network only forms once the dopant concentration exceeds a certain critical value. In principle this mechanism could extend to other highly doped materials.

6.3 Diffusion Regimes

The mobility of a native defect and/or dopant atom reflects the physical and chemical environment under which diffusion is occurring. Two types of environment arise: conditions of chemical equilibrium and those of chemical nonequilibrium. Diffusion in temperature gradients is excluded – only isothermal conditions are considered.

6.3.1 Chemical Equilibrium: Self- and Isoconcentration Diffusion

Chemical equilibrium means that the concentrations of all chemical components, including native defects, are uniform throughout the semiconductor, and where appropriate (such as in a compound material), the solid is in equilibrium with the ambient vapor of the components so that the level of nonstoichiometry is defined. Experimentally this requires diffusion to be carried out in a sealed system. Self-diffusion refers to the diffusion of the host atoms, such as Si atoms diffusing in the Si lattice. Isoconcentration diffusion describes the diffusion of dopant atoms when the same dopant concentration is uniform throughout the sample, such as for As diffusion in Si for a constant As doping level. In either case, diffusion can only be observed if some of the particular diffusing atoms are tagged, such as by using a radioisotope or an isotopically enriched diffusion source. The diffusivity of a tagged or tracer species is related to the concentration of the native defect that provides the diffusion path, and the self-diffusivity (the diffusivity of the tracer) is always significantly smaller than the associated defect diffusivity. Specific relations can be found in *Shaw* [6.5]. A diffusion flux of the tracer arises from a tracer concentration gradient, which is not to be confused with a chemical concentration gradient.

6.3.2 Chemical Diffusion (or Diffusion Under Nonequilibrium Conditions)

This category contains all of the diffusion phenomena that are of technological interest and importance. In this case, diffusion occurs due to spatial gradients in the concentrations of the chemical components in the material, which are in turn caused by departures from equilibrium: the diffusion processes are attempting to either restore or achieve equilibrium.

Chemical Self-Diffusion
Chemical self-diffusion describes the process whereby a compound semiconductor changes from one level of nonstoichiometry to another through changes in the native defect populations. These changes can arise due to a change in the ambient partial pressure of one

of the components (that of As in the case of GaAs for example), or through a change in temperature under a defined or controlled component partial pressure. Good examples are provided by II–VI semiconductors [6.6] and in particular $Hg_{0.8}Cd_{0.2}Te$ [6.7], where p- to n-type conversion is used to form p–n junctions by annealing in Hg vapor. Changes in the native defect concentrations can also lead to an increased dopant diffusivity; an increase in the vacancy concentration will enhance any diffusivity based on a vacancy mechanism for instance.

Dopant Diffusion

Suppose we have a dopant diffusing into the semiconductor from a source located at an external surface (such as a surface layer) or in an external phase (such as a gas or vapor). The existence of the dopant concentration gradient can lead to various effects which can influence the dopant flux. For example, if a dopant diffuses via a vacancy mechanism, then at any position in the diffusion region the increase in the dopant concentration requires a supply of vacancies, so that to maintain local defect equilibrium there must also be an associated vacancy flux. If the dopant controls the position of the Fermi level, then the concentrations of ionized native defects will increase or decrease, depending on their charge state, relative to their intrinsic concentrations (the Fermi-level effect). This means that the concentrations of native defects of opposite (the same) polarity to the dopant will be increased (decreased). Increases in the concentrations of ionized native defects due to Zn, Si or Te diffusion into GaAs/GaAlAs superlattice structures explain the disordering of the superlattices [6.8]. Usually a substitutional dopant atom will have a different size to that of the host atom it replaces. This size difference creates a local mechanical strain which in turn can cause changes in the local concentrations of native defects as well as to jump rates and hence dopant diffusivity [6.9]. If the strain is large enough, misfit dislocations will be generated [6.10], otherwise there will be a strain energy gradient matching the dopant concentration gradient which can enhance or retard the dopant flux [6.11]. More recently a new scenario has emerged: dopant diffusion in strained epilayers. Whether diffusion is enhanced or retarded depends on several poorly understood parameters [6.12]. However, significant effects are found, such as the decrease in the B diffusivity in strained $Si_{1-x}Ge_x$ epilayers by a factor of ≈ 10 as the strain increased from zero to 0.64 [6.13].

Compositional Interdiffusion (CID)

Compositional interdiffusion describes diffusion across the interface separating two materials of different chemical composition. Chemical composition here refers to major components; dopants and deviations from stoichiometry are excluded. CID can be exploited when making graded bandgap structures and during material preparation, such as in $Hg_{1-x}Cd_xTe$, where an alternating sequence of HgTe and CdTe epilayers of appropriate thicknesses are first grown and then interdiffused. CID can also pose problems in the fabrication of multiple quantum wells and superlattice structures when sharp boundaries are necessary. In particular, donor or acceptor dopant diffusion into GaAs-based superlattices can cause essentially complete intermixing on the cation sublattice [6.8]. This phenomenon is also known as diffusion-induced disorder. B or As doping also results in rapid intermixing at a Si/Ge interface [6.14].

Transient Enhanced Diffusion (TED)

Ion implantation is often the preferred way to achieve a doped layer. The implantation process does however create a significant amount of lattice damage so that a subsequent anneal stage is needed in order to achieve full electrical activity of the implanted dopant and recovery of the lattice damage. During the implantation process, the implant ions create collision cascades of vacancies and self-interstitials (an excess of native defects). The post-implant anneal serves to remove or reduce this excess. In addition to vacancy/interstitial recombination, the excess native defects can interact to form clusters (which also may contain implant ions as well as residual impurities) and extended defects, such as dislocation loops. At the start of the post-implant anneal, the local concentrations of vacancies and self-interstitials in the implant region can greatly exceed equilibrium values and therefore enhance the implant ion diffusivity in the implant region. As annealing proceeds the excess concentrations will diminish and will be reflected in a diminishing dopant diffusivity until values appropriate to local equilibrium are reached. This temporary enhancement in the dopant diffusivity is known as TED. The topic is a complex one to analyze quantitatively and detailed consideration of the issues involved in the case of B implants in Si can be found in the review by *Jain* et al. [6.15]. In the fabrication of shallow p–n junctions using ion implants and rapid thermal annealing (RTA), TED can determine the lower limit to junction depth. TED of B in Si can be reduced with coimplants of Si prior to RTA [6.16]. TED of Be and Si in GaAs has also been discussed [6.17]. Some workers use the term TED to describe the enhanced diffusivity of a dopant, incorporated during growth, which occurs when the structure is annealed at a higher temperature than the growth temperature, such that there is an initial supersaturation of the relevant native defects at the anneal temperature.

Segregation, Gettering, Precipitation and Clustering

A variety of important scenarios arise, involving many of the above regimes, during growth and/or thermal processing stages of materials and structures. The segregation of acceptor dopants in InP [6.18] and in III–V heterostructures [6.19, 20] has been observed and modeled. The segregation (or accumulation) due to diffusion of the dopant is in effect a partitioning process to preferred (higher solubility) regions within the layer structure. Gettering describes the segregation, or cleanup, of a fast-diffusing impurity from the active regions of a device structure. Such impurities are typically Group IB and transition metals and are incorporated either during growth or during subsequent processing. Gettering sites in Si are provided by O precipitates, self-ion implant damage layers and nanocavities [6.21]. In contrast, Group IB impurities are gettered in $Hg_{1-x}Cd_xTe$ by regions of high cation vacancy concentration [6.22].

Precipitation occurs when a species – whether native defect, dopant or impurity – becomes supersaturated, and in order to achieve its equilibrium concentration the species excess is removed by the formation of precipitates within the host lattice. A self-interstitial or vacancy excess can be removed through the nucleation of dislocation loops, stacking faults or voids, which then provide sites for the precipitation of the remaining excess. In the case of a dopant, impurity or nonstoichiometric excess, nucleation of a precipitate can be spontaneous (homogeneous) or heterogeneous. The latter occurs at the site of an impurity atom (for example, C atoms in Si serve as nucleation centers for the precipitation of O) or at dislocations, giving rise to the term "decoration". Growth of any precipitate proceeds via diffusion of the particular species from solution in the matrix to the precipitate and is generally diffusion-limited. Invariably local stress fields will be present which influence the diffusion and, if present initially, they may also play a role in the nucleation stage. The precipitation of O impurities in Si presents a unique case study because of the high [O], its technological importance and its complexity [6.23, 24]. The rather simpler case of B precipitation in Si has been described by *Solmi* et al. [6.25]. A cluster (or agglomerate) refers to a configuration of at least a few dopant atoms (with or without associated native point defects) or host species. *Solmi* and *Nobili* [6.26] have identified $(2As_{Si} - V_{Si})^0$ and $(4As_{Si} - V_{Si})^+$ clusters in heavily As-doped Si. Heavy C doping [6.27] and B implants [6.28] in Si give rise to self-interstitial clusters with C and B respectively. In Si, according to *Ortiz* et al. [6.29], if the number of self-interstitials in a cluster exceeds ≈ 10 there is a transition to a $\{113\}$ defect.

6.3.3 Recombination-Enhanced Diffusion

The local energy released in the nonradiative recombination of excess free carriers can help a diffusing species to surmount the energy barrier separating it from an adjacent lattice site – in other words, the energy barrier facing a jump is effectively reduced. This situation is important in the degradation of performance of device structures which utilize high excess minority carrier concentrations, such as light-emitting and laser diodes.

6.3.4 Surface Effects

The concentrations of native point defects within the bulk can be altered by surface processes. In the case of Si it is well known that during surface oxidation or nitridation there is injection of Si interstitials or of vacancies respectively. All diffusants can therefore be affected during the duration of the process. Ion beam milling causes the injection of Hg interstitials into $Hg_{1-x}Cd_xTe$ in sufficient quantities to effect p- to n-type conversion.

6.3.5 Short Circuit Paths

The existence of dislocations and subgrain boundaries in single-crystal materials generally provides high diffusivity routes for all atomic species relative to the surrounding matrix. Care is always needed when evaluating experimental data to ensure that bulk diffusion is not being masked by short circuit paths [6.30]. In the case of polycrystalline Si, the grain boundaries may provide high diffusivity routes, as in the cases of As and B [6.31], or retard diffusion, as for Au [6.32]. The situation is a complex one, as grain growth also occurs during any anneal. *Kaur* et al. [6.33] have provided a comprehensive account of short circuit path diffusion.

6.4 Internal Electric Fields

When the dopant concentration is large enough to make the diffusion zone electrically extrinsic, free carriers from the dopant, due to their much higher mobility, will diffuse ahead of the parent dopant atoms. This separation creates a local electric field whose direction is such as to pull the dopant atoms after the free carriers (and also to pull the free carriers back). Provided that the diffusion length $>\approx$ six Debye screening lengths (typically $\approx 10^2$ nm), the diffusion zone can be regarded as electrically neutral (the space charge density is negligible) [6.34]. In this situation the local electric field E is given by $-(kT/en)(\partial n/\partial x)$ for an ionized donor dopant diffusing parallel to the x-axis, and nondegenerate conditions apply: k, T, e and n are Boltzmann's constant, the absolute temperature, the electronic charge and the free electron concentration respectively. E exerts a force on each ionized donor (D^+) parallel to the x-axis, creating a local donor flux $-(D(D^+)[D^+]/n)(\partial n/\partial x)$ due to drift in the electric field: $D(D^+)$ is the donor diffusivity [6.35]. This drift flux adds to the diffusion flux, $-D(D^+)\partial[D^+]/\partial x$, to give the total donor flux at any position in the diffusion region, so that the donor flux in this case is increased due to E. E will also cause drift of any other charged species.

Internal electric fields can arise in other circumstances such as in depletion layers where E must be calculated from Poisson's equation, in graded bandgap structures [6.11, 36], and at the interfaces of heterostructures. Cubic II–VI and III–V strained layer heterostructures grown on the {111} direction are piezoelectric and typical strains from lattice mismatch of $\approx 1\%$ can give $E \approx 10^5$ V/cm in the absence of free carrier screening [6.37]. In wurtzite heterostructures based on the Ga, In nitrides, even higher fields are found ($E \approx 10^6$ V/cm) due to piezoelectric and spontaneous polarization [6.38]. These fields can be important in CID and chemical self-diffusion.

6.5 Measurement of Diffusion Coefficients

6.5.1 Anneal Conditions

Accurate control of sample temperature and ambient are essential if controlled and reproducible results are to be obtained in a diffusion anneal. Depending on the time spent at the anneal temperature, the warm-up and cool-down times may also be important. An appropriate choice of ambient is needed to preserve the sample surface (to avoid evaporation, surface melting or alloying with the dopant source for example). For compound semiconductors it is necessary to define the level of non-stoichiometry by controlling the ambient partial pressure of one of the components, such as As for GaAs or Hg for $Hg_{1-x}Cd_xTe$. If the dopant is in an external phase, knowledge of the phase diagram of all of the components is required [6.1, 35]. Control over partial pressure is best achieved in a sealed system, typically a fused silica ampoule. Annealing in a vertical or horizontal resistance-heated furnace requires a minimum anneal time of 30 to 60 min in order to avoid uncertainties due to warm-up and cool-down. The drive to shallow dopant profiles has been facilitated through rapid thermal annealing (RTA) techniques. These are based on radiant heating of the sample, and linear heating rates of 100–400 °C/s with cooling rates of up to 150 °C/s are available. RTA however precludes the use of a sealed system and, in this case, a popular means of preventing surface deterioration is to seal the sample with an inert, impervious capping layer, made of silicon nitride for example.

6.5.2 Diffusion Sources

Consideration is limited to planar samples with diffusion normal to a principal face. This is a common situation and diffusion of a dopant or tracer species can take place from: (i) a surrounding vapor or gas phase; (ii) a surface layer, which may be evaporated, chemically deposited (CVD) or a spun-on silicate glass, all incorporating the diffusant; (iii) epilayers containing the diffusant, which may provide the external surface or be buried within the epitaxial structure; (iv) ion implants of a dopant either directly into the sample surface or into a thin surface layer so as to avoid lattice damage. It is obviously desirable that negligible diffusion occurs prior to reaching the anneal temperature when the diffusant is incorporated into an epilayer. In self-diffusion experiments the tracer can be a radiotracer or an isotopically enriched species. A key requirement for either form of tracer is availability, and a radiotracer must have a half-life that is long enough for the experiments to be carried out.

6.5.3 Profiling Techniques

Determining the spatial distribution of a diffusant for various anneal times is fundamental to obtaining its diffusion coefficient or diffusivity. Most methods are destructive, as they generally require a bevel section through the diffusion zone or the sequential removal of layers. The two broad profiling categories are electrical and species-specific. Electrical methods are primarily the p–n junction method, spreading resistance and capacitance–voltage profiling. Limitations of the electrical methods are: (i) assumptions are needed to link the electrical data to the diffusant (for example, that the diffusant is the only electrically active center and that it is fully ionized); (ii) the assumption that the anneal temperature defect situation is "frozen-in" during cooldown. Electrical methods are the most direct means of measuring chemical self-diffusivities and can readily detect changes in host concentrations of < 1 part in 10^4. Species-specific (chemical element or isotope) profiling means that the chemical concentration of the diffusant is determined regardless of its location(s) in the lattice and of its electrical state. Profiling of the diffusant using a radiotracer has been widely used [6.39], but in the past decade or so secondary ion mass spectrometry (SIMS) has become what is essentially the standard procedure for diffusant profiling. This is because SIMS can measure diffusant concentrations within the range 10^{16} to 10^{22} cm^{-3} with spatial resolutions at best of several nanometers per decade (of concentration). Primary factors determining the resolution are progressive roughening of the eroded surface and "knock-on" effects due to the probing ion beam displacing the diffusant to greater depths. A further problem may arise when the atomic mass of a dopant is close to that of the host species.

Nondestructive profiling techniques applicable to CID in quantum well and superlattice structures utilize either high-resolution X-ray diffraction (HRXRD) [6.40] or photoluminescence (PL). The detail in the X-ray diffraction patterns reflects the CID profiles at the interfaces and can also reveal the presence of strain in the structures. The use of PL requires the presence of optically active centers in the quantum well. CID changes the shape and depth of the quantum well, which in turn changes the photon energies in the luminescence spectra. HRXRD and PL can also be combined. A particular advantage of these techniques is that they allow successive anneals to be performed on the same sample.

6.5.4 Calculating the Diffusion Coefficient

Once a planar concentration profile has been obtained, the first step is to see if the profile can be fitted to a solution of Fick's second law. The simplest solution occurs for a diffusivity D independent of the diffusant concentration (c), for a constant surface concentration c_0 and a diffusion length $\ll$ the layer or sample thickness. The solution is $c = c_0 \mathrm{erfc}[x/2\sqrt{(Dt)}]$ [6.1]. If the profile is not erfc, it may be because D varies with c, and D (as a function of c) can be obtained by a Boltzmann–Matano analysis [6.35]. It is important to recognize that the erfc or Boltzmann–Matano solutions are only valid provided c_0 does not change with time and that c/c_0 versus $x/\sqrt{t}$ for profiles at various t reduce to a single profile. More complex situations and profiles require numerical integration of the appropriate diffusion equation(s) and matching to the experimental c versus x profile; in other words a suitable model with adjustable parameters is used to simulate the observed profiles. The interpretation of HRXRD and/or PL data provides a good example of a simulation scene in which an assumed D, either c-dependent or -independent, is used to calculate the resulting CID profile and its effect on the X-ray patterns and/or PL spectra. Whereas SIMS can observe diffusivities as low as $\approx 10^{-19}$ cm^2/s, the HRXRD limit is $\approx 10^{-23}$ cm^2/s.

6.6 Hydrogen in Semiconductors

Hydrogen is a ubiquitous element in semiconductor materials and can be incorporated either by deliberate doping or inadvertently, at significant concentrations, during growth and/or in subsequent surface treatments where organic solvents, acid or plasma etching are used. H is known to passivate electrically active centers by forming complexes with dopants and native defects as well as by bonding to the dangling bonds at extended defects. Such interactions may well affect the diffusivities of the dopant and native defect. This expectation is realized in the case of O in Si, where the presence of H can enhance O diffusivity by two to three orders of magnitude [6.41]. Ab initio calculations show that, at least in the Group IV and III–V semiconductors, H is incorporated interstitially in the three charge states, H$^+$, H^0 and H$^-$, with the Fermi level controlling the relative concentrations. In addition to interactions with dopant atoms and native defects, H_2 molecules also form. *Mathiot* [6.42] has modeled H diffusion in terms of simultaneous diffusion by the three interstitial charge states with the formation of immobile neutral complexes. In polycrystalline Si, the grain boundaries retard H diffusion, so H diffuses faster in the surrounding lattice than in the grain boundary.

6.7 Diffusion in Group IV Semiconductors

Diffusants divide into one of five categories: self-, other Group IVs, slow diffusers (typically dopants from Groups III and V), intermediate diffusers and fast diffusers. The materials of interest are Ge, Si, Si/Ge alloys and SiC. A particular feature is that self-diffusion is always slower than the diffusion of other diffusants. With the exception of SiC, which has the zinc blende structure, as well as numerous polytypes (the simplest of which is the wurtzite, $2H - SiC$, form), the other members of this group have the diamond lattice structure.

6.7.1 Germanium

The evidence to date identifies the dominant native defect in Ge as the singly ionized vacancy acceptor, V_{Ge}^- [6.43], which can account for the features found in self-diffusion and in the diffusivities of dopants from groups III and V. The self-diffusivity, relative to the electrically intrinsic value, is increased in n-type Ge and decreased in p-type as expected from the dependence of $[V_{Ge}^-]$ on the Fermi level. In intrinsic Ge the best parameters for the self-diffusivity are $D_0 = 13.6 \text{ cm}^2/\text{s}$ and $Q = 3.09 \text{ eV}$, from *Werner* et al. [6.44], because of the wide temperature range covered (535–904 °C). The diffusivities of donor dopants (P, As, Sb) are very similar in magnitude, as are those for acceptor dopants (Al, In, Ga). The acceptor group diffusivities, however, are very close to the intrinsic self-diffusivity, whereas those for the donor group are 10^2 to 10^3 times larger. Li is a fast (interstitial) diffuser with a diffusivity exceeding the donor group diffusivities by factors of 10^7 to 10^5 between 600 and 900 °C, whereas Cu [6.45] and Au [6.46] are intermediate (dissociative) diffusers.

6.7.2 Silicon

Si stands alone due to the intensive investigations that have been lavished on it over the past 50 years. In the early days diffusion data yielded many perplexing features. Today the broad aspects are understood along with considerable detail, depending on the topic. Diffusion in Si covers many more topics than arise in any other semiconductor and it is still a very active area of R & D. It is now recognised that, apart from foreign purely interstitial species, self-interstitials, Si_i, and vacancies, V_{Si}, are involved in all diffusion phenomena. So far the best self-diffusion parameters obtained for intrinsic Si are $D_0 = 530 \text{ cm}^2/\text{s}$ and $Q = 4.75 \text{ eV}$ in the temperature range 855–1388 °C [6.47]. Two distinct facets of self- and dopant diffusion in Si are: (a) the diffusivity has two or three components, each with differing defect charge states; (b) the diffusivity reflects contributions from both Si_i and V_{Si} [6.2, 43, 48, 49]. Thus the Si self-diffusivity is determined by Si_i and V_{Si} mechanisms and by three separate defect charge states: neutral (0), positive (+) and negative (−). Identifying which charge state goes with which defect remains a problem. For the common dopants (B, P, As and Sb), B and P diffuse primarily via the Si_i defect, As diffuses via both Si_i and V_{Si} defects, whereas Sb diffuses primarily via V_{Si}. Two defect charge states are involved for B (0, 1+), As (0, 1−) and Sb (0, 1−), and three for P (0, 1−, 2−). The situation for Al [6.50], Ga [6.9] and In [6.51] has Si_i dominant for Al and In diffusion whereas both Si_i and V_{Si} are involved for Ga. The associated charge states are Al (0, 1+), Ga (0, 1+) and In (0). The diffusivities of the Group V donor dopants (P, As, Sb) lie close to each other and are up to a factor of ≈ 10 greater than the self-diffusivity. The acceptor dopants (B, Al, Ga, In) also form a group with diffusivities that are up to a factor of $\approx 10^2$ greater than the donor dopants. A recently observed interesting feature is that the diffusivities of B and P in intrinsic material depend on the length of the anneal time, showing an initial change until reaching a final value [6.52]. This time effect is attributed to the time needed for equilibration of the V_{Si} and Si_i concentrations at the anneal temperature.

The data presented by *Tan* and *Gösele* [6.43] show that Au, Pt and Zn are intermediate (kick-out) diffusers and that H, Li, Cu, Ni and Fe are fast interstitial diffusers. Recent evidence shows that Ir diffusion occurs via both kick-out and dissociative mechanisms [6.53]. To provide some perspective: at 1000 °C the diffusivity of H is $\approx 10^{-4} \text{ cm}^2/\text{s}$ compared to a self-diffusivity of $8 \times 10^{-17}\,\mathrm{cm^2/s}$. C and O are important impurities because, though electrically neutral, they occur in high concentrations and can affect the electrical properties. Although O occupies interstitial sites and diffuses interstitially it should be classed as an intermediate diffuser because a diffusion jump entails the breaking of two Si−O bonds. C has a diffusivity that is a little larger than those of Group III dopants: its mechanism is unresolved between the "kick-out" mode or a diffusing complex comprising a Si_i and a substitutional C.

6.7.3 $Si_{1-x}Ge_x$ Alloys

Si and Ge form a continuous range of alloys in which there is a random distribution of either element as well as a continuous variation of bandgaps. The alloys have attracted considerable interest from a device perspective and are usually prepared as epilayers on Si substrates so that the epilayer will generally be in a strained state. Diffusivity data are sparse and, in the case of dopants, limited to B, P and Sb. One might expect that the diffusivity $D(Z)$ of dopant Z would increase continuously as x goes from 0 to 1 at any given temperature below the melting point of Ge. However, in the case of B, $D(B)$ hardly varies for $x \lesssim 0.4$; even so, $D(B)$ increases by a factor $\approx 10^3$ from $\approx 10^{-15}\,\mathrm{cm^2/s}$ in traversing the composition range at $900\,°\mathrm{C}$ [6.54, 55]. $D(P)$ increases by a factor of ≈ 4 for x values between 0 and 0.24, only to show a decrease at $x = 0.40$ [6.55]. Limited data suggest that $D(Sb)$ rises continuously across the composition range, increasing by a factor $\approx 10^6$ at $900\,°\mathrm{C}$ [6.56]. Surface oxidation enhances $D(B)$ and $D(P)$, indicating that the diffusivities are dominated by a self-interstitial mechanism, whereas $D(Sb)$ is reduced by surface oxidation, pointing to a vacancy mechanism. Compressive strain retards $D(B)$ whereas tensile strain gives a marginal enhancement [6.55]. Compressive strain enhances $D(P)$ and $D(Sb)$ [6.57]. Overall, some disagreement exists between different workers about the behavior of $D(Z)$, which may well stem from difficulties with characterizing the experimental conditions. Compositional interdiffusion has been characterized at the interface between Si and layers with $x < 0.2$ [6.58].

6.7.4 Silicon Carbide

Its large bandgap, high melting point and high dielectric breakdown strength make SiC a suitable material for devices intended for operation at high temperatures and high powers. It also has potential optoelectronic applications. Characterizing the material is complicated, as SiC occurs in a range of polytypes (different stacking sequences of close packed layers). Common polytypes are the cubic zinc blende phase 3C−SiC and the hexagonal phases 2H−SiC (wurtzite), 4H−SiC and 6H−SiC. This combination of high melting point, polytypism and variations in stoichiometry makes it difficul to measure diffusivities. Typical diffusion anneal temperatures for acceptor (B, Al, Ga) and donor (N, P) dopants are in the range $1800-2100\,°\mathrm{C}$. Ab initio calculations for single vacancies and anti-sites in 4H−SiC [6.59] found the Si_C and C_{Si} anti-sites to be both neutral and therefore generally inactive (electrically and optically).The C vacancy is amphoteric with charge states ranging from $2+$ to $2-$. The Si vacancy is also amphoteric with charge states ranging from $1+$ to $3-$. Similar calculations for self-interstitials in 3C−SiC [6.60] predict divalent donor behavior for both Si and C interstitials. *Bockstedte* et al. [6.61] have calculated, using ab initio methods, the activation energies Q for self-diffusion in 3C−SiC by vacancies and self-interstitials. Generally Q is smaller for self-interstitials but the defect charge state is also an important factor. The Si vacancy is predicted to be metastable, readily transforming to the stable complex V_C-C_{Si}: the complex $V_{Si}-Si_C$ is unstable, reverting to V_C.

The Si and C self-diffusivities, $D(Si)$ and $D(C)$, respectively, were measured between 1850 and $2300\,°\mathrm{C}$ by *Hong* et al. ([6.62] and references therein) in both 3C−SiC and 6H−SiC. The ratio $D(C)/D(Si)$ was ≈ 650 in 3C−SiC and ≈ 130 in 6H−SiC. N doping increased $D(Si)$ and reduced (marginally) $D(C)$. This behavior suggests that native acceptors are important for Si self-diffusion and that native donors are only marginally involved in determining $D(C)$. Of particular interest is that, between the two polytypes, the self-diffusivities in 6H-SiC exceeded those in 3C−SiC by less than a factor of ≈ 3. This suggests that diffusivities are insensitive to the particular polytype. More recent measurements of $D(C)$, between 2100 and $2350\,°\mathrm{C}$, in 4H−SiC found diffusivities that were $\approx 10^5$ times smaller than the earlier results for 3C−SiC and 6H−SiC, mainly because of differences in D_0 [6.63]. There is currently no explanation for these huge differences and the question of the reliability of self-diffusivity data must be considered.

Earlier work by *Vodakov* et al. [6.64] found that the diffusivity of B in six different polytypes of SiC, excluding 3C-SiC, varied by $\leq 30\%$, not only for diffusion along the c-axis but also perpendicular to it. The diffusivities of some common dopants have been summarized by *Vodakov* and *Mokhov* [6.65]. B diffusion mechanisms in 4H and 6H-SiC have been discussed by *Usov* et al. [6.66]. A recent finding is that an SiO$_2$ layer on the surface of 6H-SiC greatly enhances B diffusion [6.67], yielding a diffusivity of $\approx 6 \times 10^{-16}$ cm^2/s at 900 °C. This compares to a temperature of $\approx$ 1400 °C (extrapolated) for the same diffusivity without an SiO$_2$ layer. Electric fields of $\approx 10^6$ V/cm have been found in 4H/3C/4H-SiC quantum wells due to spontaneous polarization in the 4H-SiC matrix [6.68].

6.8 Diffusion in III–V Compounds

The III–V binary compounds are formed between the cations B, Al, Ga, In and the anions N, P, As and Sb. Mutual solubility gives rise to the ternaries, such as Al$_{1-x}$Ga$_x$As, and to the quaternaries, such as In$_{1-x}$Ga$_x$As$_{1-y}$P$_y$. The B compounds offer little more than academic interest, whereas the rest of the III–V family are important materials in both electronic and optoelectronic devices. The nitrides all have the wurtzite structure, with the remaining compounds possessing the zinc blende structure. In view of the wide range of binaries, ternaries, and so on, it is not surprising that diffusivity measurements have focused mainly on those compounds relevant to devices: essentially GaAs and GaAs-based materials. An important characteristic of these compounds is the high vapor pressures of the anion components; it is the variations in these components that lead to significant changes in levels of nonstoichiometry. This means that a proper characterization, at a given temperature, of any diffusivity must specify the doping level and the ambient anion vapor pressure during the anneal: the latter determines native defect concentrations in intrinsic samples, and both factors have equal importance in controlling the concentrations under extrinsic conditions. On both the anion and the cation sublattices, the possible native point defects are the vacancy, the self-interstitial and the anti-site and all can occur in one or more charge states.

6.8.1 Self-Diffusion

Self-diffusivity data are limited to the Ga and In compounds [6.35, 69], and even here systematic measurements are restricted to GaAs [6.43, 69] and GaSb ([6.70] and references therein). For GaAs, early evidence (based largely on CID in AlGaAs structures) concluded that the Ga self-diffusivity D(Ga) was determined by the triply ionized Ga vacancy V_{Ga}^{3-} and doubly ionized Ga interstitial Ga$_i^{2+}$. More recent and direct measurements of D(Ga) in Ga isotope heterostructures identified the three vacancy charge states V_{Ga}^{2-}, V_{Ga}^{1-} and V_{Ga}^{0} as being responsible for D(Ga) in intrinsic and lightly doped GaAs; the possibility remains that V_{Ga}^{3-} and Ga$_i^{2+}$ could dominate at high doping levels. Between 800 and 1200 °C the Arrhenius parameters for D(Ga) are $D_0 = 0.64$ cm^2/s and $Q = 3.71$ eV in intrinsic GaAs under a partial As$_4$ vapor pressure of ≈ 1 atm. The situation for As self-diffusion is less clear, but the evidence points to the dominance (in the diffusion process) of the neutral As interstitial over the As vacancy (the supposedly dominant native defect, the As anti-site, is not involved). Data have been obtained for both Ga and Sb self-diffusion in intrinsic GaSb under Ga- and Sb-rich conditions. There is a conflict between the results obtained with bulk material and those from isotope heterostructures (see [6.70] and references therein). *Shaw* [6.70] concluded that the defects involved in Ga self-diffusion were the Frenkel pair Ga$_i$V$_{Ga}$ and V$_{Ga}$ even though the Ga anti-site Ga$_{Sb}$ appears to be the dominant native defect. Two parallel mechanisms were also identified for Sb self-diffusion, namely one due to the defect pair Sb$_i$V$_{Ga}$ and the second due to either to the mixed vacancy pair V$_{Ga}$V$_{Sb}$ or to the triple defect V$_{Ga}$Ga$_{Sb}$V$_{Ga}$. Reliable results for D(Ga) in intrinsic GaP under a partial vapor pressure (P$_4$) of ≈ 1 atm are also available [6.71]: between 1000 and 1190 °C the Arrhenius parameters for D(Ga) are $D_0 = 2.0$ cm^2/s and $Q = 4.5$ eV. Data on the effects of doping and changing partial pressure are lacking.

6.8.2 Dopant Diffusion

Most of the data on dopant diffusion in the III–Vs refer to GaAs [6.35], notably for Be [6.72], Cd [6.69], C, Si, S, Zn and Cr [6.43]. The singly ionized acceptors Be, Zn and Cd (which occupy Ga sites) and the singly ionized donors C and S (which occupy As sites) all diffuse via the kick-out mechanism. The native interstitials involved are Ga$_i^{2+}$ and As$_i^{0}$, apart from Be where the data are best accounted for in terms of the singly ionized interstitial Ga$_i^{1+}$. Si is an amphoteric dopant and at low concentrations it predominantly occupies Ga sites as a singly ionized donor Si$_{Ga}^{1+}$. At high concentrations compensation starts to occur due to increasing occupancy as a singly ionized acceptor on As sites. At low concentrations Si_{Ga}^{1+} diffusion is attributed to a vacancy mechanism (V_{Ga}^{3-}). Cr sits on Ga sites and is a deep-level acceptor dopant important in the growth of high resistivity GaAs. Depending on circumstances, it can diffuse by either the kick-out or the Frank–Turnbull mechanism. The creation of extended defects in the diffusion zone by Zn in-diffusion in GaAs is a well-established feature. The same feature has also been found by *Pöpping* et al. [6.73] for Zn in-diffusion in GaP. They further concluded that Zn diffuses via the kick-out process in GaP through the involvement of either Ga_i^{1+} or Ga_i^{2+}.

6.8.3 Compositional Interdiffusion

The III–V binaries, ternaries and quaternaries are the bases for the fabrication of numerous quantum well and superlattice structures. CID is clearly an issue in the integrity of such structures. The general situation in which the cation and anion sublattices in each layer can contain up to four different components with concentrations ranging from 0 to 100% presents an impossibly complex problem for characterizing diffusion behavior with any rigour. The role of strain in the layers must also be considered a parameter. As a consequence, CID studies have been limited to simpler structures, primarily GaAs-AlAs and GaAs-AlGaAs with interpretations in terms of known diffusion features in GaAs [6.43]. Doping is an important ingredient of these multilayer structures and it was soon discovered that the acceptors Be, C, Mg, Zn and the donors Si, Sn, S, Se and Te could all cause complete disorder of the structure through enhancement of the CID process on either or both sublattices [6.43, 74]. An interesting exception, however, is found in GaAs-GaAsSb, where either Si or Be reduce CID. Two generally accepted reasons for these dopant effects are: (i) the Fermi-level effect in which the dopant (acceptor/donor) concentration is high enough to make the semiconductor extrinsic so that the concentrations of native (donor/acceptor) defects are increased; (ii) if the dopant diffuses by the kick-out mechanism then in-diffusion will generate a local excess of the native self-interstitial. Clearly (i) operates for dopants incorporated during growth or by subsequent in-diffusion, whereas (ii) is restricted to in-diffusion. Either way the increase in the local native defect concentration(s) leads to a direct enhancement of CID. In the case of GaAs-GaAsSb, cited above, Si will also decrease the concentrations of native donors such as native anion vacancies, which would have a direct impact on and reduction of CID on the anion sublattice. On the other hand, Be should increase native donor concentrations and therefore give enhanced CID of the anions, contrary to observation. Overall, the general features of the dopant-induced disordering process seem to be understood but problems still remain. *Harrison* [6.74] has commented on the approximations commonly made when extracting quantitative information from CID data. The demands of III–V device technology present increasing complexity when attempting to understand the physical processes involved, so that recourse to empirical recipes is sometimes needed. This is illustrated by structures comprising GaInNAs quantum wells with GaAs barriers, all enclosed within AlAs outer layers, whose optoelectronic properties can be improved by the judicious choice of time/temperature anneals [6.75].

6.9 Diffusion in II–VI Compounds

Interest in II–VI materials pre-dates that in the III–Vs because of their luminescence properties in the visible spectrum, which, based on powder technology, resulted in the application of the bigger bandgap materials (such as ZnS) as phosphors in luminescent screens. The development of crystal growth techniques extended interest in the optoelectronic properties of the wider family of II–VI binary compounds formed between the group II cations Zn, Cd and Hg and the group VI anions S, Se and Te. As with the III–Vs, ternary and quaternary compounds are readily formed. The ternary range of compositions $Hg_{1-x}Cd_x$Te has proved to be the most important family member because of their unique properties and consequent extensive exploitation in infra-red systems. ZnS, CdS and CdSe crystallize in the wurtzite structure, whereas the remaining binaries have the zinc blende structure. The native point defects that can occur are similar to the III–Vs; namely, vacancies, self-interstitials and anti-sites for the cation and anion sublattices. Recent interest has expanded to include the cations Be, Mg and Mn, usually in ternary or quaternary systems. A distinctive feature of atomic diffusion in the II–VI compounds is the much higher diffusivities relative to those in the Group IV and III–V semiconductors. The relative ease of measurement has ensured that much more self- and dopant diffusion data are available compared to the III–Vs. A further difference is that both cation and anion equilibrium vapor pressures are significant compared to the III–Vs, where the cation vapor pressures are negligible. Unless otherwise stated, the material in the following sections is drawn from the reviews by *Shaw* [6.6, 76, 77] and by *Capper* et al. [6.78].

6.9.1 Self-Diffusion

Where the anion self-diffusivity D_A has been measured as a function of the ambient anion or cation partial pressure in undoped material (ZnSe, CdS, CdSe, CdTe and $Hg_{0.8}Cd_{0.2}Te$), a consistent pattern of behavior has emerged: in traversing the composition range from anion-rich to cation-rich, D_A is inversely proportional to the rising cation vapor pressure, P_C, until close to cation saturation, when D_A starts to increase with P_C. Strong donor doping in anion-rich CdS and CdSe had no effect on D_A. This evidence points to either a neutral anion interstitial or a neutral anion antisite/anion vacancy complex as the diffusion mechanism over most of the composition range, changing to an anion vacancy mechanism as the cation-rich limit is approached.

The situation for cation self-diffusion proves to be more complicated due to the different variations of the cation self-diffusivity D_C with P_C across the compounds. In undoped ZnSe, ZnTe, CdTe and $Hg_{0.8}Cd_{0.2}Te$ (above $\approx 300\,°C$), D_C is largely independent of P_C across the composition range. Such an independence excludes native point defect diffusion mechanisms and (excluding nondefect mechanisms) points to self-diffusion via neutral complexes such as a cation interstitial/cation vacancy or a cation vacancy/anion vacancy pair. Donor or acceptor doping increases D_C, indicating the involvement of ionized native defects or complexes. The Arrhenius parameters for Zn self-diffusion in undoped ZnSe above $760\,°C$ are $D_0 = 9.8\,cm^2/s$ $Q = 3.0\,eV$ and those for Hg in undoped $Hg_{0.8}Cd_{0.2}Te$ above $250\,°C$ are $D_0 = 3.8 \times 10^{-3}\,cm^2/s$ and $Q = 1.22\,eV$. In the case of undoped ZnS, CdS, CdSe and HgTe, D_C generally varies with P_C across the composition range. The simplest variations are found in CdSe and HgTe. In CdSe, D_C can be attributed to the parallel diffusion of singly (1+) and doubly (2+) ionized Cd self-interstitials. D_C in HgTe initially falls with P_C and then increases when crossing from anion-rich to cation-rich material, corresponding to diffusion by a singly ionized (1−) Hg vacancy and by a singly ionized (1+) Hg interstitial respectively. The behavior patterns in ZnS and CdS, however, present substantial problems in their interpretation: donor doping can also enhance D_C, pointing to the participation of an ionized native acceptor mechanism.

6.9.2 Chemical Self-Diffusion

Changes in the electrical conductivity or conductivity type caused by step changes to P_C in sample anneals have been used to characterize the change in level of nonstoichiometry through the chemical self-diffusivity, D_Δ, in CdS, CdTe and $Hg_{0.8}Cd_{0.2}Te$. D_Δ obviously describes the diffusion of one or more ionized native defects, but in itself it does not identify the defect(s). In CdS and CdTe, D_Δ is attributed to the singly ionized (1+) and/or doubly ionized (2+) Cd interstitial; in CdTe, depending on the temperature, D_Δ exceeds D_C by a factor 10^5 to 10^6. Modeling based on the simultaneous in-diffusion and out-diffusion of doubly ionized cation interstitials (2+) and vacancies (2−) gives a satisfactory quantitative account of type conversion (p → n) in $Hg_{0.8}Cd_{0.2}Te$ [6.7].

6.9.3 Dopant Diffusion

Although much information on dopant diffusion is available, it is mainly empirical and it is not uncommon for a dopant diffusivity to be independent of dopant concentration (as revealed by an *erfc* profile – a constant diffusivity for a given diffusion profile) under one set of conditions only to give profiles which cannot be characterized by single diffusivities when the conditions are changed. Equally, the variation of a dopant diffusivity with P_C may differ at different temperatures. A further difficulty when attempting to identify a diffusion mechanism is that the local electroneutrality condition is usually not known with any certainty due to significant concentrations of various ionized native defects. A good illustration of the problems encountered is provided by In diffusion in $Hg_{0.8}Cd_{0.2}Te$, where diffusion of the singly ionized (1−) pair $In_{Hg}V_{Hg}$ can account for some of the diffusion features. Some dopants, however, can present clear-cut diffusion properties which permit a well-defined interpretation. The diffusion of As in $Hg_{0.8}Cd_{0.2}Te$ is one such case [6.79]. All of the observed features of $D(As)$ are accounted for on the basis that: (i) As occupies both cation and anion lattice sites as singly ionized donors (1+) and acceptors (1−) respectively; (ii) only the ionized donor is mobile and diffuses by a vacancy mechanism on the cation sublattice; (iii) the diffusion sample is electrically intrinsic throughout, so the As concentration is always less than the intrinsic free carrier concentration.

6.9.4 Compositional Interdiffusion

Empirical information, based on bulk material, exists for CID in the following ternaries: (ZnCdHg)Te, (ZnCd)Se, (ZnCdHg)SeTe, (ZnCd)SSe, CdSeTe, ZnCdS, HgCdTe and CdMnTe. It might be expected that features evident in the binaries, such as donor doping enhancing the cation diffusivity but having no effect on that of the anion, and the anion diffusivity increasing (decreasing) with anion (cation) vapor pressure across most of the composition range, would continue to be seen. This means that in a ternary or quaternary system, donor doping will enhance CID on the cation sublattice, but not on the anion sublattice, and annealing under a high (low) anion (cation) vapor pressure will enhance CID on the anion sublattice. This effect of the anion vapor pressure has been confirmed in CdSSe and CdSeTe and more recently in ZnSSe/ZnSe superlattices [6.80]. In (donor) doping has also been found to enhance the CID of the cations in CdMnTe [6.81], as has N (acceptor) doping in ZnMgSSe/ZnSSE superlattices [6.82]. The consequences of doping on CID in the II–VIs are obviously very similar to the III–V situation.

6.10 Conclusions

The first step in a diffusion investigation is to collect empirical data, which then leads to the second step where experiments can be designed to study the effects of the Fermi level (through the background doping level), of the ambient atmosphere (such as oxidizing, inert or vapor pressure of a system component) and of the sample structure (such as an MBE layer or a quantum well). The third step is to identify the diffusion mechanism and the associated defects using the experimental results in conjunction with the results from first-principles calculations of defect formation energies and their activation energies for diffusion. Clarification of the active processes involved can be gained by numerical modeling (see *Noda* [6.83]). These data then provide the basis for the development of process simulators and defect engineering in which the concentrations and spatial distributions of host atoms, dopants and defects are organized according to requirement. Most progress towards achieving this ideal scenario has been made in Si and to a lesser extent in GaAs and $Hg_{0.8}Cd_{0.2}Te$. The reality elsewhere is that the boundaries between the steps are blurred, with the third step often being undertaken with inadequate experimental information. Much work remains to be done in order to master our understanding of diffusion processes in semiconductors.

6.11 General Reading and References

General background material for diffusion in semiconductors can be found in *Shaw* [6.5], *Tuck* [6.1], *Abdullaev* and *Dzhafarov* [6.11] and *Tan* et al. [6.8]. More specific accounts are given by *Fair* [6.9] and *Fahey* et al. [6.84] for Si, by *Frank* et al. [6.3] for Si and Ge, by *Tan* and *Gösele* [6.43] for Si, Ge and GaAs, by *Tuck* [6.35] for the III–Vs and by *Shaw* [6.6, 77] for the II–VIs. *H in Semiconductors II* (1999) ed. by N. H. Nickel (*Semiconductors and Semimetals*, 61, Academic, San Diego) provides a recent account of H in semiconductors. The volumes in the EMIS Datareviews Series (IEE, Stevenage, UK) cover all of the important semiconductors. The series *Defects and Diffusion in Semiconductors* ed. by D. J. Fisher (Trans Tech., Brandrain 6, Switzerland) offers an annual and selective retrospective of recent literature.

References

6.1 B. Tuck: *Introduction to Diffusion in Semiconductors* (Peregrinus, Stevenage 1974)
6.2 A. Ural, P. B. Griffin, J. D. Plummer: J. Appl. Phys. **85**, 6440 (1999)
6.3 W. Frank, U. Gösele, H. Mehrer, A. Seeger: In: *Diffusion in Crystalline Solids*, ed. by G. E. Murch, A. S. Nowick (Academic, Orlando 1984) Chapt.2
6.4 D. Mathiot, J. C. Pfister: J. Appl. Phys. **66**, 970 (1989)

6.5 D. Shaw: In: *Atomic Diffusion in Semiconductors*, ed. by D. Shaw (Plenum, London 1973) Chapt.1
6.6 D. Shaw: In: *Widegap II–VI Compounds for Optoelectronic Applications*, ed. by H. E. Ruda (Chapman and Hall, London 1992) Chapt.10
6.7 D. Shaw, P. Capper: J. Mater. Sci. Mater. El. **11**, 169 (2000)
6.8 T. Y. Tan, U. Gösele, S. Yu: Crit. Rev. Sol. St. Mater. Sci. **17**, 47 (1991)
6.9 R. B. Fair: In: *Impurity Doping Processes in Silicon*, ed. by F. F. Y. Wang (North-Holland, Amsterdam 1981) Chapt.7
6.10 S. M. Hu: J. Appl. Phys. **70**, R53 (1991)
6.11 G. B. Abdullaev, T. D. Dzhafarov: *Atomic Diffusion in Semiconductor Structures* (Harwood, Chur 1987)
6.12 M. Laudon, N. N. Carlson, M. P. Masquelier, M. S. Daw, W. Windl: Appl. Phys. Lett. **78**, 201 (2001)
6.13 K. Rajendran, W. Schoenmaker: J. Appl. Phys. **89**, 980 (2001)
6.14 H. Takeuchi, P. Ranada, V. Subramanian, T-J. King: Appl. Phys. Lett. **80**, 3706 (2002)
6.15 S. C. Jain, W. Schoenmaker, R. Lindsay, P. A. Stolk, S. Decoutere, M. Willander, H. E. Maes: J. Appl. Phys. **91**, 8919 (2002)
6.16 L. Shao, J. Chen, J. Zhang, D. Tang, S. Patel, J. Liu, X. Wang, W-K. Chu: J. Appl. Phys. **96**, 919 (2004)
6.17 Y. M. Haddara, J. C. Bravman: Ann. Rev. Mater. Sci. **28**, 185 (1998)
6.18 I. Lyubomirsky, V. Lyahovitskaya, D. Cahen: Appl. Phys. Lett. **70**, 613 (1997)
6.19 C. H. Chen, U. Gösele, T. Y. Tan: Appl. Phys. A **68**, 9, 19, 313 (1999)
6.20 P. N. Grillot, S. A. Stockman, J. W. Huang, H. Bracht, Y. L. Chang: J. Appl. Phys. **91**, 4891 (2002)
6.21 E. Chason, S. T. Picraux, J. M. Poate, J. O. Borland, M. I. Current, T. Diaz de la Rubia, D. J. Eaglesham, O. W. Holland, M. E. Law, C. W. Magee, J. W. Mayer, J. Melngailis, A. F. Tasch: J. Appl. Phys. **81**, 6513 (1997)
6.22 J. L. Melendez, J. Tregilgas, J. Dodge, C. R. Helms: J. Electron. Mater. **24**, 1219 (1995)
6.23 A. Borghesi, B. Pivac, A. Sassella, A. Stella: J. Appl. Phys. **77**, 4169 (1995)
6.24 K. F. Kelton, R. Falster, D. Gambaro, M. Olmo, M. Cornaro, P. F. Wei: J. Appl. Phys. **85**, 8097 (1999)
6.25 S. Solmi, E. Landi, F. Baruffaldi: J. Appl. Phys. **68**, 3250 (1990)
6.26 S. Solmi, D. Nobili: J. Appl. Phys. **83**, 2484 (1998)
6.27 B. Colombeau, N. E. B. Cowern: Semicond. Sci. Technol. **19**, 1339 (2004)
6.28 S. Mirabella, E. Bruno, F. Priolo, D. De Salvador, E. Napolitani, A. V. Drigo, A. Carnera: Appl. Phys. Lett. **83**, 680 (2003)
6.29 C. J. Ortiz, P. Pichler, T. Fühner, F. Cristiano, B. Colombeau, N. E. B. Cowern, A. Claverie: J. Appl. Phys. **96**, 4866 (2004)
6.30 D. Shaw: Semicond. Sci. Technol. **7**, 1230 (1992)
6.31 H. Puchner, S. Selberherr: IEEE Trans. Electron. Dev. **42**, 1750 (1995)
6.32 C. Poisson, A. Rolland, J. Bernardini, N. A. Stolwijk: J. Appl. Phys. **80**, 6179 (1996)
6.33 I. Kaur, Y. Mishin, W. Gust: *Fundamentals of Grain and Interphase Boundary Diffusion* (Wiley, Chichester 1995)
6.34 S. M. Hu: J. Appl. Phys. **43**, 2015 (1972)
6.35 B. Tuck: *Atomic Diffusion in III–V Semiconductors* (Adam Hilger, Bristol 1988)
6.36 L. S. Monastyrskii, B. S. Sokolovskii: Sov. Phys. Semicond. **16**, 1203 (1992)
6.37 E. A. Caridi, T. Y. Chang, K. W. Goossen, L. F. Eastman: Appl. Phys. Lett. **56**, 659 (1990)
6.38 A. Hangleiter, F. Hitzel, S. Lafmann, H. Rossow: Appl. Phys. Lett. **83**, 1169 (2003)
6.39 S. J. Rothman: In: *Diffusion in Crystalline Solids*, ed. by G. E. Murch, A. S. Nowick (Academic, Orlando 1984) Chapt.1
6.40 R. M. Fleming, D. B. McWhan, A. C. Gossard, W. Wiegmann, R. A. Logan: J. Appl. Phys. **51**, 357 (1980)
6.41 Y. L. Huang, Y. Ma, R. Job, W. R. Fahrner, E. Simeon, C. Claeys: J. Appl. Phys. **98**, 033511 (2005)
6.42 D. Mathiot: Phys. Rev. B **40**, 5867 (1989)
6.43 T. Y. Tan, U. Gösele: In: *Handbook of Semiconductor Technology*, Vol.1, ed. by K. A. Jackson, W. Schröter (Wiley-VCH, Weinheim 2000) Chapt.5
6.44 M. Werner, H. Mehrer, H. D. Hochheimer: Phys. Rev. B **37**, 3930 (1985)
6.45 N. A. Stolwijk, W. Frank, J. Hölzl, S. J. Pearton, E. E. Haller: J. Appl. Phys. **57**, 5211 (1985)
6.46 A. Strohm, S. Matics, W. Frank: Diffusion and Defect Forum **194–199**, 629 (2001)
6.47 H. Bracht, E. E. Haller, R. Clark-Phelps: Phys. Rev. Lett. **81**, 393 (1998)
6.48 A. Ural, P. B. Griffin, J. D. Plummer: Phys. Rev. Lett. **83**, 3454 (1999)
6.49 A. Ural, P. B. Griffin, J. D. Plummer: Appl. Phys. Lett. **79**, 4328 (2001)
6.50 O. Krause, H. Ryssel, P. Pichler: J. Appl. Phys **91**, 5645 (2002)
6.51 S. Solmi, A. Parisini, M. Bersani, D. Giubertoni, V. Soncini, G. Carnevale, A. Benvenuti, A. Marmiroli: J. Appl. Phys. **92**, 1361 (2002)
6.52 J. S. Christensen, H. H. Radamson, A. Yu. Kuznetsov, B. G. Svensson: Appl. Phys. Lett. **82**, 2254 (2003)
6.53 L. Lerner, N. A. Stolwijk: Appl. Phys. Lett. **86**, 011901 (2005)
6.54 N. R. Zangenberg, J. Fage-Pedersen, J. Lundsgaard Hansen, A. Nylandsted-Larsen: Defect Diffus. Forum **194–199**, 703 (2001)
6.55 N. R. Zangenberg, J. Fage-Pedersen, J. Lundsgaard Hansen, A. Nylandsted-Larsen: J. Appl. Phys **94**, 3883 (2003)
6.56 A. D. N. Paine, A. F. W. Willoughby, M. Morooka, J. M. Bonar, P. Phillips, M. G. Dowsett, G. Cooke: Defect Diffus. Forum **143–147**, 1131 (1997)
6.57 J. S. Christensen, H. H. Radamson, A. Yu. Kuznetsov, B. G. Svensson: J. Appl. Phys. **94**, 6533 (2003)

6.58 D. B. Aubertine, P. C. McIntyre: J. Appl. Phys. **97**, 013531 (2005)
6.59 L. Torpo, M. Marlo, T. E. M. Staab, R. M. Nieminen: J. Phys. Condens. Matter **13**, 6203 (2001)
6.60 J. M. Lento, L. Torpo, T. E. M. Staab, R. M. Nieminen: J. Phys. Condens. Matter **16**, 1053 (2004)
6.61 M. Bockstedte, A. Mattausch, O. Pankratov: Phys. Rev. B **68**, 205201 (2003)
6.62 J. D. Hong, R. F. Davis, D. E. Newbury: J. Mater. Sci. **16**, 2485 (1981)
6.63 M. K. Linnarsson, M. S. Janson, J. Zhang, E. Janzen, B. G. Svensson: J. Appl. Phys. **95**, 8469 (2004)
6.64 Yu. A. Vodakov, G. A. Lomakina, E. N. Mokhov, V. G. Oding: Sov. Phys. Solid State **19**, 1647 (1977)
6.65 Yu. A. Vodakov, E. N. Mokhov: In: *Silicon Carbide – 1973*, ed. by R. C. Marshall, J. W. Faust Jr, C. E. Ryan (Univ. South Carolina Press, Columbia 1973) p. 508
6.66 I. O. Usov, A. A. Suvorova, Y. A. Kudriatsev, A. V. Suvorov: J. Appl. Phys. **96**, 4960 (2004)
6.67 N. Bagraev, A. Bouravleuv, A. Gippius, L. Klyachkin, A. Malyarenko: Defect Diffus. Forum **194-199**, 679 (2001)
6.68 S. Bai, R. P. Devaty, W. J. Choyke, U. Kaiser, G. Wagner, M. F. MacMillan: Appl. Phys. Lett. **83**, 3171 (2003)
6.69 N. A. Stolwijk, G. Bösker, J. Pöpping: Defect Diffus. Forum **194-199**, 687 (2001)
6.70 D. Shaw: Semicond. Sci. Technol. **18**, 627 (2003)
6.71 L. Wang, J. A. Wolk, L. Hsu, E. E. Haller, J. W. Erickson, M. Cardona, T. Ruf, J. P. Silveira, F. Brione: Appl. Phys. Lett. **70**, 1831 (1997)
6.72 J. C. Hu, M. D. Deal, J. D. Plummer: J. Appl. Phys. **78**, 1595 (1995)
6.73 J. Pöpping, N. A. Stolwijk, G. Bösker, C. Jäger, W. Jäger, U. Södervall: Defect Diffus. Forum **194-199**, 723 (2001)
6.74 I. Harrison: J. Mater. Sci. Mater. Electron. **4**, 1 (1993)
6.75 S. Govindaraju, J. M. Reifsnider, M. M. Oye, A. L. Holmes: J. Electron. Mater. **32**, 29 (2003)
6.76 D. Shaw: J. Cryst. Growth **86**, 778 (1988)
6.77 D. Shaw: J. Electron. Mater. **24**, 587 (1995)
6.78 P. Capper, C. D. Maxey, C. L. Jones, J. E. Gower, E. S. O'Keefe, D. Shaw: J. Electron. Mater. **28**, 637 (1999)
6.79 D. Shaw: Semicond. Sci. Technol. **15**, 911 (2000)
6.80 M. Kuttler, M. Grundmann, R. Heitz, U. W. Pohl, D. Bimberg, H. Stanzel, B. Hahn, W. Gebbhart: J. Cryst. Growth **159**, 514 (1994)
6.81 A. Barcz, G. Karczewski, T. Wojtowicz, J. Kossut: J. Cryst. Growth **159**, 980 (1996)
6.82 M. Strassburg, M. Kuttler, O. Stier, U. W. Pohl, D. Bimberg, M. Behringer, D. Hommel: J. Cryst. Growth **184-185**, 465 (1998)
6.83 T. Noda: J. Appl. Phys. **94**, 6396 (2003)
6.84 P. M. Fahey, P. B. Griffin, J. D. Plummer: J. Appl. Phys. **61**, 289 (1989)

7. Photoconductivity in Materials Research

Photoconductivity is the incremental change in the electrical conductivity of a substance upon illumination. Photoconductivity is especially apparent for semiconductors and insulators, which have low conductivity in the dark. Significant information can be derived on the distribution of electronic states in the material and on carrier generation and recombination processes from the dependence of the photoconductivity on factors such as the exciting photon energy, the intensity of the illumination or the ambient temperature. These results can in turn be used to investigate optical absorption coefficients or concentrations and distributions of defects in the material. Methods involving either steady state currents under constant illumination or transient methods involving pulsed excitation can be used to study the electronic density of states as well as the recombination. The transient time-of-flight technique also allows carrier drift mobilities to be determined.

7.1 **Steady State Photoconductivity Methods** 138
    7.1.1 The Basic Single-Beam Experiment 138
    7.1.2 The Constant Photocurrent Method (CPM) 141
    7.1.3 Dual-Beam Photoconductivity (DBP) 141
    7.1.4 Modulated Photoconductivity (MPC) 141

7.2 **Transient Photoconductivity Experiments**... 142
    7.2.1 Current Relaxation from the Steady State 143
    7.2.2 Transient Photoconductivity (TPC) 143
    7.2.3 Time-of-Flight Measurements (TOF) 144
    7.2.4 Interrupted Field Time-of-Flight (IFTOF).................................... 145

References 146

Photoconductivity has traditionally played a significant role in materials research, and most notably so in the study of covalently bonded semiconductors and insulators. Indeed, since it is the incremental conductivity generated by the absorption of (optical) photons, photoconductivity can be most clearly resolved in situations where the intrinsic dark conductivity of the material is low. This conductivity in the dark, leading to "dark current", is due to the thermal equilibrium density of free carriers in the material and must be subtracted from any measured current in order to obtain the actual photocurrent. The basic processes that govern the magnitude of the photocurrent are the generation of free electrons and holes through the absorption of incident photons, the transport of those free carriers through the material under the influence of an electric field, and the recombination of the photoexcited electrons and holes. The study of any of those aspects as a function of the characteristics of the current-inducing illumination, as well as the study of their development upon changes in that illumination over time, will offer insights into the structure and electronic properties of the material under investigation. However, given the fact that three separate processes are involved in the production of a specific photocurrent, it follows that any analysis of experimental data in terms of system parameters will require a sufficiently comprehensive data set that will allow for differentiation between alternative interpretations. For instance, a low photocurrent may be the result of a low optical absorption coefficient at the given photon energy, but it may also be due to significant geminate recombination of the photogenerated electron–hole pairs, or it may reflect the formation of excitons. The combined use of different types of photoconductivity experiments is therefore often advisable, as is the combination of photoconductivity with related experiments such as photoluminescence or charge collection.

A wide variety of experimental techniques based on photoconductivity have come into general use over the years. They can be divided into two main groups, one involving steady state photoconductivity (SSPC), where the focus is on stationary photocurrent levels, and a second one involving transient effects (TPC) where the time evolution of the photocurrent is studied. We will use this division in our survey of the various methods, but we should point out that SSPC can also be measured through ac excitation. The information that can be obtained about the material under investigation is generally not specific to either the SSPC or TPC method that is used, but will depend on the wider context of the measurements. Recombination can be studied via TPC, but the temperature dependence of SSPC can also be used to identify different recombination mechanisms, while details of the electronic density of states (DOS) in the band gap of a semiconductor can be inferred either from the spectral response of the SSPC or from a proper analysis of TPC. Detailed discussions of the general principles of photoconductivity may be found in the standard monographs by *Bube* [7.1, 2], *Ryvkin* [7.3] and *Rose* [7.4].

7.1 Steady State Photoconductivity Methods

7.1.1 The Basic Single-Beam Experiment

The simplest photoconductivity experiment uses a constant monochromatic light source to generate equal excess densities of free electrons and holes, $\Delta n = \Delta p$, that lead to a change in the conductivity by

$$\Delta\sigma = \sigma_{ph} = e(\mu_n \Delta n + \mu_p \Delta p), \quad (7.1)$$

where e is the electronic charge and μ_n and μ_p are the electron and hole mobilities, respectively. The basic experimental arrangement is illustrated in Fig. 7.1a, where L and A are the length and the cross-sectional area of the sample and the photocurrent I_{ph} corresponds to $\sigma_{ph}AF$, where $F = V/L$ is the electric field applied. The end surfaces of the sample are covered by a metallic electrode. However, since materials of current interest are often used in thin film rather than bulk form, interdigitated electrodes of the type shown in Fig. 7.1b are frequently used in actual measurement geometries. In general, a fraction of the photogenerated carriers becomes immobilized by getting trapped at various defects such that not every part of Δn and Δp contributes equally to the photoconductivity in (7.1). The effect of such trapping on the photoconductivity is reflected in the use of values for the mobilities μ_n and μ_p that are lower – and not necessarily symmetrically lower – than the theoretical free-carrier mobility μ_0. In fact, for a significant number of materials with widespread practical applications, one of either the product $\mu_n \Delta n$ or the product $\mu_p \Delta p$ turns out to be much larger than the other because of strongly unequal carrier mobilities. For instance, the electron term dominates in intrinsic silicon,

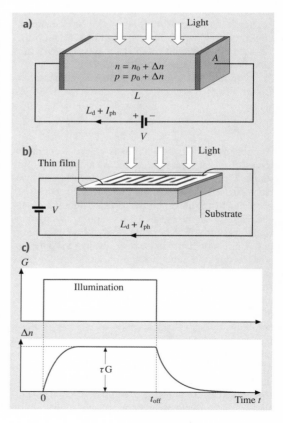

Fig. 7.1 (a) Basic arrangement for photoconductivity measurements, with V the applied voltage, L the sample length and A the cross-sectional area. I_d, n_0 and p_0 are the current and the carrier densities in the dark, and I_{ph}, Δn, Δp are the incremental values caused by the illumination. (b) Example of interdigitated electrode configuration for a thin film sample. (c) Schematic time development of the excess carrier concentration Δn in response to a period of illumination

while the photocurrent in chalcogenide glasses is carried by holes. In those instances, (7.1) effectively reduces to a one-carrier equation.

In the $\mu_n \Delta n$ or $\mu_p \Delta p$ products, the mobility μ_i is a material parameter that, in general, will depend on temperature and sample characteristics, while the excess carrier density $\Delta n = \Delta p$ is determined by a combination of material and external parameters. Phenomenologically, the excess density Δn can be written as the product $G\tau_i$, where G is the rate of generation of free electrons and holes per unit volume, and τ_i is the average lifetime of the excess carrier. Introducing these quantities into (7.1) leads to the form

$$\sigma_{ph} = eG(\mu_n \tau_n + \mu_p \tau_p),\quad (7.2)$$

which explicitly displays the mobility–lifetime products that are frequently used to characterize photoconductors. The relationship between the steady state values of Δn and G is illustrated in Fig. 7.1c, where the build-up and decay of Δn when the illumination is turned on and turned off are also shown. Those time-dependent aspects of photoconductivity will be addressed in a later section.

The generation rate G is defined by

$$G = \eta(I_0/h\nu)(1-R)[1-\exp(-\alpha d)]/d,\quad (7.3)$$

where η is the quantum efficiency of the generation process, I_0 is the incident illumination intensity (energy per unit time and unit area), $h\nu$ is the photon energy, R is the reflection coefficient of the sample, α is the optical absorption coefficient of the material, and d is the sample thickness. A quantum efficiency $\eta < 1$ signifies that, due to geminate recombination of the carriers or of exciton formation, not every absorbed photon generates a free electron and hole that will contribute to the photocurrent. The values of the parameters η, R and α depend, in general, on the wavelength of the illuminating light. Consequently, monochromatic illumination from a tunable light source can be used to obtain energy-resolved information about the sample, while illumination with white light will only offer a global average. Under many experimental circumstances, the condition $\alpha d \ll 1$ will hold over a significant energy range (when the sample thickness is small with respect to the optical absorption depth of the material). Equation (7.3) can then be simplified to

$$G \cong \eta(I_0/h\nu)(1-R)\alpha.\quad (7.4)$$

The free-carrier lifetimes of the excess electrons and holes, τ_n and τ_p, in (7.2) are governed by recombination with carriers of opposite sign. Assuming, for simplicity, the frequently encountered case of photoconductivity dominated by one type of carrier (known as the majority carrier), and assuming electrons to be the majority carrier, the recombination rate can be written as $\tau_n^{-1} = b(p_0 + \Delta p)$, where b is a recombination constant, and p_0 and Δp are the equilibrium and excess minority carrier densities. It then follows that the photoconductivity

$$\begin{aligned}\sigma_{ph} \propto \Delta n = G\tau_n &= G/b(p_0 + \Delta p)\\ &= G/b(p_0 + \Delta n).\end{aligned}\quad (7.5)$$

Equation (7.5) indicates that a linear relationship $\sigma_{ph} \propto G$ holds for $\Delta n \ll p_0$ (a low excess carrier density), while high excitation levels with $\Delta n \gg p_0$ lead to $\sigma_{ph} \propto G^{1/2}$. These linear and quadratic recombination regimes are also referred to as mono- and bimolecular recombination. For a given light source and temperature, variations in G correspond to variations in the light intensity I_0, and therefore $\sigma_{ph} \propto I_0^\gamma$ with $1/2 \leq \gamma \leq 1$. The value of γ itself will of course depend on the light intensity I_0. However, I_0 is not the only factor that determines the value of γ: intermediate γ values may indicate a $\Delta n \approx p_0$ condition, but they may equally be caused by a distribution of recombination centers, as outlined below [7.4].

From a materials characterization point of view, SSPC offers the possibility of using the above equations to determine the absorption coefficient as a function of the energy of the incoming photons, and thus explore the electronic density of states around the band gap of a semiconductor. When single-crystalline samples of materials with sufficiently well-defined energy levels are studied, maxima corresponding to specific optical transitions may be seen in the photoconductivity spectra. A recent example, involving the split valence band of a p-$CdIn_2Te_4$ crystal, may be found in You et al. [7.5]. Another example is given in Fig. 7.2, where the spectral distribution of the photocurrent is shown for optical-quality diamond films prepared by chemical vapor deposition [7.6]. The rise in photocurrent around 5.5 eV corresponds to the optical gap of diamond, while the shoulders at ≈ 1.5 eV and ≈ 3.5 eV signal the presence of defect distributions in the gap. The data in Fig. 7.2 were obtained under ac conditions using chopped light and a lock-in amplifier. The changes in the observed phase shift can then also be used to locate the energies at which transitions to specific features of the density of states (DOS) become of importance. The use of ac excitation and lock-in detection has the added advantage of strongly reducing uncorrelated noise, but

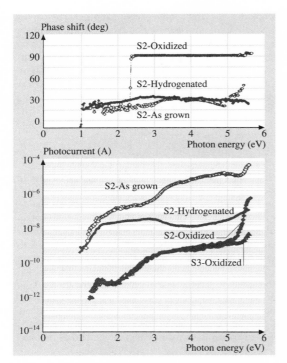

Fig. 7.2 Room temperature ac photocurrent spectra, measured at 7 Hz, after various treatments of CVD diamond layers deposited at 920 °C (S2) and 820 °C (S3) (after [7.6])

cess, depend on the temperature through the approximate Fermi–Dirac occupation probability function $\exp[(E - E_F)/kT]$, thus making recombination a temperature-dependent process. In photoconductors, recombination is mediated by carrier traps in the bandgap. The presence of discrete trapping levels leads to thermally activated photocurrents, with the activation energy indicating the energetic positions of the traps. *Main* and *Owen* [7.8] and *Simmons* and *Taylor* [7.9] showed that the positive photocurrent activation energy in the monomolecular recombination regime corresponds to the distance above the Fermi level of a donor-like center, while a negative activation energy value in the bimolecular region refers to the energy position above the valence band edge of an acceptor-like center. Figure 7.3 illustrates this photocurrent behavior for amorphous As_2Se_3 [7.7]. The above pattern is characteristic of chalcogenide glasses, where the intrinsic charged defects with negative effective correlation energy act as recombination centers [7.10]. SSPC measurements can thus determine the recombination levels of those defects.

In highly photosensitive materials, such as selenium or hydrogenated amorphous silicon (a-Si:H), measurements in the monomolecular region are hindered by the problem of satisfying the $\Delta n \ll p_0$ condition. In addition, the SSPC temperature dependence in a-Si:H does not exhibit a definite activation energy due to the presence of a more distributed and complex set of traps that even induce regions of superlinear dependence on light intensity [7.11]. This illustrates that SSPC analysis is not necessarily straightforward.

Whenever the electronic density of states in the band gap of a photoconductor consists of a distribution of traps (as is the case in amorphous materials), a quasi-Fermi level $E_{qF} = E_F - kT \ln(1 + \Delta n/n)$, linked to the excess carrier density, can be defined. This quasi-Fermi level will – to a first approximation – correspond to the de-

care must be taken to ensure that the ac frequency remains lower than the response rate of the investigated system over the spectral range of interest.

The equilibrium free-carrier densities n_0 and p_0, which play a role in the recombination pro-

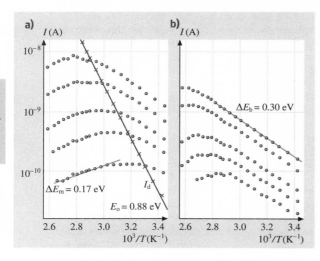

Fig. 7.3a,b Temperature dependence of the steady state dark and photocurrents in an a-As_2Se_3 bulk sample, illuminated at 1.55 eV with intensities of 0.84, 3.5, 9.8, 38 and 120×10^{12} photons/cm²s (**a**), and illuminated at 1.85 eV with intensities of 0.56, 1.7, 4.6, 27 and 77×10^{12} photons/cm²s (**b**). ΔE_m and ΔE_b represent the photocurrent activation energies in the monomolecular and bimolecular recombination regimes respectively, and E_σ is the activation energy of the dark current I_d (after [7.7])

marcation level that divides the DOS into a shallower part where carriers will be trapped and subsequently re-emitted and a deeper part where traps have become recombination centers. In other words, varying the light intensity influences both carrier generation and recombination rates. When several trapping centers with quite different characteristics are present in the photoconductor, shifts in the positions of the quasi-Fermi levels can then produce unexpected results. Instances of $\sigma_{ph} \propto I_0^\gamma$, with $\gamma > 1$ (as referred to above) will be observed for some materials, while combinations that actually produce *negative* photoconductivity, $\sigma_{ph} < 0$, have also been encountered [7.4].

7.1.2 The Constant Photocurrent Method (CPM)

The constant photocurrent method (CPM) has been used by *Vaněček* and coworkers [7.12, 13] to determine the optical absorption coefficient as a function of photon energy, $\alpha(E)$, via (7.2–7.4). In CPM, the photocurrent is kept constant by continually adjusting the light intensity I_0 while the photon energy is scanned across the spectrum. The constant photocurrent implies that the quasi-Fermi levels have immobile positions and thus that the free-carrier lifetime is a constant, τ. It then follows that in

$$\sigma_{ph} = e\mu\tau(I_0/h\nu)(1-R)\eta\alpha \tag{7.6}$$

the product $(I_0/h\nu)\alpha$ will remain constant, and that α can be determined from it, provided that any energy dependencies for the parameters μ, R and η of (7.6) are negligible. The value at which the photocurrent is fixed can be chosen freely, but will in practice be dictated by the low-absorption region of the sample. However, since even low-level photocurrents can still be measured with high precision, the method is especially useful at low values of optical absorption where standard transmission measurements lose their accuracy.

In 'absolute' CPM, the optical transmission through the film is measured at the same time as the photocurrent, and the data from the two measurements are combined in order to remove optical interference fringes from the data and to fix the value of the proportionality constant [7.12]. The experimental arrangement used in such absolute CPM measurements is shown schematically in Fig. 7.4. The CPM experiment can be operated with either dc or ac illumination, but the absorption spectra retrieved will not be identical. AC illumination can be obtained using a mechanical chopper (as suggested in Fig. 7.4), but also, for instance, from an ac-driven light-emitting

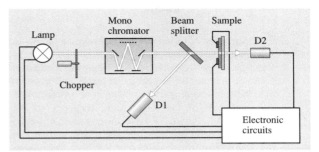

Fig. 7.4 Schematic diagram of an 'absolute' CPM set-up. Photodetector D1 is used to regulate the intensity of the lamp, while detector D2 measures the transmitted light (after [7.12])

diode. *Main* et al. [7.14] showed that, in the dc mode, transitions involving initially unoccupied DOS levels raise the absorption above the value that is seen with the ac technique. Systematic comparison of *dc* and *ac* results allows us, therefore, to distinguish between occupied states below the operative Fermi level and unoccupied ones above it. In cases where the quantum efficiency of carrier generation η can be taken as unity, CPM gives $\alpha(E)$ directly as $1/I_0$, and this method is widely used, for example for hydrogenated amorphous silicon. However, for materials such as chalcogenide glasses or organic semiconductors where η itself is energy-dependent, it is only the product $\eta\alpha$ that is readily obtained.

7.1.3 Dual-Beam Photoconductivity (DBP)

Like the CPM discussed above, the dual-beam photoconductivity (DBP) technique is used to determine the sub-bandgap optical absorption in a photoconductor. A constant, uniformly absorbed illumination I_0 is used to establish a constant excess carrier density in the material, and hence a constant free-carrier lifetime τ. The chopped signal $I'(E)$ of a low-intensity, tunable light source is added to this background to generate variations in photoconductivity $\delta\sigma_{ph}(E)$. Synchronous lock-in detection of the small ac signal then provides the information needed to deduce $\alpha(E)$. By carrying out measurements at different values of the background illumination intensity, DBP allows the photoconductor absorption to be tested for changing quasi-Fermi level positions. Changes in the resolved $\alpha(E)$ curves can then be used to obtain information on the DOS distribution in the sample. An example of this use of DBP can be found in *Günes* et al. [7.15], where differences in absorption between annealed and light-soaked hydrogenated amorphous silicon samples are studied.

7.1.4 Modulated Photoconductivity (MPC)

The experimental technique that has become known as modulated photoconductivity (MPC) is used to determine the energetic distribution of states in the bandgap of a photoconductor by analyzing the phase shift between ac photoexcitation and the ensuing ac photocurrent as a function of the modulation frequency of the light [7.17, 18]. Figure 7.5 shows the essential parts of an MPC set-up, and illustrates the phase difference between the illumination and the photocurrent. Two modulation frequency ranges with distinct characteristics are identified. In the high-frequency region, from a few Hz up to the kHz range, the signal is dominated by carrier release from traps, with a release rate that matches the modulation frequency. The usual assumption, that the release probability decreases exponentially with the trap depth according to $r \propto \exp(-E/kT)$, gives the link between the measured phase shift and the DOS of the material. The relationship between the two is expressed by

$$g(E) \propto \sin(\Phi)/I_{ac}, \quad E = k_B T \ln(\nu_0/\omega), \quad (7.7)$$

where Φ and I_{ac} are the phase and intensity of the ac photocurrent, k_B is the Boltzmann constant, T the temperature, ν_0 the attempt-to-escape frequency and ω the modulation frequency. At the low-frequency end, recombination and trapping in deep states determine the phase shifts and the DOS varies according to $\tan(\Phi)/\omega$. The transition between the two regions is tied to the position of the quasi-Fermi levels and can, therefore, be shifted by changing the illumination intensity. MPC works best with photoconductors where one carrier type dominates the current, and therefore only one side of the bandgap needs be taken into account in the analysis. Examples of MPC-determined DOS profiles are given in Fig. 7.6 [7.16]. The figure shows a profile for the conduction band side of the bandgap of an as-deposited polymorphous silicon sample, as well as those for the sample following light soaking and after subsequent annealing.

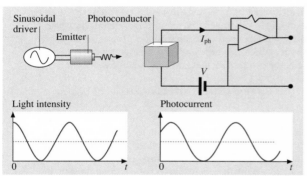

Fig. 7.5 Schematic diagram of an MPC set-up (*upper frames*), and of the phase relationship between the exciting light intensity and the resulting photocurrent (*lower frames*)

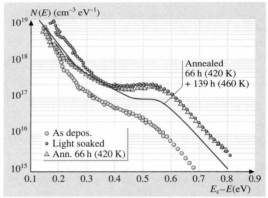

Fig. 7.6 DOS below the CB edge of a polymorphous silicon sample deposited at 423 K and measured (by MPC) as-deposited, after light soaking, and after two stages of subsequent annealing (after [7.16])

7.2 Transient Photoconductivity Experiments

The study of transient aspects of photoconductivity can relate to either the build-up or the relaxation of steady state photocurrents, or to a material's response to pulsed excitation. While the SSPC turn-on transient reflects the interplay between generation and recombination of carriers (an interplay that often leads to a current overshoot at high excitation levels), the SSPC relaxation upon turn-off only involves recombination and is therefore easier to analyze. Nevertheless, a simple exponential decay of the photocurrent, as sketched in Fig. 7.1c, will only be observed when a unique recombination path is followed, a situation that is the exception rather than the rule. Transient photoconductivity (TPC) caused by pulsed excitation is generally simpler to analyze. Indeed, whereas a quasi-equilibrium distribution of trapped photogenerated carriers will build up or be present in the photoconductor's bandgap under SSPC, the TPC experiments can be analyzed against the background of the thermal equilibrium distribution of carriers in the material.

7.2.1 Current Relaxation from the Steady State

Upon termination of steady state illumination, the generation term drops out of the rate equation that describes the nonequilibrium carrier distribution, but the carrier density itself and the operative recombination process are not altered. Consequently, the initial photocurrent decay will be governed by whatever recombination mode existed under SSPC conditions. Spectroscopic analysis of the relaxation current in terms of the distribution of states in the bandgap can be readily achieved in the case of monomolecular recombination [7.19], with the product of photocurrent and time being proportional to the DOS:

$$I_{\text{ph}}(t)t \propto g(E), \quad E = k_\text{B} T \ln(\nu_0 t). \tag{7.8}$$

In (7.8), $k_\text{B}T$ is the Boltzmann energy and ν_0 is the attempt-to-escape frequency. When, on the other hand, bimolecular recombination dominates, the link between the current and the distribution of recombination centers is much less direct and spectroscopic analysis is difficult. Unfortunately, bimolecular recombination is dominant in good photoconductors.

In spite of the above, relaxation of the steady state current has often been used to obtain a first-order estimate for free-carrier lifetimes, even when this had to be done on a purely phenomenological basis due to a lack of sufficient information on the recombination mechanisms involved. An exponential fit to the initial part of the decay is then often used to make the estimate. In cases where more than one – sometimes vastly different – recombination mechanisms are operative, this initial decay does not necessarily represent the most significant proportion of carriers. This is certainly the case whenever so-called persistent photoconductivity is observed; one of the relaxation times involved is then longer than the observation time.

7.2.2 Transient Photoconductivity (TPC)

In the standard transient photoconductivity (TPC) experiment, free carriers are excited into the transport band at time $t = 0$ by a short light pulse. They are then moved along by the electric field until their eventual disappearance through recombination, but before this happens they will have been immobilized a number of times by various traps that are present in the material. Since the carrier distributions are in thermal equilibrium at the start of the experiment, both the trapping sites for electrons above the Fermi level and the hole trapping sites below E_F are empty, such that the newly created carriers are not excluded from any of those trapping sites. Given that carrier release from a trap is a thermally activated process with the trap depth being the activation energy, deeper traps immobilize carriers for longer times and lead to lower values for the transient current. As shallower states release trapped carriers sooner, retrapping of those carriers will lead to increased occupation of the deeper states and further reduction of the current level. To allow this thermalization of the excited carriers to run its full course until recombination sets in, the experiments are traditionally carried out in the so-called secondary photocurrent mode, whereby the sample is supplied with ohmic electrical contacts and carrier loss is by recombination only. Coplanar electrode geometries (gap cells) are mostly used. Expressions that link the transient current to the distribution of localized states can be derived [7.20], but they are difficult to invert in the general case. Nevertheless, as long as recombination can be neglected, the relationship $g(E) \propto [I(t)t]^{-1}$ can be used as a first-order estimate.

For the special case of an exponential DOS, the solution is straightforward: a $g(E) \propto \exp(-E/E_0)$ distribution of trapping levels leads to a power law for the transient current $I(t) \propto t^{-(1-\alpha)}$ with $\alpha = k_\text{B}T/E_0$. In other words, the width of the exponential distribution E_0 can be deduced from the slope of the power law decay of the current. Essentially exponential distributions were found to dominate the valence band tail of equilibrated amorphous As_2Se_3 samples over a wide energy range [7.7], but no other examples have emerged.

An elegant way to circumvent the difficulties posed by a time domain analysis of the transient current is to transpose the current decay into the frequency domain by a Fourier transform [7.21]. Since the TPC current decay is the photoconductor's response to an impulse excitation, its Fourier transform gives the frequency response $I(\omega)$ of that photoconductor. In fact, this $I(\omega)$ corresponds to the photocurrent intensity I_{ac} as used in the MPC method, and the same procedures can thus be used to extract the information on density and energy distribution of localized states in the band gap. Not just Fourier transform but also Laplace transform techniques have been applied to the conversion of TPC signals into DOS information. A comparison and discussion of the results may be found in [7.22]. Examples of Fourier

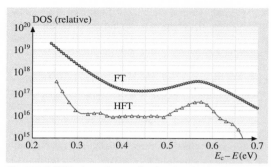

Fig. 7.7 DOS below the conduction band edge in a-Si:H, obtained through Fourier transforms of the transient photocurrent; HFT: the high-resolution analysis of [7.22], FT: the earlier analysis according to [7.21]

transform TPC analysis, as originally proposed and as developed since, are shown in Fig. 7.7 for an a-Si:H sample. Whereas the energy range that can be probed is limited in MPC by the frequency range of the lock-in amplifier, it is the smallest resolution time of the detection system that limits the range in the case of TPC, the latter one being generally more advantageous.

7.2.3 Time-of-Flight Measurements (TOF)

The time-of-flight (TOF) experiment, originally designed to determine the drift mobility of free carriers in high-mobility materials, has been highly successfully adapted to low-mobility materials such as organic or amorphous semiconductors [7.24], where it has been used for drift mobility measurements but also as an alternative TPC technique to study the energy distribution of localized states. While majority carriers will dominate photocurrents in traditional TPC, TOF allows independent measurements with majority and minority carriers, and thus independent examinations of the valence band side and conduction band side of the band gap.

For TOF measurements, the sample consists of a layer of the photoconducting material sandwiched between two electrodes that are blocking carrier injection into the sample. At least one of the electrodes must be semitransparent to permit the photoexcitation of free carriers in the material just beyond the illuminated contact by a strongly absorbed light flash. Depending on the polarity of the electric field applied across the sample, either electrons or holes will then be drifted through the sample. At their arrival at the back contact, the current will drop since the blocking contact ensures that only the primary photocurrent is measured. From the transit time

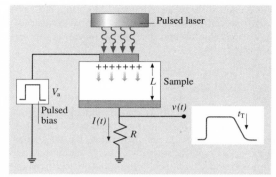

Fig. 7.8 TOF measurement set-up shown for the case of holes being drifted through the length L of the sample by a positive applied voltage. Choosing the resistance R that generates an output voltage to be low minimizes RC distortion of the signal at short times; choosing it to be high enhances the detectability of weak signals at the expense of time resolution

t_T (the time needed for the charge sheet to cross the sample), the drift mobility μ_d can be calculated according to $\mu_d = L/t_T F$, where L is the sample length and F the applied electric field. The essential elements of a TOF

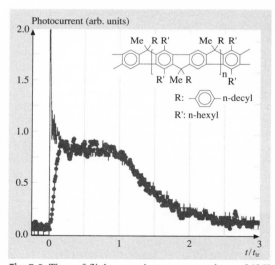

Fig. 7.9 Time-of-flight transients measured at 243 K in methyl-substituted ladder-type poly(para)phenylene (MeLPPP) with 60 kV/cm (line) and 300 kV/cm (•) applied, and normalized to a transit time set to 90% of the pre-transit current. The *inset* shows the chemical structure of MeLPPP (after [7.23])

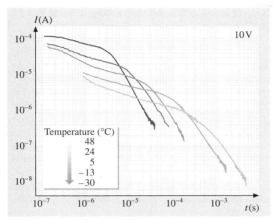

Fig. 7.10 Example of TOF hole transients measured at several temperatures, as indicated, with 10 V applied across a 5.6 μm-thick a-Si:H sample grown in an expanding thermal plasma at 0.85 nm/s and 250 °C substrate temperature, and sandwiched between Mo contacts (after [7.25])

measuring circuit are displayed in Fig. 7.8. The transit time can be measured directly on the current trace, in which case it is variously defined as the time at which the current has dropped by values ranging from 10% to 50% (the latter one being most commonly used), or it can be obtained by integrating the current and using the time at which the collected charge saturates. Obtaining a true value of μ_d requires that the field F be uniform and constant during the carrier transit, which means that F should only be applied a short time before the optical excitation and that the transit time should be short with respect to the dielectric relaxation time in the material. Figure 7.9 shows TOF transients in a conjugated polymer whereby a 10% drop is used to define the transit time.

In materials with a wide distribution of localized gap states, as is generally the case in disordered photoconductors, the drifting charge package spreads out along the length of the sample, and a representative transit time can only be discerned as a change of slope in a double-logarithmic plot of current versus time. The curves in Fig. 7.10 (from [7.25]) illustrate such behavior. Measurements at different temperatures and applied fields are then needed to ascertain that the observed feature marks an actual carrier transit rather than deep trapping of the photogenerated charge. In the materials that exhibit this anomalously dispersive transport, the pre-transit current will have the characteristics of the TPC described in the previous section,

and the information about the distribution of gap states $g(E)$ that is contained in the current transient can be extracted in the same ways. Both pre-transit current transients and measured drift mobility values have been employed in the past to estimate the DOS in the band tails of disordered semiconductors. In the latter case, specific $g(E)$ functions are explored through trap-controlled transport modeling to reproduce the experimental dependence of μ_d on the temperature and the electric field. This technique has since been replaced by the more direct procedures described in preceding sections.

At times longer than the TOF transit time, a steeper current decay testifies to the fact that carriers are leaving the sample. The post-transit current that is then observed is increasingly due to the emission of carriers that were trapped in states deep in the bandgap. Provided that the conditions are such that the probability of subsequent deep retrapping of the same carriers is negligible, a proper analysis of these post-transit TOF current transients permits the elucidation of the distribution of localized states deeper in the gap [7.27] with, as in (7.8), $g(E) \propto I(t)t$ expressing the correspondence.

7.2.4 Interrupted Field Time-of-Flight (IFTOF)

The interrupted field time-of-flight (IFTOF) experiment differs from the time-of-flight experiment described

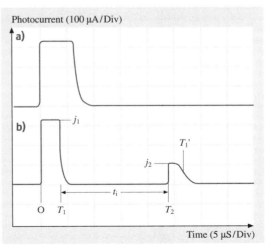

Fig. 7.11a,b Comparison of current traces in TOF (**a**) and IFTOF (**b**) experiments. The applied electric field is turned off in case (**b**) for a length of time t_i (after [7.26])

in the previous section in that the applied field that drives the photogenerated carrier packet through the sample is turned off for some period of time before the carriers have completed their transit. As illustrated in Fig. 7.11, a lower current intensity is measured when the field is turned on again, signalling that some of the drifting carriers have become immobilized in deep traps [7.28]. By studying the drop in current as a function of the interruption time t_i, the deep-trapping lifetime of the carriers can be evaluated. Recombination can be routinely neglected in TOF experiments since only one type of carrier drifts through the sample, but by charging a sample with carriers of one polarity before performing an IFTOF experiment that drifts carriers of the opposite polarity through the sample, recombination parameters can be studied too [7.29].

Another interesting method for studying the recombination process is – just like IFTOF – based on a simple modification of the TOF experiment: after generating free carriers through one contact and drifting the slower type of carrier into the sample, a second light pulse through the other contact sends a sheet of oppositely charged carriers towards the first one. The two carrier packages will cross and some electrons and holes will recombine during that crossing, thereby affecting the observed current levels and providing a way to study the recombination process. An elegant example of the application of this technique to amorphous selenium can be found in *Haugen* and *Kasap* [7.30].

References

7.1   R. H. Bube: *Photoconductivity of Solids* (Wiley, New York 1960)
7.2   R. H. Bube: *Photoelectronic Properties of Semiconductors* (Cambridge Univ. Press, Cambridge 1992)
7.3   S. M. Ryvkin: *Photoelectric Effects in Semiconductors* (Consultants Bureau, New York 1964)
7.4   A. Rose: *Concepts in Photoconductivity and Allied Problems* (Krieger, Huntington 1978)
7.5   S. H. You, K. J. Hong, T. S. Jeong, C. J. Youn, J. S. Park, D. C. Shin, J. D. Moon: J. Appl. Phys. **95**, 4042 (2004)
7.6   M. Nesládek, L. M. Stals, A. Stesmans, K. Iakoubovskii, G. J. Adriaenssens, J. Rosa, M. Vaněček: Appl. Phys. Lett. **72**, 3306 (1998)
7.7   G. J. Adriaenssens: Philos. Mag. B **62**, 79 (1990) and references therein
7.8   C. Main, A. E. Owen: In: *Electronic and Structural Properties of Amorphous Semiconductors*, ed. by P. G. Le Comber, J. Mort (Academic, London 1973) p. 527
7.9   J. G. Simmons, G. W. Taylor: J. Phys. C **7**, 3051 (1974)
7.10 G. J. Adriaenssens, N. Qamhieh: J. Mater. Sci. Mater. El. **14**, 605 (2003)
7.11 H. Fritzsche, B.-G. Yoon, D.-Z. Chi, M. Q. Tran: J. Non-Cryst. Solids **141**, 123 (1992)
7.12 M. Vaněček, J. Kočka, A. Poruba, A. Fejfar: J. Appl. Phys. **78**, 6203 (1995)
7.13 M. Vaněček, J. Kočka, J. Stuchlík, A. Tříska: Solid State Commun. **39**, 1199 (1981)
7.14 C. Main, S. Reynolds, I. Zrinščak, A. Merazga: Mater. Res. Soc. Symp. Proc. **808**, 103 (2004)
7.15 M. Günes, C. Wronski, T. J. McMahon: J. Appl. Phys. **76**, 2260 (1994)
7.16 C. Longeaud, D. Roy, O. Saadane: Phys. Rev. B **65**, 85206 (2002)
7.17 H. Oheda: J. Appl. Phys. **52**, 6693 (1981)
7.18 R. Brüggemann, C. Main, J. Berkin, S. Reynolds: Philos. Mag. B **62**, 29 (1990)
7.19 M. S. Iovu, I. A. Vasiliev, E. P. Colomeico, E. V. Emelianova, V. I. Arkhipov, G. J. Adriaenssens: J. Phys. Condens. Mat. **16**, 2949 (2004)
7.20 A. I. Rudenko, V. I. Arkhipov: Philos. Mag. B **45**, 209 (1982)
7.21 C. Main, R. Brüggemann, D. P. Webb, S. Reynolds: Solid State Commun. **83**, 401 (1992)
7.22 C. Main: J. Non-Cryst. Solids **299**, 525 (2002)
7.23 D. Hertel, A. Ochse, V. I. Arkhipov, H. Bässler: J. Imag. Sci. Technol. **43**, 220 (1999)
7.24 W. E. Spear: J. Non-Cryst. Solids **1**, 197 (1969)
7.25 M. Brinza, E. V. Emelianova, G. J. Adriaenssens: Phys. Rev. B **71**, 115209 (2005)
7.26 S. Kasap, B. Polishuk, D. Dodds, S. Yannacopoulos: J. Non-Cryst. Solids **114**, 106 (1989)
7.27 G. F. Seynhaeve, R. P. Barclay, G. J. Adriaenssens, J. M. Marshall: Phys. Rev. B **39**, 10196 (1989)
7.28 S. Kasap, B. Polishuk, D. Dodds: Rev. Sci. Instrum. **61**, 2080 (1990)
7.29 S. Kasap, B. Fogal, M. Z. Kabir, R. E. Johanson, S. K. O'Leary: Appl. Phys. Lett. **84**, 1991 (2004)
7.30 C. Haugen, S. O. Kasap: Philos. Mag. B **71**, 91 (1995)

8. Electronic Properties of Semiconductor Interfaces

In this chapter we investigate the electronic properties of semiconductor interfaces. Semiconductor devices contain metal–semiconductor, insulator–semiconductor, insulator–metal and/or semiconductor–semiconductor interfaces. The electronic properties of these interfaces determine the characteristics of the device. The band structure lineup at all these interfaces is determined by one unifying concept, the continuum of interface-induced gap states (IFIGS). These intrinsic interface states are the wavefunction tails of electron states that overlap the fundamental band gap of a semiconductor at the interface; in other words they are caused by the quantum-mechanical tunneling effect. IFIGS theory quantitatively explains the experimental barrier heights of well-characterized metal–semiconductor or Schottky contacts as well as the valence-band offsets of semiconductor–semiconductor interfaces or semiconductor heterostructures. Insulators are viewed as semiconductors with wide band gaps.

8.1	Experimental Database	149
	8.1.1 Barrier Heights of Laterally Homogeneous Schottky Contacts	149
	8.1.2 Band Offsets of Semiconductor Heterostructures	152
8.2	IFIGS-and-Electronegativity Theory	153
8.3	Comparison of Experiment and Theory	155
	8.3.1 Barrier Heights of Schottky Contacts	155
	8.3.2 Band Offsets of Semiconductor Heterostructures	156
	8.3.3 Band-Structure Lineup at Insulator Interfaces	158
8.4	Final Remarks	159
References		159

In his pioneering article entitled *Semiconductor Theory of the Blocking Layer*, *Schottky* [8.1] finally explained the rectifying properties of metal–semiconductor contacts, which had first been described by *Braun* [8.2], as being due to a depletion of the majority carriers on the semiconductor side of the interface. This new depletion-layer concept immediately triggered a search for a physical explanation of the barrier heights observed in metal–semiconductor interfaces, or Schottky contacts as they are also called in order to honor Schottky's many basic contributions to this field.

The early Schottky–Mott rule [8.3, 4] proposed that n-type (p-type) barrier heights were equal to the difference between the work function of the metal and the electron affinity (ionization energy) of the semiconductor. A plot of the experimental barrier heights of various metal–selenium rectifiers versus the work functions of the corresponding metals did indeed reveal a linear correlation, but the slope parameter was much smaller than unity [8.4]. To resolve the failure of the very simple and therefore attractive Schottky–Mott rule, *Bardeen* [8.5] proposed that electronic interface states in the semiconductor band gap play an essential role in the charge balance at metal–semiconductor interfaces.

Heine [8.6] considered the quantum-mechanical tunneling effect at metal–semiconductor interfaces and noted that *for energies in the semiconductor band gap, the volume states of the metal have tails in the semiconductor*. *Tejedor* and *Flores* [8.7] applied this same idea to semiconductor heterostructures where, for energies in the band-edge discontinuities, the volume states of one semiconductor tunnel into the other. The continua of interface-induced gap states (IFIGS), as these evanescent states were later called, are an intrinsic property of semiconductors and they are the *fundamental* physical mechanism that determines the band-structure lineup at both metal–semiconductor contacts and semiconductor heterostructures: in other words, at all semiconductor interfaces. Insulator interfaces are also included in this, since insulators may be described as wide-gap semi-

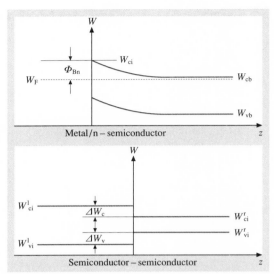

Fig. 8.1 Schematic energy-band diagrams of metal–semiconductor contacts and semiconductor heterostructures. W_F: Fermi level; Φ_{Bn}: barrier height; W_v and W_c: valence-band maximum and conduction-band minimum, respectively; ΔW_v and ΔW_c: valence- and conduction-band offset, respectively; i and b: values at the interface and in the bulk, respectively; r and l: right and left side, respectively

conductors. Figure 8.1 shows schematic band diagrams of an n-type Schottky contact and a semiconductor heterostructure.

The IFIGS continua derive from both the valence- and the conduction-band states of the semiconductor. The energy at which their predominant character changes from valence-band-like to conduction-band-like is called their branch point. The position of the Fermi level relative to this branch point then determines the sign and the amount of the net charge in the IFIGS. Hence, the IFIGS give rise to intrinsic interface dipoles. Both the barrier heights of Schottky contacts and the band offsets of heterostructures thus divide up into a zero-charge-transfer term and an electric-dipole contribution.

From a more chemical point of view, these interface dipoles may be attributed to the partial ionic character of the covalent bonds between atoms right at the interface. Generalizing *Pauling's* [8.8] electronegativity concept, the difference in the electronegativities of the atoms involved in the interfacial bonds also describes the charge transfer at semiconductor interfaces. Combining the physical IFIGS and the chemical electronegativity concept, the electric-dipole contributions of Schottky barrier heights as well as those of heterostructure band offsets vary proportional to the difference in the electronegativities of the metal and the semiconductor and of the two semiconductors, respectively. The electronegativities of the Group IV elemental and the IV–IV, III–V, and II–VI compound semiconductors are almost equal, since the elements that constitute these semiconductors are all placed in the middle of the Periodic Table. Hence, the IFIGS dipole terms of the respective semiconductor heterostructures will be small and may be neglected [8.9]. The valence-band offsets of nonpolar, of lattice-matched and of metamorphic heterostructures should thus equal the difference between the branch-point energies of the semiconductors in contact.

The theoreticians appreciated *Heine*'s IFIGS concept at once. The initial reluctance of most experimentalists was motivated by the observation that the predictions of the IFIGS theory only marked upper limits for the barrier heights observed with *real* Schottky contacts [8.10]. *Schmitsdorf* et al. [8.11] finally resolved this dilemma. They found a linear decrease in the effective barrier height with increasing ideality factors for their Ag/n-Si(111) diodes. Such behavior has been observed for all of the Schottky contacts investigated so far. *Schmitsdorf* et al. attributed this correlation to patches of decreased barrier heights and lateral dimensions smaller than the depletion layer width [8.12]. Consequently, they extrapolated their plots of effective barrier height versus ideality factor to the ideality factor determined by the image-force or Schottky effect [8.13] alone; in this way, they obtained the barrier heights of the laterally homogeneous contacts. The barrier heights of laterally uniform contacts can also be determined from capacitance–voltage measurements (C/V) and by applying ballistic-electron-emission microscopy (BEEM) and internal photoemission yield spectroscopy (IPEYS). The I/V, C/V, BEEM, and IPEYS data agree within the margins of experimental error.

Mönch [8.14] found that the barrier heights of laterally homogeneous Schottky contacts as well as the experimentally observed valence band offsets of semiconductor heterostructures agree excellently with the predictions of the IFIGS-and-electronegativity theory.

8.1 Experimental Database

8.1.1 Barrier Heights of Laterally Homogeneous Schottky Contacts

I/V Characteristics

The current transport in real Schottky contacts occurs via thermionic emission over the barrier provided the doping level of the semiconductor is not too high [8.15]. For doping levels larger than approximately 10^{18} per cm^3, the depletion layer becomes so narrow that tunnel or field emission through the depletion layer prevails. The current–voltage characteristics then become ohmic rather than rectifying.

For thermionic emission over the barrier, the current–voltage characteristics may be written as (see, for example, [8.14])

$$I_{\text{te}} = A A_R^* T^2 \exp\left(-\Phi_{\text{Bn}}^{\text{eff}}/k_B T\right) \exp(e_0 V_c / n k_B T) \\ \times [1 - \exp(-e_0 V_c / k_B T)] , \quad (8.1)$$

where A is the diode area, A_R^* is the effective Richardson constant of the semiconductor, and k_B, T, and e_0 are Boltzmann's constant, the temperature, and the electronic charge, respectively. The effective Richardson constant is defined as

$$A_R^* = \frac{4\pi e_0 k_B m_n^*}{h^3} = A_R \frac{m_n^*}{m_0} , \quad (8.2)$$

where $A_R = 120\,\text{A cm}^{-2}\,\text{K}^{-2}$ is the Richardson constant for thermionic emission of nearly free electrons into vacuum, h is Planck's constant, and m_0 and m_n^* are the vacuum and the effective conduction-band mass of electrons, respectively. The externally applied bias V_a divides up into a voltage drop V_c across the depletion layer of the Schottky contact and an IR drop at the series resistance R_s of the diode, so that $V_c = V_a - IR_s$. For *ideal* (intimate, abrupt, defect-free, and, above all, laterally homogeneous) Schottky contacts, the effective zero-bias barrier height $\Phi_{\text{Bn}}^{\text{eff}}$ equals the difference $\Phi_{\text{Bn}}^{\text{hom}} - \delta\Phi_{\text{if}}^0$ between the homogeneous barrier height and the zero-bias image-force lowering (see [8.14])

$$\delta\Phi_{\text{if}}^0 = e_0 \left[\frac{2e_0^2 N_d}{(4\pi)^2 \varepsilon_\infty^2 \varepsilon_b \varepsilon_0^3}\left(e_0 \left|V_i^0\right| - k_B T\right)\right]^{1/4} , \quad (8.3)$$

where N_d is the donor density, $e_0|V_i^0|$ is the zero-bias band bending, ε_∞ and ε_b are the optical and the bulk dielectric constant, respectively, and ε_0 is the permittivity of vacuum. The ideality factor n describes the voltage dependence of the barrier height and is defined by

$$1 - 1/n = \partial \Phi_{\text{Bn}}^{\text{eff}} / \partial e_0 V_c . \quad (8.4)$$

For *real* diodes, the ideality factors n are generally found to be larger than the ideality factor

$$n_{\text{if}} = \left(1 - \frac{\delta\Phi_{\text{if}}^0}{4e_0|V_i^0|}\right)^{-1} , \quad (8.5)$$

which is determined by the image-force effect only.

The effective barrier heights and the ideality factors of real Schottky diodes fabricated under experimentally identical conditions vary from one specimen to the next. However, the variations of both quantities are correlated, and the $\Phi_{\text{Bn}}^{\text{eff}}$ values become smaller as the ideality factors increase. As an example, Fig. 8.2 displays $\Phi_{\text{Bn}}^{\text{eff}}$ versus n data for Ag/n-Si(111) contacts with $(1 \times 1)^i$- unreconstructed and $(7 \times 7)^i$-reconstructed interfaces [8.11]. The dashed and dash-dotted lines are the linear least-squares fits to the data points. The linear dependence of the effective barrier height on the ideality factor may be written as

$$\Phi_{\text{Bn}}^{\text{eff}} = \Phi_{\text{Bn}}^{\text{nif}} - \varphi_p(n - n_{\text{if}}) , \quad (8.6)$$

where $\Phi_{\text{Bn}}^{\text{nif}}$ is the barrier height at the ideality factor n_{if}. Several conclusions may be drawn from this relation. First, the $\Phi_{\text{Bn}}^{\text{eff}} - n$ correlation shows that more than

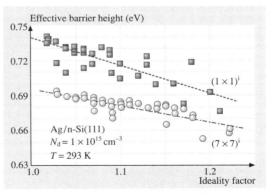

Fig. 8.2 Effective barrier heights versus ideality factors determined from I/V characteristics of Ag/n-Si(111)-$(7 \times 7)^i$ and -$(1 \times 1)^i$ contacts at room temperature. The *dashed* and *dash-dotted* lines are the linear least-squares fits to the data. After [8.11]

one physical mechanism determines the barrier heights of *real* Schottky contacts. Second, the extrapolation of Φ_{Bn}^{eff} versus n curves to n_{if} removes all mechanisms that cause a larger bias dependence of the barrier height than the image-force effect itself from consideration. Third, the extrapolated barrier heights Φ_{Bn}^{nif} are equal to the zero-bias barrier height $\Phi_{Bn}^{hom} - \delta\Phi_{if}^{0}$ of the laterally homogeneous contact.

The laterally homogeneous barrier heights obtained from Φ_{Bn}^{eff} versus n curves to n_{if} are not necessarily characteristic of the corresponding *ideal* contacts. This is illustrated by the two data sets displayed in Fig. 8.2, which differ in the interface structures of the respective diodes. Quite generally, structural rearrangements such as the $(7 \times 7)^i$ reconstruction are connected with a redistribution of the valence charge. The bonds in perfectly ordered bulk silicon, the example considered here, are purely covalent, and so reconstructions are accompanied by electric $Si^{+\Delta q} - Si^{-\Delta q}$ dipoles. The Si(111)-(7×7) reconstruction is characterized by a stacking fault in one half of its unit mesh [8.16]. *Schmitsdorf* et al. [8.11] quantitatively explained the experimentally observed reduction in the laterally homogeneous barrier height of the $(7 \times 7)^i$ with regard to the $(1 \times 1)^i$ diodes by the electric dipole associated with the stacking fault of the Si(111)-7×7 reconstruction.

Patches of reduced barrier height with lateral dimensions smaller than the depletion layer width that are embedded in large areas of laterally homogeneous barrier height is the only known model that explains a lowering of effective barrier heights with increasing ideality factors. In their phenomenological studies of such patchy Schottky contacts, *Freeouf* et al. [8.12] found that the potential distribution exhibits a saddle point in front of such nanometer-size patches of reduced barrier height. Figure 8.4 explains this behavior. The saddle-point barrier height strongly depends on the voltage drop V_c across the depletion layer. *Freeouf* et al. simulated the current transport in such patchy Schottky contacts and found a reduction in the effective barrier height and a correlated increase in the ideality factor as they reduced the lateral dimensions of the patches. However, they overlooked the fact that the barrier heights of the laterally homogeneous contacts may be obtained from Φ_{Bn}^{eff} versus n plots, by extrapolating to n_{if}.

C/V Characteristics

Both the space charge and the width of the depletion layers at metal–semiconductor contacts vary as a function of the externally applied voltage. The space-charge the-

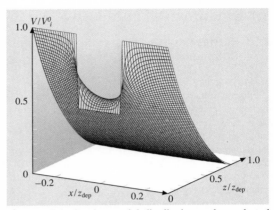

Fig. 8.3 Calculated potential distribution underneath and around a patch of reduced interface potential embedded in a region of larger interface band-bending. The lateral dimension and the interface potential reduction of the patch are set to two tenths of the depletion layer width z_{dep} and one half of the interface potential of the surrounding region

ory gives the variation in the depletion layer capacitance per unit area as (see [8.14])

$$C_{dep} = \{e_0^2 \varepsilon_b \varepsilon_0 N_d / 2 [e_0(|V_i^0| - V_c) - k_B T]\}^{1/2} . \tag{8.7}$$

The current through a Schottky diode biased in the reverse direction is small, so the *IR* drop due to the series resistance of the diode may be neglected. Consequently, the extrapolated intercepts on the abscissa of $1/C_{dep}^2$ versus V_a plots give the band bending $e_0|V_i^0|$ at the interface, and together with the energy distance $W_n = W_F - W_{cb}$ from the Fermi level to the conduction band minimum in the bulk, one obtains the flat-band barrier height $\Phi_{Bn}^{fb} \equiv \Phi_{Bn}^{hom} = e_0|V_i^0| + W_n$ which equals the laterally homogeneous barrier height of the contact.

As an example, Fig. 8.4 displays the flat-band barrier heights of the same Ag/n-Si(111) diodes that are discussed in Fig. 8.2. The dashed and dash-dotted lines are the Gaussian least-squares fits to the data from the diodes with $(1 \times 1)^i$ and $(7 \times 7)^i$ interface structures, respectively. Within the margins of experimental error the peak C/V values agree with the laterally homogeneous barrier heights obtained from the extrapolations of the I/V data shown in Fig. 8.2. These data clearly demonstrate that barrier heights characteristic of laterally homogeneous Schottky contacts can be only obtained from I/V or C/V data from many diodes fabricated under identical conditions rather than from a single diode. However, the

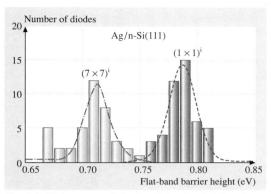
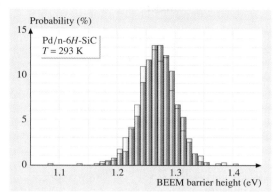

Fig. 8.4 Histograms of flat-band barrier heights determined from C/V characteristics of Ag/n-Si(111)-$(7\times 7)^i$ and -$(1\times 1)^i$ contacts at room temperature. The data were obtained with the same diodes discussed in Fig. 8.2. The *dashed* and *dash-dotted lines* are the Gaussian least-squares fits to the data. After [8.11]

Fig. 8.5 Histograms of local BEEM barrier heights of two Pd/n-6H-SiC(0001) diodes with ideality factors of 1.06 (*gray solid bars*) and 1.49 (*empty bars*). The data were obtained by fitting the square law (8.8) to 800 BEEM $I_{\text{coll}}/V_{\text{tip}}$ spectra each. Data from *Im* et al. [8.17]

effective barrier heights and the ideality factors vary as a function of the diode temperature. Hence, effective barrier heights and ideality factors evaluated from the I/V characteristics for one and the same diode recorded at different temperatures are also suitable for determining the corresponding laterally homogeneous barrier height (see [8.14]).

Ballistic-Electron-Emission Microscopy

In ballistic-electron-emission microscopy (BEEM) [8.18], a tip injects almost monoenergetic electrons into the metal film of a Schottky diode. These tunnel-injected electrons reach the semiconductor as ballistic electrons provided that they lose no energy on their way through the metal. Hence, the collector current I_{coll} is expected to set in when the ballistic electrons surpass the metal–semiconductor barrier; in other words, if the voltage V_{tip} applied between tip and metal film exceeds the *local* potential barrier $\Phi_{\text{Bn}}^{\text{loc}}(z)/e_0$. *Bell* and *Kaiser* [8.19] derived the square law

$$I_{\text{coll}}(z) = R^* I_{\text{tip}} \left[e_0 V_{\text{tip}} - \Phi_{\text{Bn}}^{\text{loc}}(z) \right]^2 \quad (8.8)$$

for the BEEM $I_{\text{coll}}/V_{\text{tip}}$ characteristics, where I_{tip} is the injected tunnel current. BEEM measures *local* barrier heights; most specifically, the saddle-point barrier heights in front of nanometer-sized patches rather than their lower barrier heights right at the interface.

BEEM is *the* experimental tool for measuring spatial variations in the barrier height on the nanometer-scale.

The local barrier heights are determined by fitting relation (8.8) to measured $I_{\text{coll}}/V_{\text{tip}}$ characteristics recorded at successive tip positions along lateral line scans. Figure 8.5 displays histograms of the local BEEM barrier heights of two Pd/n-6H-SiC(0001) diodes [8.17]. The diodes differ in their ideality factors, 1.06 and 1.49, which are close to and much larger, respectively, than the value $n_{\text{if}} = 1.01$ determined solely by the image-force effect. Obviously, the *nanometer-scale* BEEM histograms of the two diodes are identical although their *macroscopic* ideality factors and therefore their patchinesses differ. Two important conclusions were drawn from these findings. First, these data suggest the existence of two different types of patches, intrinsic and extrinsic ones. The intrinsic patches might be correlated with the random distributions of the ionized donors and acceptors which cause nanometer-scale lateral fluctuations in the interface potential. A few gross interface defects of extrinsic origin, which escape BEEM observations, are then responsible for the variations in the ideality factors. Second, Gaussian least-squares fits to the histograms of the local BEEM barrier heights yield peak barrier heights of 1.27 ± 0.03 eV. Within the margins of experimental error, this value agrees with the laterally homogeneous value of 1.24 ± 0.09 eV which was obtained by extrapolation of the linear least-squares fit to a $\Phi_{\text{Bn}}^{\text{eff}}$ versus n plot to n_{if}. The nanometer-scale BEEM histograms and the macroscopic I/V characteristics thus provide identical barrier heights of laterally homogeneous Schottky contacts.

Internal Photoemission Yield Spectroscopy

Metal-semiconductor contacts show a photoelectric response to optical radiation with photon energies smaller than the width of the bulk band gap. This effect is caused by photoexcitation of electrons from the metal over the interfacial barrier into the conduction band of the semiconductor. Experimentally, the internal photoemission yield, which is defined as the ratio of the photoinjected electron flux across the barrier into the semiconductor to the flux of the electrons excited in the metal, is measured as a function of the energy of the incident photons. Consequently, this technique is called internal photoemission yield spectroscopy (IPEYS). Cohen et al. [8.21] derived that the internal photoemission yield varies as a function of the photon energy $\hbar\omega$ as

$$Y(\hbar\omega) \propto \left(\hbar\omega - \Phi_{Bn}^{IPEYS}\right)^2 / \hbar\omega \,. \tag{8.9}$$

Patches only cover a small portion of the metal–semiconductor interface, so the threshold energy Φ_{Bn}^{IPEYS} will equal the barrier height Φ_{Bn}^{hom} of the laterally homogeneous part of the contact minus the zero-bias image-force lowering $\delta\Phi_{if}^0$.

In Fig. 8.6, experimental $[Y(\hbar\omega) \cdot \hbar\omega]^{1/2}$ data for a Pt/p-Si(001) diode [8.20] are plotted versus the energy of the exciting photons. The dashed line is the linear least-squares fit to the data. The deviation of the experimental $[Y(\hbar\omega) \cdot \hbar\omega]^{1/2}$ data towards larger values slightly below and above the threshold is caused by the shape of the Fermi–Dirac distribution function at finite temperatures and by the existence of patches with barrier heights smaller and larger than Φ_{Bn}^{hom}.

8.1.2 Band Offsets of Semiconductor Heterostructures

Semiconductors generally grow layer-by-layer, at least initially. Hence, core-level photoemission spectroscopy (PES) is a very reliable tool and the one most widely used to determine the band-structure lineup at semiconductor heterostructures. The valence-band offset may be obtained from the energy positions of core-level lines in X-ray photoelectron spectra recorded with bulk samples of the semiconductors in contact and with the interface itself [8.22]. Since the escape depths of the photoelectrons are on the order of just 2 nm, one of the two semiconductors must be sufficiently thin. This condition is easily met when heterostructures are grown by molecular beam epitaxy (MBE) and PE spectra are recorded during growth interrupts.

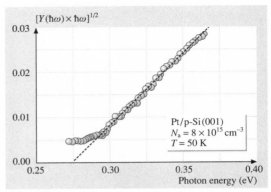

Fig. 8.6 Spectral dependence of the internal photoemission yield $\sqrt{Y(\hbar\omega) \cdot \hbar\omega}$ of a Pt/p-Si(001) diode versus the photon energy of the exciting light. The *dashed line* is the linear least-squares fit to the data for photon energies larger than 0.3 eV. Data from Turan et al. [8.20]

The valence-band discontinuity is then given by (see Fig. 8.7)

$$\Delta W_v = W_{vir} - W_{vil} = W_i(n_r l_r) - W_i(n_l l_l) \\ + [W_{vbr} - W_b(n_r l_r)] - [W_{vbl} - W_b(n_l l_l)] \,, \tag{8.10}$$

where $n_r l_r$ and $n_l l_l$ denote the core levels of the semiconductors on the right (r) and the left (l) side of the interface, respectively. The subscripts i and b characterize interface and bulk properties, respectively. The

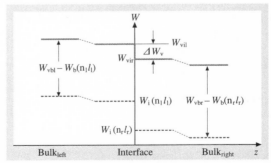

Fig. 8.7 Schematic energy band diagram at semiconductor heterostructures. W_{vb} and W_{vi} are the valence-band maxima and $W_b(nl)$ and $W_i(nl)$ are the core levels in the bulk and at the interface, respectively. The subscripts l and r denote the semiconductors on the right and the on the left side of the interface. ΔW_v is the valence-band offset. The *thin dashed lines* account for possible band-bending from space-charge layers

energy difference $W_i(n_r l_r) - W_i(n_l l_l)$ between the core levels of the two semiconductors at the interface is determined from energy distribution curves of photoelectrons recorded during MBE growth of the heterostructure. The energy positions $W_{vbr} - W_b(n_r l_r)$ and $W_{vbl} - W_b(n_l l_l)$ of the core levels relative to the valence-band maxima in the bulk of the two semiconductors are evaluated separately.

Another widely used technique for determining band offsets in heterostructures is internal photoemission yield spectroscopy. The procedure for evaluating the IPEYS signals is the same as described in Sect. 8.1.1.

8.2 IFIGS-and-Electronegativity Theory

Because of the quantum-mechanical tunneling effect, the wavefunctions of bulk electrons decay exponentially into vacuum at surfaces or, more generally speaking, at solid–vacuum interfaces. A similar behavior occurs at interfaces between two solids [8.6, 7]. In energy regions of Schottky contacts and semiconductor heterostructures where occupied band states overlap a band gap, the wavefunctions of these electrons will tail across the interface. The only difference to solid–vacuum interfaces is that the wavefunction tails oscillate at solid–solid interfaces. Figure 8.8 schematically explains the tailing effects at surfaces and semiconductor interfaces. For the band-structure lineup at semiconductor interfaces, only the tailing states within the gap between the top valence and the lowest conduction band are of any real importance since the energy position of the Fermi level determines their charging state. These wavefunction tails or interface-induced gap states (IFIGS) derive from the continuum of the virtual gap states (ViGS) of the complex semiconductor band structure. Hence, the IFIGS are an intrinsic property of the semiconductor.

The IFIGS are made up of valence-band and conduction-band states of the semiconductor. Their net charge depends on the energy position of the Fermi level relative to their branch point, where their character changes from predominantly donor- or valence band-like to mostly acceptor- or conduction band-like. The band-structure lineup at semiconductor interfaces is thus described by a zero-charge-transfer term and an electric dipole contribution.

In a more chemical approach, the charge transfer at semiconductor interfaces may be related to the partly ionic character of the covalent bonds at interfaces. *Pauling* [8.8] described the ionicity of single bonds in diatomic molecules by the difference between the electronegativities of the atoms involved. The binding energies of core-level electrons are known to depend on the chemical environment of the atoms or, in other words, on the ionicity of their chemical bonds. Figure 8.9 displays experimentally observed chemical shifts for Si(2p) and Ge(3d) core levels induced by metal adatoms

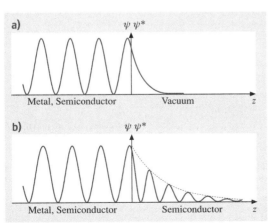

Fig. 8.8a,b Wavefunctions at clean surfaces (**a**) and at metal–semiconductor and semiconductor–semiconductor interfaces (**b**) (schematically)

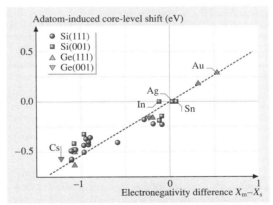

Fig. 8.9 Chemical shifts of Si(2p) and Ge(3d) core levels induced by metal adatoms on silicon and germanium surfaces, respectively, as a function of the difference $X_m - X_s$ in the metal and the semiconductor electronegativities in Pauling units. After [8.14]

on silicon and germanium surfaces as a function of the difference $X_m - X_s$ between the Pauling atomic electronegativity of the metal and that of the semiconductor atoms. The covalent bonds between metal and substrate atoms still persist at metal–semiconductor interfaces, as ab-initio calculations [8.23] have demonstrated for the example of Al/GaAs(110) contacts. The pronounced linear correlation of the data displayed in Fig. 8.9 thus justifies the application of *Pauling's* electronegativity concept to semiconductor interfaces.

The combination of the physical IFIGS and the chemical electronegativity concept yields the barrier heights of ideal p-type Schottky contacts and the valence-band offsets of ideal semiconductor heterostructures as

$$\Phi_{Bp} = \Phi_{bp}^p - S_X(X_m - X_s) \quad (8.11)$$

and

$$\Delta W_v = \Phi_{bpr}^p - \Phi_{bpl}^p + D_X(X_{sr} - X_{sl}) , \quad (8.12)$$

respectively, where $\Phi_{bp}^p = W_{bp} - W_v(\Gamma)$ is the energy distance from the valence-band maximum to the branch point of the IFIGS or the p-type branch-point energy. It has the physical meaning of a zero-charge-transfer barrier height. The slope parameters S_X and D_X are explained at the end of this section.

The IFIGS derive from the virtual gap states of the complex band structure of the semiconductor. Their branch point is an average property of the semiconductor. *Tersoff* [8.24, 27] calculated the branch-point energies Φ_{bp}^p of Si, Ge, and 13 of the III–V and II–VI compound semiconductors. He used a linearized augmented plane-wave method and the local density approximation. Such extensive computations may be avoided. *Mönch* [8.28] applied *Baldereschi's* concept [8.29] of mean-value k-points to calculate the branch-point energies of zincblende-structure compound semiconductors. He first demonstrated that the quasi-particle band gaps of diamond, silicon, germanium, $3C$-SiC, GaAs and CdS at the mean-value k-point equal their average or dielectric band gaps [8.30]

$$W_{dg} = \hbar\omega_p/\sqrt{\varepsilon_\infty - 1} , \quad (8.13)$$

where $\hbar\omega_p$ is the plasmon energy of the bulk valence electrons. *Mönch* then used *Tersoff's* Φ_{bp}^p values, calculated the energy dispersion $W_v(\Gamma) - W_v(\underline{k}_{mv})$ of the topmost valence band in the empirical tight-binding approximation (ETB), and plotted the resulting branch-point energies $W_{bp} - W_v(\underline{k}_{mv}) = \Phi_{bp}^p + [W_v(\Gamma) - W_v(\underline{k}_{mv})]_{ETB}$ at the mean-value k-point $\underline{k}_{mv}$ versus the widths of the dielectric band gaps W_{dg}. The linear least-squares

Table 8.1 Optical dielectric constants, widths of the dielectric band gap, and branch-point energies of diamond-, zincblende- and chalcopyrite-structure semiconductors and of some insulators

Semiconductor	ε_∞	W_{dg}(eV)	Φ_{bp}^p(eV)
C	5.70	14.40	1.77
Si	11.90	5.04	0.36[a]
Ge	16.20	4.02	0.18[a]
$3C$-SiC	6.38	9.84	1.44
$3C$-AlN	4.84	11.92	2.97
AlP	7.54	6.45	1.13
AlAs	8.16	5.81	0.92
AlSb	10.24	4.51	0.53
$3C$-GaN	5.80	10.80	2.37
GaP	9.11	5.81	0.83
GaAs	10.90	4.97	0.52
GaSb	14.44	3.8	0.16
$3C$-InN	–	6.48	1.51
InP	9.61	5.04	0.86
InAs	12.25	4.20	0.50
InSb	15.68	3.33	0.22
$2H$-ZnO	3.72	12.94	3.04[b]
ZnS	5.14	8.12	2.05
ZnSe	5.70	7.06	1.48
ZnTe	7.28	5.55	1.00
CdS	5.27	7.06	1.93
CdSe	6.10	6.16	1.53
CdTe	7.21	5.11	1.12
CuGaS$_2$	6.15	7.46	1.43
CuInS$_2$	6.3*	7.02	1.47
CuAlSe$_2$	6.3*	6.85	1.25
CuGaSe$_2$	7.3*	6.29	0.93
CuInSe$_2$	9.00	5.34	0.75
CuGaTe$_2$	8.0*	5.39	0.61
CuInTe$_2$	9.20	4.78	0.55
AgGaSe$_2$	6.80	5.96	1.09
AgInSe$_2$	7.20	5.60	1.11
SiO$_2$	2.10		3.99[c]
Si$_3$N$_4$	3.80		1.93[c]
Al$_2$O$_3$	3.13		3.23[c]
ZrO$_2$	4.84		≈ 3.2[c]
HfO$_2$	4.00		2.62[c]

*$\varepsilon_\infty = n^2$, [a] [8.24], [b] [8.25], [c] [8.26]

fit to the data of the zincblende-structure compound semiconductors [8.28]

$$\Phi_{bp}^p = 0.449 \cdot W_{dg} - [W_v(\Gamma) - W_v(\underline{k}_{mv})]_{ETB} , \quad (8.14)$$

indicates that the branch points of these semiconductors lie 5% below the middle of the energy gap at the mean-value k-point. Table 8.1 displays the p-type branch-point energies of the Group IV elemental semiconductors, of SiC, and of III–V and II–VI compound semiconductors, as well as of some insulators.

A simple phenomenological model of Schottky contacts with a continuum of interface states and a constant density of states D_{is} across the semiconductor band gap yields the slope parameter [8.31, 32]

$$S_X = A_X / \left[1 + \left(e_0^2 / \varepsilon_i \varepsilon_0 \right) D_{is} \delta_{is} \right], \tag{8.15}$$

where ε_i is an interface dielectric constant. The parameter A_X depends on the electronegativity scale chosen and amounts to $0.86\,\mathrm{eV}$/Miedema-unit and $1.79\,\mathrm{eV}$/Pauling-unit. For $D_{is} \to 0$, relation (8.15) yields $S_X \to 1$ or, in other words, if no interface-induced gap states were present at the metal–semiconductor interfaces one would obtain the Schottky–Mott rule. The extension δ_{is} of the interface states may be approximated by their charge decay length $1/2q_{is}$. Mönch [8.32] used theoretical D_{gs}^{mi} and q_{gs}^{mi} data for metal-induced gap states (MIGS), as the IFIGS in Schottky contacts are traditionally called, and plotted the $(e_0^2/\varepsilon_0) D_{gs}^{mi}/2q_{gs}^{mi}$ values versus the optical susceptibility $\varepsilon_\infty - 1$. The linear least-squares fit to the data points yielded [8.32]

$$A_X / S_X - 1 = 0.1 \cdot (\varepsilon_\infty - 1)^2, \tag{8.16}$$

where the reasonable assumption $\varepsilon_i \approx 3$ was made.

To a first approximation, the slope parameter D_X of heterostructure band offsets may be equated with the slope parameter S_X of Schottky contacts, since the IFIGS determine the intrinsic electric-dipole contributions to both the valence-band offsets and the barrier heights. Furthermore, the Group IV semiconductors and the elements constituting the III–V and II–VI compound semiconductors are all placed in the center columns of the Periodic Table and their electronegativities thus only differ by up to 10%. Consequently, the electric-dipole term $D_X \cdot (X_{sr} - X_{sl})$ may be neglected [8.9], so that (8.12) reduces to

$$\Delta W_v \cong \Phi_{bpr}^p - \Phi_{bpl}^p \tag{8.17}$$

for practical purposes.

8.3 Comparison of Experiment and Theory

8.3.1 Barrier Heights of Schottky Contacts

Experimental barrier heights of intimate, abrupt, clean and (above all) laterally homogeneous Schottky contacts on n-Si and n-GaAs as well as n-GaN, and the three SiC polytypes $3C$, $6H$ and $4H$ are plotted in Figs. 8.10 and 8.11, respectively, versus the difference in the Miedema electronegativities of the metals and the semiconductors. *Miedema's* electronegativities [8.33, 34] are preferred since they were derived from properties of metal alloys and intermetallic compounds, while *Pauling* [8.8] considered covalent bonds in small molecules. The p- and n-type branch-point energies, $\Phi_{bp}^p = W_{bp} - W_v(\Gamma)$ and $\Phi_{bp}^n = W_c - W_{bp}$, respectively, add up to the fundamental band-gap energy $W_g = W_c - W_v(\Gamma)$. Hence, the barrier heights of n-type Schottky contacts are

$$\Phi_{Bn}^{hom} = \Phi_{bp}^n + S_X (X_m - X_s). \tag{8.18}$$

The electronegativity of a compound is taken as the geometric mean of the electronegativities of its constituent atoms.

First off all, the experimental data plotted in Figs. 8.10 and 8.11 clearly demonstrate that the different experimental techniques, I/V, BEEM, IPEYS and PES, yield barrier heights of laterally homogeneous Schottky contacts which agree within the margins of experimental error.

Second, all experimental data are quantitatively explained by the branch-point energies (8.14) and the slope parameters (8.16) of the IFIGS-and-electronegativity theory. As was already mentioned in Sect. 8.1.1, the stacking fault, which is part of the interfacial Si(111)-$(7 \times 7)^i$ reconstruction, causes an *extrinsic* electric dipole in addition to the *intrinsic* IFIGS electric dipole. The latter one is present irrespective of whether the interface structure is reconstructed or $(1 \times 1)^i$-unreconstructed. The extrinsic stacking fault-induced electric dipole quantitatively explains the experimentally observed barrier height lowering of 76 ± 2 meV.

Third, the IFIGS lines in Figs. 8.11a and 8.11b were drawn using the branch-point energies calculated for *cubic* 3C-GaN and 3C-SiC, respectively, since relation (8.12) was derived for zincblende-structure compounds only. However, the Schottky contacts were prepared on wurtzite-structure 2H-GaN and not just on cubic 3C-SiC but also on its hexagonal polytypes $4H$ and $6H$. The good agreement between the experimen-

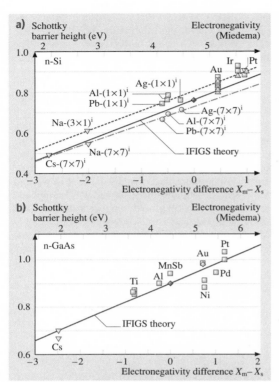

Fig. 8.10a,b Barrier heights of laterally homogeneous n-type silicon (**a**) and GaAs Schottky contacts (**b**) versus the difference in the Miedema electronegativities of the metals and the semiconductors. The ○ and □, ◇, △, and ▽ symbols differentiate the data from I/V, BEEM, IPEYS, and PES measurements, respectively. The *dashed* and the *dash-dotted lines* are the linear least-squares fits to the data from diodes with $(1 \times 1)^i$-unreconstructed and $(7 \times 7)^i$-reconstructed interfaces, respectively. The *solid* IFIGS *lines* are drawn with $S_X = 0.101$ eV/Miedema-unit and $\Phi_{bp}^p = 0.36$ eV for silicon (**a**) and with $S_X = 0.08$ eV/Miedema-unit and $\Phi_{bp}^p = 0.5$ eV for GaAs (**b**). After [8.14]

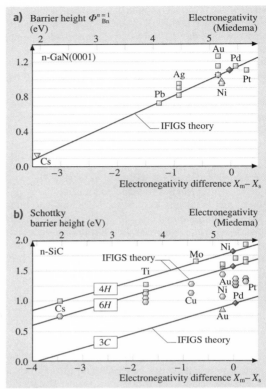

Fig. 8.11a,b Barrier heights of laterally homogeneous n-type GaN(0001) (**a**) and 3*C*-, 4*H*-, and 6*H*-SiC Schottky contacts (**b**) versus the difference in the Miedema electronegativities of the metals and the semiconductors. (**a**): The □ ◇, △, and ▽ symbols differentiate the data from I/V, BEEM, IPEYS, and PES measurements, respectively. The *solid* IFIGS *line* is drawn with $S_X = 0.29$ eV/Miedema-unit and $\Phi_{bp}^p = 2.37$ eV. (**b**): The □, ◇, and ○ *symbols* differentiate data of 4*H*-, 6*H*- and 3*C*-SiC Schottky contacts, respectively. The *solid* IFIGS *lines* are drawn with the band gaps of the polytypes minus $\Phi_{bp}^p = 1.44$ eV of cubic 3*C*-SiC and $S_X = 0.24$ eV/Miedema-unit. After [8.14]

tal data and the IFIGS lines indicates that the p-type branch-point energies are rather insensitive to the specific bulk lattice structure of the semiconductor. This conclusion is further justified by the band-edge discontinuities of the semiconductor heterostructures, which were experimentally observed and are discussed in Sect. 8.3.2, and by the band-edge offsets of 3*C*/2*H* homostructures that were calculated for various semiconductors [8.35–39].

8.3.2 Band Offsets of Semiconductor Heterostructures

In the bulk, and at interfaces of sp^3-coordinated semiconductors, the chemical bonds are covalent. The simplest semiconductor–semiconductor interfaces are *lattice-matched* heterostructures. However, if the bond lengths of the two semiconductors differ then the interface will respond with tetragonal lattice distortions. Such

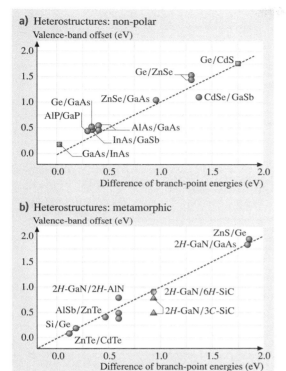

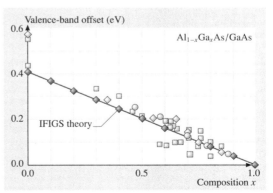

the other hand, causes extrinsic electric dipoles. Their components normal to the interface will add an extrinsic electric-dipole contribution to the valence-band offset. In the following, only nonpolar, lattice-matched isovalent, and metamorphic heterostructures will be discussed.

The valence-band offsets at nonpolar, in other words (110)-oriented, heterostructures of compound semiconductors should equal the difference in the branch-point energies of the two semiconductors in contact provided the intrinsic IFIGS electric-dipole contribution can be neglected, see relation (8.17). Figure 8.12a displays respective experimental results for diamond- and zincblende-structure semiconductors as a function of the difference in the branch-point energies given in Table 8.1. The dashed line clearly demonstrates that the experimental data are execellently explained by the theoretical branch-point energies or, in other words, by the IFIGS theory.

As an example of lattice-matched and isovalent heterostructures, Fig. 8.13 shows valence-band offsets for $Al_{1-x}Ga_xAs/GaAs$ heterostructures as a function of the alloy composition x. The IFIGS branch-point energies of the alloys were calculated assuming virtual $Al_{1-x}Ga_x$ cations [8.28], and were found to vary linearly as a function of composition between the values of AlAs and GaAs. More refined first-principles calculations yielded identical results [8.41, 42]. Figure 8.13 reveals that the theoretical IFIGS valence-band offsets fit the experimental data excellently.

Figure 8.12b displays valence-band offsets for metamorphic heterostructures versus the difference in the branch-point energies of the two semiconductors. The

Fig. 8.12a,b Valence band offsets at nonpolar (110)-oriented (**a**) and metamorphic semiconductor heterostructures (**b**) versus the difference between the p-type branch-point energies of the semiconductors in contact. After [8.14]

pseudomorphic interfaces are under tensile or compressive stress. If the strain energy becomes too large then it is energetically more favorable to release the stress by the formation of misfit dislocations. Such *metamorphic* interfaces are almost relaxed.

In contrast to isovalent heterostructures, the chemical bonds at heterovalent interfaces require special attention, since interfacial donor- and acceptor-type bonds may cause interfacial electric dipoles [8.40]. No such extrinsic electric dipoles will exist normal to nonpolar (110) interfaces. However, polar (001) interfaces behave quite differently. Acceptor bonds or donor bonds normal to the interface would exist at abrupt heterostructures. But, for reasons of charge neutrality, they have to be compensated by a corresponding density of donor bonds and acceptor bonds, respectively. This may be achieved by an intermixing at the interface which, on

Fig. 8.13 Valence band offsets of lattice-matched and isovalent $Al_{1-x}Ga_xAs/GaAs$ heterostructures as a function of alloy composition x. After [8.14]

dashed line indicates that the experimental results are again excellently described by the theoretical IFIGS data. This is true not only for heterostructures between cubic zincblende- and hexagonal wurtzite-structure compounds but also for wurtzite-structure Group III nitrides grown on both cubic 3C- and hexagonal 6H-SiC substrates. These observations suggest the following conclusions. First, all of the heterostructures considered in Fig. 8.12b are only slightly (if at all) strained, although their lattice parameters differ by up to 19.8%. Second, the calculations of the IFIGS branch-point energies assumed zincblende-structure semiconductors. These values, on the other hand, reproduce the experimental valence band offsets irrespective of whether the semiconductors have zincblende, wurtzite or, as in the case of 6H-SiC, another hexagonal-polytype structure. These findings again support the conclusion drawn from the GaN and SiC Schottky barrier heights in the previous section, that the IFIGS branch-point energies are rather insensitive to the specific semiconductor bulk lattice structure.

8.3.3 Band-Structure Lineup at Insulator Interfaces

The continuing miniaturization of complementary metal–oxide–semiconductor (CMOS) devices requires gate insulators where the dielectric constants (κ) are larger than the value of the silicon dioxide conventionally used. At present, the high-κ insulators Al_2O_3, ZrO_2, and HfO_2 are being intensively studied. Insulators may be considered as wide-gap semiconductors. Hence, relations (8.11) and (8.12) also apply to insulator Schottky contacts and heterostructures. Unfortunately, the branch-point energies of these insulators cannot be obtained from relation (8.14) since it is valid for zincblende-structure compound semiconductors only. However, the experimental band offsets reported for SiO_2, Si_3N_4, Al_2O_3, and HfO_2 heterostructures may be plotted as a function of the branch-point energies of the respective semiconductors [8.26]. Figure 8.14a reveals that the valence-band offsets become smaller with increasing branch-point energy of the semiconductors. Moreover, the data points reported for the many different SiO_2 heterostructures studied indicate a linear dependence for the valence-band offsets on the branch-point energy of the semiconductors, which may be written as

$$\Delta W_v = \varphi_{vbo} \left[\Phi_{bp}^p(ins) - \Phi_{bp}^p(sem) \right], \quad (8.19)$$

since the valence band offsets of insulator homostructures will definitely vanish. Such a linear relationship

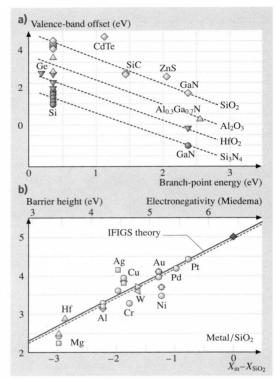

Fig. 8.14 (a) Valence band offsets of SiO_2, Si_3N_4, Al_2O_3 and HfO_2 heterostructures versus the p-type branch-point energies of the respective semiconductors. (b) n-type barrier heights of SiO_2 Schottky contacts versus the difference between electronegativities of the metal and SiO_2. The *dashed line* is the linear least-squares fit to the data points. The *solid* IFIGS *line* is drawn with $\Phi_{bp}^n = 5$ eV ($W_g = 9$ eV) and $S_X = 0.77$ eV/Miedema-unit ($\varepsilon_\infty = 2.1$). After [8.25]

can also be adopted for the Al_2O_3, HfO_2 and Si_3N_4 heterostructures, where less experimental results are available. Hence, the data displayed in Fig. 8.14a provide a means of determining the branch-point energies $\Phi_{bp}^p(ins)$ of SiO_2, Si_3N_4, and the high-κ oxides Al_2O_3 and HfO_2. The dashed lines in Fig. 8.14a are the linear least-squares fits to the respective data points. The experimental slope parameters φ_{vbo} range from 1.16 to 1.23 for HfO_2 and SiO_2 heterostructures, respectively, while relation (8.12) predicts $\varphi_{vbo} = 1$ provided that the electric dipole term $D_X \cdot (X_{sr} - X_{sl})$ vanishes. However, as well-established as this simplifying assumption is for the classical semiconductor heterostructures discussed in Sect. 8.3.2, it has questionable validity for the insulators considered here since they are much more ionic. Hence, the difference $\varphi_{vbo} - 1$ may be attributed to intrinsic electric-dipole layers at these insulator–semiconductor interfaces. The p-type branch-point energies Φ_{bp}^p of the insulators obtained from the linear least-squares fits are displayed in Table 8.1.

The reliability of these branch-point energies may be checked by, for example, analyzing barrier heights of respective insulator Schottky contacts. Such data are only available for SiO$_2$. Figure 8.14b displays the barrier heights of SiO$_2$ Schottky contacts as a function of the electronegativity difference $X_m - X_{SiO_2}$, where the electronegativity of SiO$_2$ is estimated as 6.42 Miedema-units. The linear least-squares fit

$$\Phi_{Bn} = (4.95 \pm 0.19) + (0.77 \pm 0.10) \\ \times (X_m - X_{SiO_2})[eV] \quad (8.20)$$

to the experimental data agrees excellently with the prediction from the IFIGS-and-electronegativity theory.

8.4 Final Remarks

The local density approximation to density functional theory (LDA-DFT) is the most powerful and widely used tool in theoretical studies of the ground-state properties of solids. However, excitation energies such as the width of the energy gaps between the valence and conduction bands of semiconductors cannot be correctly obtained from such calculations. The fundamental band gaps of the elemental semiconductors C, Si and Ge as well as of the III–V and II–VI compounds are notoriously underestimated by 25 to 50%. However, it became possible to compute quasi-particle energies and band gaps of semiconductors from first principles using the so-called GW approximation for the electron self-energy [8.43, 44]. The resulting band gap energies agree to within 0.1 to 0.3 eV with experimental values.

For some specific metal–semiconductor contacts, the band-structure lineup was also studied by state-of-the-art ab-initio LDA-DFT calculations. The resulting LDA-DFT barrier heights were then subjected to a-posteriori corrections which consider quasi-particle effects and, if necessary, spin-orbit interactions and semicore-orbital effects. However, comparison of the theoretical results with experimental data gives an inconsistent picture. The *mean* values of the barrier heights of Al- and Zn/p-ZnSe contacts, which were calculated for different interface configurations using ab-initio LDA-DF theory and a-posteriori spin-orbit and quasi-particle corrections [8.45, 46], agree with the experimental data to within the margins of experimental error. The same conclusion was reached for Al/Al$_{1-x}$Ga$_x$As Schottky contacts [8.47]. However, ab-initio LDA-DFT barrier heights of Al-, Ag-, and Au/p-GaN contacts [8.48,49], as well as of Al- and Ti/3C-SiC(001) interfaces [8.50,51], strongly deviate from the experimental results.

As already mentioned, ab-initio LDF-DFT valenc band offsets of Al$_{1-x}$Ga$_x$As/GaAs heterostructures [8.41, 42] reproduce the experimental results well. The same holds for *mean* values of LDF-DFT valence-band offsets computed for different interface configurations of GaN- and AlN/SiC heterostructures [8.52–56].

The main difficulty which the otherwise extremely successful ab-initio LDF-DFT calculations encounter when describing semiconductor interfaces is not the precise exchange-correlation potential, which may be estimated in the GW approximation, but their remarkable sensitivity to the geometrical and compositional structure right at the interface. This aspect is more serious at metal–semiconductor interfaces than at heterostructures between two sp^3-bonded semiconductors. The more conceptual IFIGS-and-electronegativity theory, on the other hand, quantitatively explains not only the barrier heights of ideal Schottky contacts but also the valence-band offsets of semiconductor heterostructures. Here again, the Schottky contacts are the more important case, since their zero-charge-transfer barrier heights equal the branch-point energies of the semiconductors, while the valence-band offsets are determined by the differences in the branch-point energies of the semiconductors in contact.

References

8.1 W. Schottky: Naturwissenschaften **26**, 843 (1938)
8.2 F. Braun: Pogg. Ann. Physik Chemie **153**, 556 (1874)
8.3 N. F. Mott: Proc. Camb. Philos. Soc. **34**, 568 (1938)
8.4 W. Schottky: Phys. Zeitschr. **41**, 570 (1940)
8.5 J. Bardeen: Phys. Rev. **71**, 717 (1947)
8.6 V. Heine: Phys. Rev. **138**, A1689 (1965)

8.7 C. Tejedor, F. Flores: J. Phys. C **11**, L19 (1978)
8.8 L. N. Pauling: *The Nature of the Chemical Bond* (Cornell Univ. Press, Ithaca 1939)
8.9 W. Mönch: On the Present Understanding of Schottky Contacts. In: *Festkörperprobleme*, Vol. 26, ed. by P. Grosse (Vieweg, Braunschweig 1986) p. 67
8.10 W. Mönch: Phys. Rev. B **37**, 7129 (1988)
8.11 R. Schmitsdorf, T. U. Kampen, W. Mönch: Surf. Sci. **324**, 249 (1995)
8.12 J. L. Freeouf, T. N. Jackson, S. E. Laux, J. M. Woodall: Appl. Phys. Lett. **40**, 634 (1982)
8.13 W. Schottky: Phys. Zeitschr. **15**, 872 (1914)
8.14 W. Mönch: *Electronic Properties of Semiconductor Interfaces* (Springer, Berlin, Heidelberg 2004)
8.15 H. A. Bethe: MIT Radiation Lab. Rep. 43-12 (1942)
8.16 K. Takayanagi, Y. Tanishiro, M. Takahashi, S. Takahashi: Surf. Sci. **164**, 367 (1985)
8.17 H.-J. Im, Y. Ding, J. P. Pelz, W. J. Choyke: Phys. Rev. B **64**, 075310 (2001)
8.18 W. J. Kaiser, L. D. Bell: Phys. Rev. Lett. **60**, 1406 (1988)
8.19 L. D. Bell, W. J. Kaiser: Phys. Rev. Lett. **61**, 2368 (1988)
8.20 R. Turan, B. Aslan, O. Nur, M. Y. A. Yousif, M. Willander: Appl. Phys. A **72**, 587 (2001)
8.21 J. Cohen, J. Vilms, R.J. Archer: Hewlett-Packard R&D Report AFCRL-69-0287 (1969)
8.22 R. W. Grant, J. R. Waldrop, E. A. Kraut: Phys. Rev. Lett. **40**, 656 (1978)
8.23 S. B. Zhang, M. L. Cohen, S. G. Louie: Phys. Rev. B **34**, 768 (1986)
8.24 J. Tersoff: J. Vac. Sci. Technol. B **4**, 1066 (1986)
8.25 W. Mönch: Appl. Phys. Lett. **86**, 162101 (2005)
8.26 W. Mönch: Appl. Phys. Lett. **86**, 122101 (2005)
8.27 J. Tersoff: Phys. Rev. Lett. **52**, 465 (1984)
8.28 W. Mönch: J. Appl. Phys. **80**, 5076 (1996)
8.29 A. Baldereschi: Phys. Rev. B **7**, 5212 (1973)
8.30 D. R. Penn: Phys. Rev. **128**, 2093 (1962)
8.31 A. M. Cowley, S. M. Sze: J. Appl. Phys. **36**, 3212 (1965)
8.32 W. Mönch: Appl. Surf. Sci. **92**, 367 (1996)
8.33 A. R. Miedema, F. R. de Boer, P. F. de Châtel: J. Phys. F **3**, 1558 (1973)
8.34 A. R. Miedema, P. F. de Châtel, F. R. de Boer: Physica **100B**, 1 (1980)
8.35 A. Qteish, V. Heine, R. J. Needs: Phys. Rev. B **45**, 6534 (1992)
8.36 P. Käckell, B. Wenzien, F. Bechstedt: Phys. Rev. B **50**, 10761 (1994)
8.37 S. Ke, K. Zhang, X. Xie: J. Phys. Condens. Mat. **8**, 10209 (1996)
8.38 J. A. Majewski, P. Vogl: MRS Internet J. Nitride Semicond. Res. **3**, 21 (1998)
8.39 S.-H. Wei, S. B. Zhang: Phys. Rev. B **62**, 6944 (2000)
8.40 W. A. Harrison, E. A. Kraut, J. R. Waldrop, R. W. Grant: Phys. Rev. B **18**, 4402 (1978)
8.41 J. S. Nelson, A. F. Wright, C. Y. Fong: Phys. Rev. B **43**, 4908 (1991)
8.42 S. B. Zhang, M. L. Cohen, S. G. Louie, D. Tománek, M. S. Hybertsen: Phys. Rev. B **41**, 10058 (1990)
8.43 M. S. Hybertsen, S. G. Louie: Phys. Rev. B **34**, 5390 (1986)
8.44 R. W. Godby, M. Schlüter, L. J. Sham: Phys. Rev. B **37**, 10159 (1988)
8.45 M. Lazzarino, G. Scarel, S. Rubini, G. Bratina, L. Sorba, A. Franciosi, C. Berthod, N. Binggeli, A. Baldereschi: Phys. Rev. B **57**, R9431 (1998)
8.46 S. Rubini, E. Pellucchi, M. Lazzarino, D. Kumar, A. Franciosi, C. Berthod, N. Binggeli, A. Baldereschi: Phys. Rev. B **63**, 235307 (2001)
8.47 J. Bardi, N. Binggeli, A. Baldereschi: Phys. Rev. B **54**, R11102 (1996)
8.48 S. Picozzi, A. Continenza, G. Satta, S. Massidda, A. J. Freeman: Phys. Rev. B **61**, 16736 (2000)
8.49 S. Picozzi, G. Profeta, A. Continenza, S. Massidda, A. J. Freeman: Phys. Rev. B **65**, 165316 (2002)
8.50 J. Hoekstra, M. Kohyama: Phys. Rev. B **57**, 2334 (1998)
8.51 M. Kohyama, J. Hoekstra: Phys. Rev. B **61**, 2672 (2000)
8.52 M. Städele, A. J. Majewski, P. Vogl: Phys. Rev. B **56**, 6911 (1997)
8.53 J. A. Majewski, M. Städele, P. Vogl: Mater. Res. Soc. Symp. Proc. **449**, 917 (1997)
8.54 N. Binggeli, P. Ferrara, A. Baldereschi: Phys. Rev. B **63**, 245306 (2001)
8.55 B. K. Agrawal, S. Agrawal, R. Srivastava, P. Srivastava: Physica E **11**, 27 (2001)
8.56 M. R. Laridjani, P. Masri, J. A. Majewski: Mater. Res. Soc. Symp. Proc. **639**, G11.34 (2001)

9. Charge Transport in Disordered Materials

This chapter surveys general theoretical concepts developed to qualitatively understand and to quantitatively describe the electrical conduction properties of disordered organic and inorganic materials. In particular, these concepts are applied to describe charge transport in amorphous and microcrystalline semiconductors and in conjugated and molecularly doped polymers. Electrical conduction in such systems is achieved through incoherent transitions of charge carriers between spatially localized states. Basic theoretical ideas developed to describe this type of electrical conduction are considered in detail. Particular attention is given to the way the kinetic coefficients depend on temperature, the concentration of localized states, the strength of the applied electric field, and the charge carrier localization length. Charge transport via delocalized states in disordered systems and the relationships between kinetic coefficients under the nonequilibrium conditions are also briefly reviewed.

9.1 General Remarks on Charge Transport in Disordered Materials 163

9.2 Charge Transport in Disordered Materials via Extended States 167

9.3 Hopping Charge Transport in Disordered Materials via Localized States 169
    9.3.1 Nearest-Neighbor Hopping 170
    9.3.2 Variable-Range Hopping 172
    9.3.3 Description of Charge-Carrier Energy Relaxation and Hopping Conduction in Inorganic Noncrystalline Materials 173
    9.3.4 Description of Charge Carrier Energy Relaxation and Hopping Conduction in Organic Noncrystalline Materials 180

9.4 Concluding Remarks 184

References 185

Many characteristics of charge transport in disordered materials differ markedly from those in perfect crystalline systems. The term "disordered materials" usually refers to noncrystalline solid materials without perfect order in the spatial arrangement of atoms. One should distinguish between disordered materials with ionic conduction and those with electronic conduction. Disordered materials with ionic conduction include various glasses consisting of a "network-formers" such as SiO_2, B_2O_3 and Al_2O_3, and of "network-modifiers" such as Na_2O, K_2O and Li_2O. When an external voltage is applied, ions can drift by hopping over potential barriers in the glass matrix, contributing to the electrical conduction of the material. Several fascinating effects have been observed for this kind of electrical conduction. One is the extremely nonlinear dependence of the conductivity on the concentration of ions in the material. Another beautiful phenomenon is the so-called "mixed alkali effect": mixing two different modifiers in one glass leads to an enormous drop in the conductivity in comparison to that of a single modifier with the same total concentration of ions. A comprehensive description of these effects can be found in the review article of *Bunde* et al. [9.1]. Although these effects sometimes appear puzzling, they can be naturally and rather trivially explained using routine classical percolation theory [9.2]. The description of ionic conduction in glasses is much simplified by the inability of ions to tunnel over large distances in the glass matrix in single transitions. Every transition occurs over a rather small interatomic distance, and it is relatively easy to describe such electrical conductivity theoretically [9.2]. On the other hand, disordered systems with electronic conduction have a much more complicated theoretical description. Transition probabilities of electrons between spatially different regions in the material significantly depend not only on the energy parameters (as in the case of ions), but also on spatial factors such as the tunnelling distance, which can be rather large. The interplay between the energy and spatial factors in the transition probabilities of electrons makes the development of a theory of electronic conduction in disordered systems challenging. Since the description of electronic conduction is less clear than that of ionic conduction, and since disordered electronic materials are widely used for various device applications, in this chapter we concentrate on disordered materials with the electronic type of electrical conduction.

Semiconductor glasses form one class of such materials. This class includes amorphous selenium, a-Se and other chalcogenide glasses, such as a-As_2Se_3. These materials are usually obtained by quenching from the melt. Another broad class of disordered materials, inorganic amorphous semiconductors, includes amorphous silicon a-Si, amorphous germanium a-Ge, and their alloys. These materials are usually prepared as thin films by the deposition of atomic or molecular species. Hydrogenated amorphous silicon, a-Si:H, has attracted much attention from researchers, since incorporation of hydrogen significantly improves conduction, making it favorable for use in amorphous semiconductor devices. Many other disordered materials, such as hydrogenated amorphous carbon (a-C:H) and its alloys, polycrystalline and microcrystalline silicon are similar to a-Si:H in terms of their charge transport properties. Some crystalline materials can also be considered to be disordered systems. This is the case for doped crystals if transport phenomena within them are determined by randomly distributed impurities, and for mixed crystals with disordered arrangements of various types of atoms in the crystalline lattice. In recent years much research has also been devoted to the study of organic disordered materials, such as conjugated and molecularly doped polymers and organic glasses, since these systems has been shown to possess electronic properties similar to those of inorganic disordered materials, while they are easier to manufacture than the latter systems.

There are two reasons for the great interest of researchers in the conducting properties of disordered materials. On the one hand, disordered systems represent a challenging field in a purely academic sense. For many years the theory of how semiconductors perform charge transport was mostly confined to crystalline systems where the constituent atoms are in regular arrays. The discovery of how to make solid amorphous materials and alloys led to an explosion in measurements of the electronic properties of these new materials. However, the concepts often used in textbooks to describe charge carrier transport in crystalline semiconductors are based on an assumption of long-range order, and so they cannot be applied to electronic transport in disordered materials. It was (and still is) a highly challenging task to develop a consistent theory of charge transport in such systems. On the other hand, the explosion in research into charge transport in disordered materials is related to the various current and potential device applications of such systems. These include the application of disordered inorganic and organic materials in photovoltaics (the functioning material in solar cells), in electrophotography, in large-area displays (they are used in thin film transistors), in electrical switching threshold and memory devices, in light-emitting diodes, in linear image sensors, and in optical recording devices. Readers interested in the device applications of disordered materials should be aware that there are numerous monographs on this topic: the literature on this field is very rich. Several books are recommended (see [9.3–12]), as are numerous review articles referred to in these books.

In this chapter we focus on disordered semiconductor materials, ignoring the broad class of disordered metals. In order to describe electronic transport in disordered metals, one can more or less successfully apply extended and modified conventional theoretical concepts developed for electron transport in ordered crystalline materials, such as the Boltzmann kinetic equation. Therefore, we do not describe electronic transport in disordered metals here. We can recommend a comprehensive monograph to interested readers (see [9.13]), in which modern concepts about conduction in disordered metals are presented beautifully.

Several nice monographs on charge transport in disordered semiconductors are also available. Although many of them were published several years ago (some even decades ago), we can recommend them to the interested reader as a source of information on important experimental results. These results have permitted researchers the present level of understanding of transport phenomena in disordered inorganic and organic materials. A comprehensive collection of experimental data for noncrystalline materials from the books specified above would allow one to obtain a picture of the modern state of experimental research in the field.

We will focus in this chapter on the theoretical description of charge transport in disordered materials, introducing some basic concepts developed to describe electrical conduction. Several excellent books already exist in which a theoretical description of charge transport in disordered materials is the main topic. Among others we can recommend the books of *Shklovskii* and *Efros* [9.14], *Zvyagin* [9.15], *Böttger* and *Bryksin* [9.16], and *Overhof* and *Thomas* [9.17]. There appears to be a time gap in which comprehensive monographs on the theoretical description of electrical conduction in disordered materials were not published. During this period some new and rather powerful theoretical concepts were developed. We present these concepts below, along with some more traditional ones.

9.1 General Remarks on Charge Transport in Disordered Materials

Although the literature on transport phenomena in disordered materials is enormously rich, there are still many open questions in this field due to various problems specific to such materials. In contrast to ordered crystalline semiconductors with well-defined electronic energy structures consisting of energy bands and energy gaps, the electronic energy spectra of disordered materials can be treated as quasi-continuous. Instead of bands and gaps, one can distinguish between extended and localized states in disordered materials. In an extended state, the charge carrier wavefunction is spread over the whole volume of a sample, while the wavefunction of a charge carrier is localized in a spatially restricted region in a localized state, and a charge carrier present in such a state cannot spread out in a plane wave as in ordered materials. Actually, localized electron states are known in ordered systems too. Electrons and holes can be spatially localized when they occupy donors or acceptors or some other impurity states or structural defects in ordered crystalline materials. However, the localized states usually appear as δ-like discrete energy levels in the energy spectra of such materials. In disordered semiconductors, on the other hand, energy levels related to spatially localized states usually fill the energy spectrum continuously. The energy that separates the extended states from the localized ones in disordered materials is called the mobility edge. To be precise, we will mostly consider the energy states for electrons in the following. In this case, the states above the mobility edge are extended and the states below the edge are localized. The localized states lie energetically above the extended states for holes. The energy region between the mobility edges for holes and electrons is called the mobility gap. The latter is analogous to the band gap in ordered systems, although the mobility gap contains energy states, namely the spatially localized states. Since the density of states (DOS), defined as the number of states per unit energy per unit volume, usually decreases when the energy moves from the mobility edges toward the center of the mobility gap, the energy regions of localized states in the vicinity of the mobility edges are called band tails. We would like to emphasize that the charge transport properties depend significantly on the energy spectrum in the vicinity and below the mobility edge (in the band tails). Unfortunately this energy spectrum is not known for almost all disordered materials. A whole variety of optical and electrical investigation techniques have proven unable to determine this spectrum. Since the experimental information on this spectrum is rather vague, it is difficult to develop a consistent theoretical description for charge transport ab initio. The absence of reliable information on the energy spectrum and on the structures of the wavefunctions in the vicinity and below the mobility edges can be considered to be the main problem for researchers attempting to quantitatively describe the charge transport properties of disordered materials.

An overview of the energy spectrum in a disordered semiconductor is shown in Fig. 9.1. The energy levels ε_v and ε_c denote the mobility edges for the valence and conduction bands, respectively. Electron states in the mobility gap between these energies are spatially localized. The states below ε_v and above ε_c can be occupied by delocalized holes and electrons. Some peaks in the DOS are shown in the mobility gap, which can be created by some defects with particularly high concentrations. Although there is a consensus between researchers on the general view of the DOS in disordered materials,

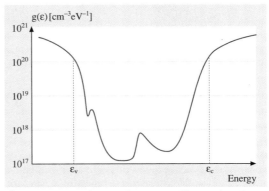

Fig. 9.1 Density of states of a noncrystalline semiconductor (schematic); ε_v and ε_c correspond to mobility edges in the conduction band and the valence band, respectively

the particular structure of the energy spectrum is not known for most disordered systems. From a theoretical point of view, it is enormously difficult to calculate this spectrum.

There are several additional problems that make the study of charge transport in disordered materials more difficult than in ordered crystalline semiconductors. The particular spatial arrangements of atoms and molecules in different samples with the same chemical composition can differ from each other depending on the preparation conditions. Hence, when discussing electrical conduction in disordered materials one often should specify the preparation conditions. Another problem is related to the long-time relaxation processes in disordered systems. Usually these systems are not in thermodynamic equilibrium and the slow relaxation of the atoms toward the equilibrium arrangement can lead to some changes in electrical conduction properties. In some disordered materials a long-time electronic relaxation can affect the charge transport properties too, particularly at low temperatures, when electronic spatial rearrangements can be very slow. At low temperatures, when tunneling electron transitions between localized states dominate electrical conduction, this long-time electron relaxation can significantly affect the charge transport properties.

It is fortunate that, despite these problems, some general transport properties of disordered semiconductors have been established. Particular attention is usually paid to the temperature dependence of the electrical conductivity, since this dependence can indicate the underlying transport mechanism. Over a broad temperature range, the direct current (DC) conductivity in disordered materials takes the form

$$\sigma = \sigma_0 \exp\left[-\left(\frac{\Delta(T)}{k_B T}\right)^\beta\right], \quad (9.1)$$

where the pre-exponential factor σ_0 depends on the underlying system and the power exponent β depends on the material and also sometimes on the temperature range over which the conductivity is studied; $\Delta(T)$ is the activation energy. In many disordered materials, like vitreous and amorphous semiconductors, σ_0 is of the order of 10^2–10^4 Ω^{-1}cm^{-1}. In such materials the power exponent β is close to unity at temperatures close to and higher than the room temperature, while at lower temperatures β can be significantly smaller than unity. In organic disordered materials, values of β that are larger than unity also have been reported. For such systems the value $\beta \approx 2$ is usually considered to be appropriate [9.18].

Another important characteristic of the electrical properties of a disordered material is its alternating current (AC) conductivity measured when an external alternating electric field with some frequency ω is applied. It has been established in numerous experimental studies that the real part of the AC conductivity in most disordered semiconductors depends on the frequency according to the power law

$$\mathrm{Re}\,\sigma(\omega) = C\omega^s, \quad (9.2)$$

where C is constant and the power s is usually smaller than unity. This power law has been observed in numerous materials at different temperatures over a wide frequency range. This frequency dependence differs drastically from that predicted by the standard kinetic theory developed for quasi-free charge carriers in crystalline systems. In the latter case, the real part of the AC conductivity has the frequency dependence

$$\mathrm{Re}\,\sigma(\omega) = \frac{ne^2}{m}\frac{\tau}{1+\omega^2\tau^2}, \quad (9.3)$$

where n is the concentration of charge carriers, e is the elementary charge, m is the effective mass and τ is the momentum relaxation time. Since the band electrons in crystalline semiconductors usually have rather short momentum relaxation times, $\tau \approx 10^{-14}$ s, the contribution of charge carriers in delocalized states to the AC conductivity usually does not depend on frequency at $\omega \ll \tau^{-1}$. Therefore, the observed frequency dependence described by (9.2) should be ascribed to the contribution of charge carriers in localized states.

One of the most powerful tools used to study the concentrations of charge carriers and their mobilities in crystalline semiconductors is the provided by measurements of the Hall constant, R_H. Such measurements also provide direct and reliable information about the sign of the charge carriers in crystalline materials. Unfortunately, this is not the case for disordered materials. Moreover, several anomalies have been established for Hall measurements in the latter systems. For example, the sign of the Hall constant in disordered materials sometimes differs from that of the thermoelectric power, α. This anomaly has not been observed in crystalline materials. The anomaly has been observed in liquid and solid noncrystalline semiconductors. Also, in some materials, like amorphous arsenic, a-As, $R_H > 0$, $\alpha < 0$, while in many other materials other combinations with different signs of R_H and α have been experimentally established.

In order to develop a theoretical picture of the transport properties of any material, the first issues to clarify are the spectrum of the energy states for charge carriers and the spatial structure of such states. Since these two central issues are yet to be answered properly for noncrystalline materials, the theory of charge transport in disordered systems should be considered to be still in its embryonic stage.

The problem of deducing electron properties in a random field is very complicated, and the solutions obtained so far only apply to some very simple models. One of them is the famous Anderson model that illustrates the localization phenomenon caused by random disorder [9.19]. In this model, one considers a regular system of rectangular potential wells with randomly varying depths, as shown schematically in Fig. 9.2. The ground state energies of the wells are assumed to be randomly distributed over the range with a width of W. First, one considers the ordered version of the model, with W equal to zero. According to conventional band theory, a narrow band arises in the ordered system where the energy width depends on the overlap integral I between the electron wavefunctions in the adjusting wells. The eigenstates in such a model are delocalized with wavefunctions of the Bloch type. This is trivial. The problem is to find the solution for a finite degree of disorder ($W \neq 0$). The result from the Anderson model for such a case is described as follows. At some particular value for the ratio $W/(zI)$, where z is the coordination number of the lattice, all electron states of the system are spatially localized. At smaller values of $W/(zI)$ some states in the outer regions of the DOS are localized and other states in the middle of the DOS energy distribution are spatially extended, as shown schematically in Fig. 9.3. This is one of the most famous results in the transport theory of disordered systems. When considering this result, one should note the following points. (i) It was obtained using a single-electron picture without taking into account long-range many-particle interactions. However, in disordered systems with localized electrons such interactions can lead to the localization of charge carriers and they often drastically influence the energy spectrum [9.14]. Therefore the applicability of the single-electron Anderson result to real systems is questionable. (ii) Furthermore, the energy structure of the Anderson model shown in Fig. 9.3 strongly contradicts that observed in real disordered materials. In real systems, the mobility gap is located between the mobility edges, as shown in Fig. 9.1, while in the Anderson model the energy region between the mobility edges is filled with delocalized states. Moreover, in one-dimensional and in some two-dimensional systems, the Anderson model predicts that all states are localized at any amount of disorder. These results are of little help when attempting to interpret the DOS scheme in Fig. 9.1.

A different approach to the localization problem is to try to impose a random potential $V(x)$ onto the band structure obtained in the frame of a traditional band theory. Assuming a classical smoothly varying (in space)

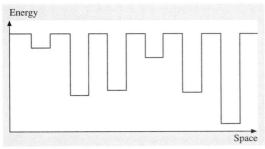

Fig. 9.2 Anderson model of disorder potential

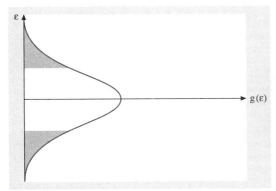

Fig. 9.3 Density of states in the Anderson model. *Hatched regions* in the tails correspond to spatially localized states

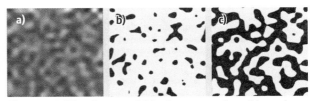

Fig. 9.4a–c Disorder potential landscape experienced by a charge carrier (**a**). Regions with energies below some given energy level E_c are *colored black*. In *frame* (**b**) this level is very low and there is no connected path through the system via *black regions*. In *frame* (**c**) the level E_c corresponds to the classical percolation level

potential $V(x)$ with a Gaussian distribution function

$$F(V) = \frac{1}{\varepsilon_0 \sqrt{2\pi}} \exp\left(-\frac{V^2}{2\varepsilon_0^2}\right), \quad (9.4)$$

one can solve the localization problem using the classical percolation theory illustrated in Fig. 9.4. In Fig. 9.4a, an example of a disorder potential experienced by electrons is shown schematically. In Fig. 9.4b and Fig. 9.4c the regions below a given energy level E_c are colored black. In Fig. 9.4b this level is positioned very low, so that regions with energies below E_c do not provide a connected path through the system. In Fig. 9.4c an infinite percolation cluster consisting only of black regions exists. The E_c that corresponds to the first appearance of such a connected path is called the classical percolation level [9.14]. Mathematically soluving the percolation problem shows that the mobility edge identified with the classical percolation level in the potential $V(x)$ is shifted with respect to the band edge of the ordered system by an amount $\xi \varepsilon_0$, where $\xi \approx 0.96$ towards the center of the bandgap [9.15]. A similar result, though with a different constant ξ, can be obtained via a quantum-mechanical treatment of a short-range potential $V(x)$ of white-noise type [9.20]. As the amplitude ε_0 of the random potential increases the band gap narrows, while the conduction and valence bands become broader. Although this result is provided by both limiting models – by the classical one with a long-range smoothly varying potential $V(x)$ and by the quantum-mechanical one with a short-range white-noise potential $V(x)$ – none of the existing theories can reliably describe the energy spectrum of a disordered material and the properties of the charge carrier wavefunctions in the vicinity of the mobility edges, in other words in the energy range which is most important for charge transport.

The DC conductivity can generally be represented in the form

$$\sigma = e \int \mu(\varepsilon) n(\varepsilon) \, d\varepsilon, \quad (9.5)$$

where e is the elementary charge, $n(\varepsilon) d\varepsilon$ is the concentration of electrons in the energy range between ε and $\varepsilon + d\varepsilon$ and $\mu(\varepsilon)$ is the mobility of these electrons. The integration is carried out over all energies ε. Under equilibrium conditions, the concentration of electrons $n(\varepsilon) d\varepsilon$ is determined by the density of states $g(\varepsilon)$ and the Fermi function $f(\varepsilon)$, which depends on the position of the Fermi energy ε_F (or a quasi-Fermi energy in the case of the stationary excitation of electrons):

$$n(\varepsilon) = g(\varepsilon) f(\varepsilon), \quad (9.6)$$

where

$$f(\varepsilon) = \frac{1}{1 + \exp\left(\frac{\varepsilon - \varepsilon_F}{k_B T}\right)}. \quad (9.7)$$

Here T is the temperature and k_B is the Boltzmann constant.

The Fermi level in almost all known disordered semiconductors under real conditions is situated in the mobility gap – in the energy range which corresponds to spatially localized electron states. The charge carrier mobility $\mu(\varepsilon)$ in the localized states below the mobility edge is much less than that in the extended states above the mobility edge. Therefore, at high temperatures, when a considerable fraction of electrons can be found in the delocalized states above the mobility edge, these states dominate the electrical conductivity of the system. The corresponding transport mechanism under such conditions is similar to that in ordered crystalline semiconductors. Electrons in the states within the energy range of the width, of the order $k_B T$ above the mobility edge, dominate the conductivity. In such a case the conductivity can be estimated as

$$\sigma \approx e \mu_c n(\varepsilon_c) k_B T, \quad (9.8)$$

where μ_c is the electron mobility in the states above the mobility edge ε_c, and $n(\varepsilon_c) k_B T$ is their concentration. This equation is valid under the assumption that the typical energy scale of the DOS function $g(\varepsilon)$ above the mobility edge is larger than $k_B T$. The position of the Fermi level in disordered materials usually depends on temperature only slightly. Combining (9.6)–(9.8), one obtains the temperature dependence of the DC conductivity in the form

$$\sigma = \sigma_0 \exp\left(-\frac{\Delta}{k_B T}\right), \quad (9.9)$$

described by (9.1) with $\beta = 1$ and constant activation energy, which is observed in most disordered semiconductors at high temperatures.

In order to obtain the numerical value of the conductivity in this high-temperature regime, one needs to know the density of states in the vicinity of the mobility edge $g(\varepsilon_c)$, and also the magnitude of the electron mobility μ_c in the delocalized states above ε_c. While the magnitude of $g(\varepsilon_c)$ is usually believed to be close to the DOS value in the vicinity of the band edge in crystalline semiconductors, there is no consensus among researchers on the magnitude of μ_c. In amorphous semiconductors μ_c is usually estimated to be in the range of $1 \text{ cm}^2/\text{V s}$ to $10 \text{ cm}^2/\text{V s}$. Unfortunately, there are no reliable theoretical calculations of this quantity for most disordered materials. The only exception is provided by so-called mixed crystals, which are also sometimes called crystalline solid solutions. In the next section we describe the theoretical method which allows one to estimate μ_c in such systems. This method can be extended to other disordered materials, provided the statistical properties of the disorder potential, essential for electron scattering, are known.

9.2 Charge Transport in Disordered Materials via Extended States

The states with energies below ε_v and above ε_c in disordered materials are believed to possess similar properties to those of extended states in crystals. Various experimental data suggest that these states in disordered materials are delocalized states. However, traditional band theory is largely dependent upon the system having translational symmetry. It is the periodic atomic structure of crystals that allows one to describe electrons and holes within such a theory as quasi-particles that exhibit behavior similar to that of free particles in vacuum, albeit with a renormalized mass (the so-called "effective mass"). The energy states of such quasi-particles can be described by their momentum values. The wavefunctions of electrons in these states (the so-called Bloch functions) are delocalized. This means that the probability of finding an electron with a given momentum is equal at corresponding points of all elementary cells of the crystal, independent on the distance between the cells.

Strictly speaking, the traditional band theory fails in the absence of translational symmetry – for disordered systems. Nevertheless, one still assumes that the charge carriers present in delocalized states in disordered materials can be approximately described by wavefunctions with a spatially homogeneous probability of finding a charge carrier with a given quasi-momentum. As for crystals, one starts from the quasi-free particle picture and considers the scattering effects in a perturbation approach following the Boltzmann kinetic description. This description is valid if the de Broglie wavelength of the charge carrier $\lambda = \hbar/p$ is much less than the mean free path $l = v\tau$, where τ is the momentum relaxation time and p and v are the characteristic values of the momentum and velocity, respectively. This validity condition for the description based on the kinetic Boltzmann equation can also be expressed as $\hbar/\tau \ll \varepsilon$, where ε is the characteristic kinetic energy of the charge carriers, which is equal to $k_B T$ for a nondegenerate electron gas and to the Fermi energy in the degenerate case. While this description seems valid for delocalized states far from the mobility edges, it fails for energy states in the vicinity of the mobility edges. So far, there has been no consensus between the theorists on how to describe charge carrier transport in the latter states. Moreover, it is not clear whether the energy at which the carrier mobility drops coincides with the mobility edge or whether it is located above the edge in the extended states. Numerous discussions of this question, mostly based on the scaling theory of localization, can be found in special review papers. For the rest of this section, we skip this rather complicated subject and instead we focus on the description of charge carrier transport in a semiconductor with a short-range random disorder potential of white-noise type. This seems to be the only disordered system where a reliable theory exists for charge carrier mobility via extended states above the mobility edge. Semiconductor solid solutions provide an example of a system with this kind of random disorder [9.20–25].

Semiconductor solid solutions $A_x B_{1-x}$ (mixed crystals) are crystalline semiconductors in which the sites of the crystalline sublattice can be occupied by atoms of two different types, A and B. Each site can be occupied by either an A or a B atom with some given probability x between zero and unity. The value x is often called the composition of the material. Due to the random spatial distributions of the A and B atoms, local statistical fluctuations in the composition inside the sample are unavoidable, meaning that mixed crystals are disordered systems. Since the position of the band edge depends on the composition x, these fluctuations in local x values lead to the disorder potential for electrons and holes within the crystal. To be precise, we will consider the influence of the random potential on a conduction band electron. Let $E_c(x)$ be the conduction band minimum for a crystal with composition x. In Fig. 9.5 a possible schematic dependence $E_c(x)$ is shown. If the average composition for the whole sample is x_0, the local positions of the band edge $E_c(x)$ fluctuate around the average value $E_c(x_0)$ according to the fluctuations of the composition x around x_0. For small deviations in composition Δx from the average value, one can use the linear relation

$$E_c(x_0 + \Delta x) = E_c(x_0) + \alpha \Delta x , \qquad (9.10)$$

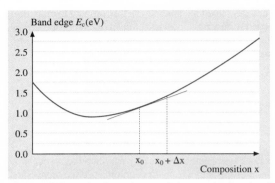

Fig. 9.5 Schematic dependence of the conduction band edge ε_c on composition x in a mixed crystal A_xB_{1-x}

where

$$\alpha = \left.\frac{dE_c(x)}{dx}\right|_{x=x_0} . \tag{9.11}$$

If the deviation of the concentration of A atoms from its mean value in some region of a sample is $\xi(r)$ and the total concentration of (sub)lattice sites is N, the deviation of the composition in this region is $\Delta x = \xi(r)/N$, and the potential energy of an electron at the bottom of the conduction band is

$$V(r) = \alpha \frac{\xi(r)}{N} . \tag{9.12}$$

Although one calls the disorder in such systems a "short-range" disorder, it should be noted that the consideration is valid only for fluctuations that are much larger than the lattice constant of the material. The term "short-range" is due to the assumption that the statistical properties of the disorder are absolutely uncorrelated. This means that potential amplitudes in the adjusting spatial points are completely uncorrelated to each other. Indeed, it is usually assumed that the correlation function of the disorder in mixed crystals can be approximated by a white-noise correlation function of the form

$$\langle \xi(r)\xi(r')\rangle = x(1-x)N\delta(r-r') . \tag{9.13}$$

The random potential caused by such compositional fluctuations is then described by the correlation function [9.20]

$$\langle V(r)V(r')\rangle = \gamma \delta(r-r') \tag{9.14}$$

with

$$\gamma = \frac{\alpha^2}{N}x(1-x) . \tag{9.15}$$

Charge carriers in mixed crystals are scattered by compositional fluctuations. As is usual in kinetic descriptions of free electrons, the fluctuations on the spatial scale of the order of the electron wavelength are most efficient. Following *Shlimak* et al. [9.23], consider an isotropic quadratic energy spectrum

$$\varepsilon_p = \frac{p^2}{2m} , \tag{9.16}$$

where p and m are the quasi-momentum and the effective mass of an electron, respectively. The scattering rate for such an electron is

$$\nu_p = \frac{2\pi}{\hbar} \sum_q \langle |V_q|^2\rangle (1-\cos\vartheta_q)\, \delta\left(\varepsilon_p - \varepsilon_{p-q}\right) , \tag{9.17}$$

where ϑ_q is the scattering angle and

$$\langle |V_q|^2\rangle = \frac{1}{\Omega}\int d^3r \exp(iqr)\langle V(r)V(0)\rangle . \tag{9.18}$$

The quantity Ω in this formula is the normalization volume. Using the correlation function (9.14), one obtains the relation

$$\langle |V_q|^2\rangle = \frac{\alpha^2 x(1-x)}{\Omega N} , \tag{9.19}$$

which shows that the scattering by compositional fluctuations is equivalent to that by a short-range potential [9.23]. Substituting (9.19) into (9.17) one obtains the following expression for the scattering rate [9.20]

$$\nu_p = \frac{\alpha^2 x(1-x)mp}{\pi \hbar^4 N} . \tag{9.20}$$

This formula leads to an electron mobility of the following form in the framework of the standard Drude approach [9.20, 23]

$$\mu_C = \frac{\pi^{3/2}}{2\sqrt{2}}\frac{e\hbar^4 N}{\alpha^2 x(1-x)m^{5/2}(k_B T)^{1/2}} . \tag{9.21}$$

Very similar formulae can be found in many recent publications (see for example *Fahy* and *O'Reily* [9.26]). It has also been modified and applied to two-dimensional systems [9.27] and to disordered diluted magnetic semiconductors [9.28].

It would not be difficult to apply this theoretical description to other disordered systems, provided the correlation function of the disorder potential takes the form of (9.14) with known amplitude γ. However, it is worth emphasizing that the short-range disorder of white-noise type considered here is a rather simple model that cannot be applied to most disordered materials. Therefore, we can conclude that the problem of theoretically describing charge carrier mobility via delocalized states in disordered materials is still waiting to be solved.

In the following section we present the general concepts developed to describe electrical conduction in disordered solids at temperatures where tunneling transitions of electrons between localized states significantly contribute to charge transport.

9.3 Hopping Charge Transport in Disordered Materials via Localized States

Electron transport via delocalized states above the mobility edge dominates the electrical conduction of disordered materials only at temperatures high enough to cause a significant fraction of the charge carriers fill these states. As the temperature decreases, the concentration of the electrons described by (9.9) decreases exponentially and so their contribution to electrical conductivity diminishes. Under these circumstances, tunneling transitions of electrons between localized states in the band tails dominate the charge transport in disordered semiconductors. This transport regime is called hopping conduction, since the incoherent sequence of tunneling transitions of charge carriers resembles a series of their hops between randomly distributed sites. Each site in this picture provides a spatially localized electron state with some energy ε. In the following we will assume that the localized states for electrons (concentration N_0) are randomly distributed in space and their energy distribution is described by the DOS function $g(\varepsilon)$:

$$g(\varepsilon) = \frac{N_0}{\varepsilon_0} G\left(\frac{\varepsilon}{\varepsilon_0}\right), \quad (9.22)$$

where ε_0 is the energy scale of the DOS distribution.

The tunneling transition probability of an electron from a localized state i to a localized state j that is lower in energy depends on the spatial separation r_{ij} between the sites i and j as

$$\nu_{ij}(r) = \nu_0 \exp\left(-\frac{2r_{ij}}{\alpha}\right), \quad (9.23)$$

where α is the localization length, which we assume to be equal for sites i and j. This length determines the exponential decay of the electron wavefunction in the localized states, as shown in Fig. 9.6. The pre-exponential factor ν_0 in (9.23) depends on the electron interaction mechanism that causes the transition. Usually it is assumed that electron transitions contributing to charge transport in disordered materials are caused by interactions of electrons with phonons. Often the coefficient ν_0 is simply assumed to be of the order of the phonon

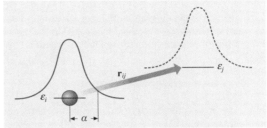

Fig. 9.6 Hopping transition between two localized states i and j with energies of ε_i and ε_j, respectively. The *solid* and *dashed lines* depict the carrier wavefunctions at sites i and j, respectively; α is the localization radius

frequency ($\approx 10^{13}\,\text{s}^{-1}$), although a more rigorous approach is in fact necessary to determine ν_0. This should take into account the particular structure of the electron localized states and also the details of the interaction mechanism [9.29, 30].

When an electron transits from a localized state i to a localized state j that is higher in energy, the transition rate depends on the energy difference between the states. This difference is compensated for by absorbing a phonon with the corresponding energy [9.31]:

$$\nu_{ij}(r, \varepsilon_i, \varepsilon_j) = \nu_0 \exp\left(-\frac{2r_{ij}}{a}\right)$$
$$\times \exp\left(-\frac{\varepsilon_j - \varepsilon_i + |\varepsilon_j - \varepsilon_i|}{2k_\text{B}T}\right).$$
(9.24)

Equations (9.23) and (9.24) were written for the case in which the electron occupies site i whereas site j is empty. If the system is in thermal equilibrium, the occupation probabilities of sites with different energies are determined by Fermi statistics. This effect can be taken into account by modifying (9.24) and adding terms that account for the relative energy positions of sites i and

j with respect to the Fermi energy ε_F. Taking into account these occupation probabilities, one can write the transition rate between sites i and j in the form [9.31]

$$v_{ij} = v_0 \exp\left(-\frac{2r_{ij}}{a}\right)$$
$$\times \exp\left(-\frac{|\varepsilon_i - \varepsilon_F| + |\varepsilon_j - \varepsilon_F| + |\varepsilon_j - \varepsilon_i|}{2k_B T}\right). \quad (9.25)$$

Using these formulae, the theoretical description of hopping conduction is easily formulated. One has to calculate the conductivity provided by transition events (the rates of which are described by (9.25)) in the manifold of localized states (where the DOS is described by (9.22)).

9.3.1 Nearest-Neighbor Hopping

Before presenting the correct solution to the hopping problem we would like to emphasize the following. The style of the theory for electron transport in disordered materials via localized states significantly differs from that used for theories of electron transport in ordered crystalline materials. While the description is usually based on various averaging procedures in crystalline systems, in disordered systems these averaging procedures can lead to extremely erroneous results. We believe that it is instructive to analyze some of these approaches in order to illustrate the differences between the descriptions of charge transport in ordered and disordered materials. To treat the scattering rates of electrons in ordered crystalline materials, one usually proceeds by averaging the scattering rates over the ensemble of scattering events. A similar procedure is often attempted for disordered systems too, although various textbooks (see, for instance, *Shklovskii* and *Efros* [9.14]) illustrate how erroneous such an approach can be in the case of disordered materials.

Let us consider the simplest example of hopping processes, namely the hopping of an electron through a system of isoenergetic sites randomly distributed in space with some concentration N_0. It will be always assumed in this chapter that electron states are strongly localized and the strong inequality $N_0 \alpha^3 \ll 1$ is fulfilled. In such a case the electrons prefer to hop between the spatially nearest sites and therefore this transport regime is often called nearest-neighbor hopping (NNH). This type of hopping transport takes place in many real systems at temperatures where the thermal energy $k_B T$ is larger than the energy scale of the DOS. In such situations the energy-dependent terms in (9.24) and (9.25) do not play any significant role and the hopping rates are determined solely by the spatial terms. The rate of transition of an electron between two sites i and j is described in this case by (9.23). The average transition rate is usually obtained by weighting this expression with the probability of finding the nearest neighbor at some particular distance r_{ij}, and by integrating over all possible distances:

$$\langle v \rangle = \int_0^\infty dr\, v_0$$
$$\times \exp\left(-\frac{2r}{\alpha}\right) 4\pi r^2 N_0 \exp\left(-\frac{4\pi}{3}r^3 N_0\right)$$
$$\approx \pi v_0 N_0 \alpha^3 . \quad (9.26)$$

Assuming that this average hopping rate describes the mobility, diffusivity and conductivity of charge carriers, one apparently comes to the conclusion that these quantities are linearly proportional to the density of localized states N_0. However, experiments evidence an exponential dependence of the transport coefficients on N_0.

Let us look therefore at the correct solution to the problem. This solution is provided in the case considered here, $N_0 \alpha^3 \ll 1$, by percolation theory (see, for instance, *Shklovskii* and *Efros* [9.14]). In order to find the transport path, one connects each pair of sites if the relative separation between the sites is smaller than some given distance R, and checks whether there is a continuous path through the system via such sites. If such a path is absent, the magnitude of R is increased and the procedure is repeated. At some particular value $R = R_c$, a continuous path through the infinite system via sites with relative separations $R < R_c$ arises. Various mathematical considerations give the following relation for R_c [9.14]:

$$\frac{4\pi}{3} N_0 R_c^3 = B_c , \quad (9.27)$$

where $B_c = 2.7 \pm 0.1$ is the average number of neighboring sites available within a distance of less than R_c. The corresponding value of R_c should be inserted into (9.23) in order to determine kinetic coefficients such as the mobility, diffusivity and conductivity. The idea behind this procedure is as follows. Due to the exponential dependence of the transition rates on the distances between the sites, the rates for electron transitions over distances $r < R_c$ are much larger than those over distances R_c. Such fast transitions do not play any significant role as a limiting factor in electron transport and so they can

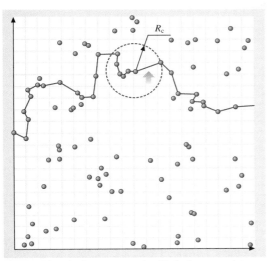

Fig. 9.7 A typical transport path with the lowest resistance. *Circles* depict localized states. The *arrow* points out the most "difficult" transition, with length R_c

be neglected in calculations of the resistivity of the system. Transitions over distances R_c are the slowest among those that are necessary for DC transport and hence such transitions determine the conductivity. The structure of the percolation cluster responsible for charge transport is shown schematically in Fig. 9.7. The transport path consists of quasi-one-dimensional segments, each containing a "difficult" transition over the distance $\approx R_c$. Using (9.23) and (9.27), one obtains the dependence of the conductivity on the concentration of localization sites in the form

$$\sigma = \sigma_0 \exp\left(-\frac{\gamma}{\alpha N_0^{1/3}}\right), \quad (9.28)$$

where σ_0 is the concentration-independent pre-exponential factor and $\gamma = 1.73 \pm 0.03$. Such arguments do not allow one to determine the exponent in the kinetic coefficients with an accuracy better than a number of the order of unity [9.14]. One should note that the quantity in the exponent in (9.28) is much larger than unity for a system with strongly localized states when the inequality $N_0 \alpha^3 \ll 1$ is valid. This inequality justifies the above derivation. The dependence described by (9.28) has been confirmed in numerous experimental studies of the hopping conductivity via randomly placed impurity atoms in doped crystalline semiconductors [9.14]. The drastic difference between this correct result and the erroneous one based on (9.26) is apparent. Unfortunately, the belief of many researchers in the validity of the procedure based on the averaging of hopping rates is so strong that the agreement between (9.28) and experimental data is often called occasional. We would like to emphasize once more that the ensemble averaging of hopping rates leads to erroneous results. The magnitude of the average rate in (9.26) is dominated by rare configurations of very close pairs of sites with separations of the order of the localization length α. Of course, such pairs allow very fast electron transitions, but electrons cannot move over considerable distances using only these close pairs. Therefore the magnitude of the average transition rate is irrelevant for calculations of the hopping conductivity. The correct concentration dependence of the conductivity is given by (9.28). This result was obtained under the assumption that only spatial factors determine transition rates of electrons via localized states. This regime is valid at reasonably high temperatures.

If the temperature is not as high and the thermal energy $k_B T$ is smaller than the energy spread of the localized states involved in the charge transport process, the problem of calculating the hopping conductivity becomes much more complicated. In this case, the interplay between the energy-dependent and the distance-dependent terms in (9.24) and (9.25) determines the conductivity. The lower the temperature, the more important the energy-dependent terms in the expressions for transition probabilities of electrons in (9.24) and (9.25) become. If the spatially nearest-neighboring sites have very different energies, as shown in Fig. 9.8, the probability of an upward electron transition between these sites can be so low that it would be more favorable for this electron to hop to a more distant site at a closer energy. Hence the typical lengths of

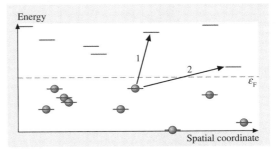

Fig. 9.8 Two alternative hopping transitions between occupied states (*filled circles*) and unoccupied states. The *dashed line* depicts the position of the Fermi level. Transitions (1) and (2) correspond to nearest-neighbor hopping and variable-range hopping regimes, respectively

electron transitions increase with decreasing temperature. This transport regime was termed "variable-range hopping". Next we describe several useful concepts developed to describe this transport regime.

9.3.2 Variable-Range Hopping

The concept of variable-range hopping (VRH) was put forward by Mott (see *Mott* and *Davis* [9.32]) who considered electron transport via a system of randomly distributed localized states at low temperatures. We start by presenting Mott's arguments. At low temperatures, electron transitions between states with energies in the vicinity of the Fermi level are most efficient for transport since filled and empty states with close energies can only be found in this energy range. Consider the hopping conductivity resulting from energy levels within a narrow energy strip with width $2\Delta\varepsilon$ symmetric to the Fermi level shown in Fig. 9.9. The energy width of the strip useful for electron transport can be determined from the relation

$$g(\varepsilon_\text{F}) \cdot \Delta\varepsilon \cdot r^3(\Delta\varepsilon) \approx 1 \, . \tag{9.29}$$

This criterion is similar to that used in (9.27), although we do not care about numerical coefficients here. Here we have to consider the percolation problem in four-dimensional space since in addition to the spatial terms considered in Sect. 9.3.1 we now have to consider the energy too. The corresponding percolation problem for the transition rates described by (9.25) has not yet been solved precisely. In (9.29) it is assumed that the energy width $2\Delta\varepsilon$ is rather small and that the DOS function $g(\varepsilon)$ is almost constant in the range $\varepsilon_\text{F} \pm \Delta\varepsilon$. One can obtain

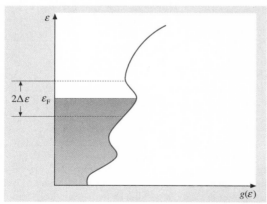

Fig. 9.9 Effective region in the vicinity of the Fermi level where charge transport takes place at low temperatures

the typical hopping distance from (9.29) as a function of the energy width $\Delta\varepsilon$ in the form

$$r(\Delta\varepsilon) \approx [g(\varepsilon_\text{F})\Delta\varepsilon]^{-1/3} \, , \tag{9.30}$$

and substitute it into (9.24) in order to express the typical hopping rate

$$\nu = \nu_0 \exp\left(-\frac{2[g(\varepsilon_\text{F})\Delta\varepsilon]^{-1/3}}{\alpha} - \frac{\Delta\varepsilon}{k_\text{B}T}\right) . \tag{9.31}$$

The optimal energy width $\Delta\varepsilon$ that provides the maximum hopping rate can be determined from the condition $\text{d}\nu/\text{d}\Delta\varepsilon = 0$. The result reads

$$\Delta\varepsilon = \left(\frac{2k_\text{B}T}{3g^{1/3}(\varepsilon_\text{F})}\right)^{3/4} . \tag{9.32}$$

After substitution of (9.32) into (9.31) one obtains Mott's famous formula for temperature-dependent conductivity in the VRH regime

$$\sigma = \sigma_0 \exp\left[-\left(\frac{T_0}{T}\right)^{1/4}\right] , \tag{9.33}$$

where T_0 is the characteristic temperature:

$$T_0 = \frac{\beta}{k_\text{B} g(\varepsilon_\text{F}) \alpha^3} . \tag{9.34}$$

Mott gave only a semi-quantitative derivation of (9.33), from which the exact value of the numerical constant β cannot be determined. Various theoretical studies in 3-D systems suggest values for β in the range $\beta = 10.0$ to $\beta = 37.8$. According to our computer simulations, the appropriate value is close to $\beta = 17.6$.

Mott's law implies that the density of states in the vicinity of the Fermi level is energy-independent. However, it is known that long-range electron–electron interactions in a system of localized electrons cause a gap (the so-called Coulomb gap) in the DOS in the vicinity of the Fermi energy [9.33, 34]. The gap is shown schematically in Fig. 9.10. Using simple semiquantitative arguments, *Efros* and *Shklovskii* [9.33] suggested a parabolic shape for the DOS function

$$g(\varepsilon) = \frac{\eta \kappa^3}{e^6} (\varepsilon - \varepsilon_\text{F})^2 \, , \tag{9.35}$$

where κ is the dielectric constant, e is the elementary charge and η is a numerical coefficient. This result was later confirmed by numerous computer simulations (see, for example, *Baranovskii* et al. [9.35]). At low temperatures, the density of states near the Fermi level has a parabolic shape, and it vanishes exactly at the Fermi energy. As the temperature rises, the gap disappears (see, for example, *Shlimak* et al. [9.36]).

As we have seen above, localized states in the vicinity of the Fermi energy are the most useful for transport at low temperatures. Therefore the Coulomb gap essentially modifies the temperature dependence of the hopping conductivity in the VRH regime at low temperatures. The formal analysis of the T-dependence of the conductivity in the presence of the Coulomb gap is similar to that for the Mott's law discussed above. Using the parabolic energy dependence of the DOS function, one arrives at the result

$$\sigma = \sigma_0 \exp\left[-\left(\frac{\tilde{T}_0}{T}\right)^{1/2}\right] \qquad (9.36)$$

with $\tilde{T}_0 = \tilde{\beta} e^2/(\kappa \alpha k_B)$, where $\tilde{\beta}$ is a numerical coefficient.

Equations (9.33) and (9.36) belong to the most famous theoretical results in the field of variable-range hopping conduction. However these formulae are usually of little help to researchers working with essentially noncrystalline materials, such as amorphous, vitreous or organic semiconductors. The reason is as follows. The above formulae were derived for the cases of either constant DOS (9.33) or a parabolic DOS (9.36) in the energy range associated with hopping conduction. These conditions can usually be met in the impurity band of a lightly doped crystalline semiconductor. In the most disordered materials, however, the energy distribution of the localized states is described by a DOS function that is very strongly energy-dependent. In amorphous, vitreous and microcrystalline semiconductors, the energy dependence of the DOS function is believed to be exponential, while in organic materials it is usually assumed to be Gaussian. In these cases, new concepts are needed in order to describe the hopping conduction. In the next section we present these new concepts and calculate the way the conductivity depends on temperature and on the concentration of localized states in various significantly noncrystalline materials.

9.3.3 Description of Charge-Carrier Energy Relaxation and Hopping Conduction in Inorganic Noncrystalline Materials

In most inorganic noncrystalline materials, such as vitreous, amorphous and polycrystalline semiconductors, the localized states for electrons are distributed over a rather broad energy range with a width of the order of an electronvolt. The DOS function that describes this energy distribution in such systems is believed to have a purely exponential shape

$$g(\varepsilon) = \frac{N_0}{\varepsilon_0} \exp\left(-\frac{\varepsilon}{\varepsilon_0}\right), \qquad (9.37)$$

where the energy ε is counted positive from the mobility edge towards the center of the mobility gap, N_0 is the total concentration of localized states in the band tail, and ε_0 determines the energy scale of the tail. To be precise, we consider that electrons are the charge carriers here. The result for holes can be obtained in an analogous way. Values of ε_0 in inorganic noncrystalline materials are believed to vary between 0.025 eV and 0.05 eV, depending on the system under consideration.

It is worth noting that arguments in favor of a purely exponential shape for the DOS in the band tails of inorganic noncrystalline materials described by (9.37) cannot be considered to be well justified. They are usually based on a rather ambiguous interpretation of experimental data. One of the strongest arguments in favor of (9.37) is the experimental observation of the exponential decay of the light absorption coefficient for photons with an energy deficit ε with respect to the energy width of the mobility gap (see, for example, *Mott* and *Davis* [9.32]). One should mention that this argument is valid only under the assumption that the energy dependence of the absorption coefficient is determined solely by the energy dependence of the DOS. However, in many cases the matrix element for electron excitation by a photon in noncrystalline materials also strongly depends on energy [9.14, 37]. Hence any argument for the shape of the DOS based on the energy dependence of the light absorption coefficient should be taken very cautiously. Another argument in favor of (9.37) comes

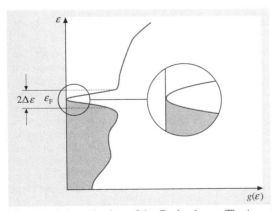

Fig. 9.10 Schematic view of the Coulomb gap. The *insert* shows the parabolic shape of the DOS near the Fermi level

from the measurements of dispersive transport in time-of-flight experiments. In order to interpret the observed time dependence of the mobility of charge carriers, one usually assumes that the DOS for the band tail takes the form of (9.37) (see, for example, *Orenstein* and *Kastner* [9.38]). One of the main reasons for such an assumption is probably the ability to solve the problem analytically without elaborate computer work.

In the following we start our consideration of the problem by also assuming that the DOS in a band tail of a noncrystalline material has an energy dependence that is described by (9.37). This simple function will allow us to introduce some valuable concepts that have been developed to describe dynamic effects in noncrystalline materials in the most transparent analytical form. We first present the concept of the so-called transport energy, which, in our view, provides the most transparent description of the charge transport and energy relaxation of electrons in noncrystalline materials.

The Concept of the Transport Energy

The crucial role of a particular energy level in the hopping transport of electrons via localized band-tail states with the DOS described by (9.37) was first recognized by *Grünewald* and *Thomas* [9.39] in their numerical analysis of equilibrium variable-range hopping conductivity. This problem was later considered by *Shapiro* and *Adler* [9.40], who came to the same conclusion as Grünewald and Thomas, namely that the vicinity of one particular energy level dominates the hopping transport of electrons in the band tails. In addition, they achieved an analytical formula for this level and showed that its position does not depend on the Fermi energy.

Independently, the rather different problem of nonequilibrium energy relaxation of electrons by hopping through the band tail with the DOS described by (9.37) was solved at the same time by *Monroe* [9.41]. He showed that, starting from the mobility edge, an electron most likely makes a series of hops downward in energy. The manner of the relaxation process changes at some particular energy ε_t, which Monroe called the transport energy (TE). The hopping process near and below TE resembles a multiple-trapping type of relaxation, with the TE playing a role similar to the mobility edge. In the multiple-trapping relaxation process [9.38], only electron transitions between delocalized states above the mobility edge and the localized band-tail states are allowed, while hopping transitions between the localized tail states are neglected. Hence, every second transition brings the electron to the mobility edge. The TE of *Monroe* [9.41] coincides exactly with the energy level discovered by *Grünewald* and *Thomas* [9.39] and by *Shapiro* and *Adler* [9.40] for equilibrium hopping transport.

Shklovskii et al. [9.42] have shown that the same energy level ε_t also determines the recombination and transport of electrons in the nonequilibrium steady state under continuous photogeneration in a system with the DOS described by (9.37).

It is clear, then, that the TE determines both equilibrium and nonequilibrium and both transient and steady-state transport phenomena. The question then arises as to why this energy level is so universal that electron hopping in its vicinity dominates all transport phenomena. Below we derive the TE by considering a single hopping event for an electron localized deep in the band tail. It is the transport energy that maximizes the hopping rate as a final electron energy in the hop, independent of its initial energy [9.43]. All derivations below are carried out for the case $k_B T < \varepsilon_0$.

Consider an electron in a tail state with energy ε_i. According to (9.24), the typical rate of downward hopping of such an electron to a neighboring localized state deeper in the tail with energy $\varepsilon_j \geq \varepsilon_i$ is

$$\nu_\downarrow(\varepsilon_i) = \nu_0 \exp\left(-\frac{2r(\varepsilon_i)}{\alpha}\right), \quad (9.38)$$

where

$$r(\varepsilon) \approx \left[\frac{4\pi}{3} \int_{\varepsilon_i}^{\infty} g(x)\,\mathrm{d}x\right]^{-1/3}. \quad (9.39)$$

The typical rate of upward hopping for such an electron to a state less deep in the tail with energy $\varepsilon_j \leq \varepsilon_i$ is

$$\nu_\uparrow(\varepsilon_i, \delta) = \nu_0 \exp\left[-\frac{2r(\varepsilon_i - \delta)}{\alpha} - \frac{\delta}{k_B T}\right], \quad (9.40)$$

where $\delta = \varepsilon_i - \varepsilon_j \geq 0$. This expression is not exact. The average nearest-neighbor distance, $r(\varepsilon_i - \delta)$, is based on all states deeper than $\varepsilon_i - \delta$. For the exponential tail, this is equivalent to considering a slice of energy with a width of the order ε_0. This works for a DOS that varies slowly compared with $k_B T$, but not in general. It is also assumed for simplicity that the localization length, α, does not depend on energy. The latter assumption can be easily jettisoned at the cost of somewhat more complicated forms of the following equations.

We will analyze these hopping rates at a given temperature T, and try to find the energy difference δ that provides the fastest typical hopping rate for an electron placed initially at energy ε_i. The corresponding energy difference, δ, is determined by the condition

$$\frac{\mathrm{d}\nu_\uparrow(\varepsilon_i, \delta)}{\mathrm{d}\delta} = 0 . \tag{9.41}$$

Using (9.37), (9.39) and (9.40), we find that the hopping rate in (9.40) has its maximum at

$$\delta = \varepsilon_i - 3\varepsilon_0 \ln \frac{3\varepsilon_0 (4\pi/3)^{1/3} N_0^{1/3} \alpha}{2k_\mathrm{B} T} . \tag{9.42}$$

The second term in the right-hand side of (9.42) is called the transport energy ε_t after *Monroe* [9.41]:

$$\varepsilon_\mathrm{t} = 3\varepsilon_0 \ln \frac{3\varepsilon_0 (4\pi/3)^{1/3} N_0^{1/3} \alpha}{2k_\mathrm{B} T} . \tag{9.43}$$

We see from (9.42) that the fastest hop occurs to the state in the vicinity of the TE, independent of the initial energy ε_i, provided that ε_i is deeper in the tail than ε_t; in other words, if $\delta \geq 0$. This result coincides with that of *Monroe* [9.41]. At low temperatures, the TE ε_t is situated deep in the band tail, and as the temperature rises it moves upward towards the mobility edge. At some temperature T_c, the TE merges with the mobility edge. At higher temperatures, $T > T_\mathrm{c}$, the hopping exchange of electrons between localized band tail states becomes inefficient and the dynamic behavior of electrons is described by the well-known multiple-trapping model (see, for instance, *Orenstein* and *Kastner* [9.38]). At low temperatures, $T < T_\mathrm{c}$, the TE replaces the mobility edge in the multiple-trapping process [9.41], as shown in Fig. 9.11. The width, W, of the maximum of the hopping rate is determined by the requirement that near ε_t the hopping rate, $\nu_\uparrow(\varepsilon_i, \delta)$, differs by less than a factor of e from the value $\nu_\uparrow(\varepsilon_i, \varepsilon_i - \varepsilon_\mathrm{t})$. One finds [9.42]

$$W = \sqrt{6\varepsilon_0 k_\mathrm{B} T} . \tag{9.44}$$

For shallow states with $\varepsilon_i \leq \varepsilon_\mathrm{t}$, the fastest hop (on average) is a downward hop to the nearest spatially localized state in the band tail, with the rate determined by (9.38) and (9.39). We recall that the energies of electron states are counted positive downward from the mobility edge towards the center of the mobility gap. This means that electrons in the shallow states with $\varepsilon_i \leq \varepsilon_\mathrm{t}$ normally hop into deeper states with $\varepsilon > \varepsilon_i$, whereas electrons in the deep states with $\varepsilon_i > \varepsilon_\mathrm{t}$ usually hop upward in energy into states near ε_t in the energy interval W, determined by (9.44).

This shows that ε_t must play a crucial role in those phenomena, which are determined by electron hopping in the band tails. This is indeed the case, as shown in numerous review articles where comprehensive theories based on the concept of the TE can be found (see,

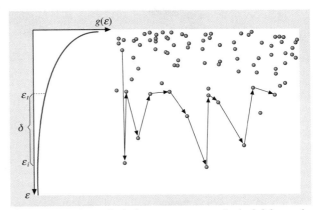

Fig. 9.11 Hopping path via the transport energy. In the *left frame*, the exponential DOS is shown schematically. The *right frame* depicts the transport path constructed from upward and downward hops. The upward transitions bring the charge carrier to sites with energies close to the transport energy ε_t

for instance, *Shklovskii* et al. [9.42]). We will consider only one phenomenon here for illustration, namely the hopping energy relaxation of electrons in a system with the DOS described by (9.37). This problem was studied initially by *Monroe* [9.41].

Consider an electron in some localized shallow energy state close to the mobility edge. Let the temperature be low, $T < T_\mathrm{c}$, so that the TE, ε_t, lies well below the mobility edge, which has been chosen here as a reference energy, $\varepsilon = 0$. The aim is to find the typical energy, $\varepsilon_\mathrm{d}(t)$, of our electron as a function of time, t. At early times, as long as $\varepsilon_\mathrm{d}(t) < \varepsilon_\mathrm{t}$, the relaxation is governed by (9.38) and (9.39). The depth $\varepsilon_\mathrm{d}(t)$ of an electron in the band tail is determined by the condition

$$\nu_\downarrow [\varepsilon_\mathrm{d}(t)] t \approx 1 . \tag{9.45}$$

This leads to the double logarithmical dependence $\varepsilon_\mathrm{d}(t) \propto \varepsilon_0 \ln[\ln(\nu_0 t)] + C$, where constant C depends on $\varepsilon_0, N_0, \alpha$ in line with (9.38) and (9.39). Indeed, (9.38) and (9.45) prescribe the logarithmic form of the time dependence of the hopping distance, $r(t)$, and (9.37) and (9.39) then lead to another logarithmic dependence $\varepsilon_\mathrm{d}[r(t)]$ [9.41]. At the time

$$t_\mathrm{C} \approx \nu_0^{-1} \exp\left(\frac{3\varepsilon_0}{k_\mathrm{B} T}\right) \tag{9.46}$$

the typical electron energy, $\varepsilon_\mathrm{d}(t)$, approaches the TE ε_t, and the style of the relaxation process changes. At $t > t_\mathrm{c}$, every second hop brings the electron into states in the vicinity of the TE ε_t from where it falls downward in energy to the nearest (in space) localization site. In the latter relaxation process, the typical electron energy is determined by the condition [9.41]

$$\nu_\uparrow [\varepsilon_d(t), \varepsilon_t] t \approx 1 , \qquad (9.47)$$

where $\nu_\uparrow [\varepsilon_d(t), \varepsilon_t]$ is the typical rate of electron hopping upward in energy toward the TE [9.41]. This condition leads to a typical energy position of the relaxing electron at time t of

$$\varepsilon_d(t) \approx 3\varepsilon_0 \ln [\ln (\nu_0 t)] - \varepsilon_0 \left[8/\left(N_0 \alpha^3\right) \right] . \qquad (9.48)$$

This is a very important result, which shows that in a system where the DOS has a pure exponential energy dependence, described by (9.37), the typical energy of a set of independently relaxing electrons would drop deeper and deeper into the mobility gap with time. This result is valid as long as the electrons do not interact with each other, meaning that the occupation probabilities of the electron energy levels are not taken into account. This condition is usually met in experimental studies of transient processes, in which electrons are excited by short (in time) pulses, which are typical of time-of-flight studies of the electron mobility in various disordered materials. In this case, only a small number of electrons are present in the band tail states. Taking into account the huge number of localized band tail states in most disordered materials, one can assume that most of the states are empty and so the above formulae for the hopping rates and electron energies can be used. In this case the electron mobility is a time-dependent quantity [9.41]. A transport regime in which mobility of charge carriers is time-dependent is usually called dispersive transport (see, for example, *Mott* and *Davis* [9.32], *Orenstein* and *Kastner* [9.38], *Monroe* [9.41]). Hence we have to conclude that the transient electron mobility in inorganic noncrystalline materials with the DOS in the band tails as described by (9.37) is a time-dependent quantity and the transient electrical conductivity has dispersive character. This is due to the nonequilibrium behavior of the charge carriers. They continuously drop in energy during the course of the relaxation process.

In some theoretical studies based on the Fokker–Planck equation it has been claimed that the maximum of the energy distribution of electrons coincides with the TE ε_t and hence it is independent of time. This statement contradicts the above result where the maximum of the distribution is at energy $\varepsilon_d(t)$, given by (9.48). The Fokker–Planck approach presumes the diffusion of charge carriers over energy. Hence it is invalid for describing the energy relaxation in the exponential tails, in which electron can move over the full energy width of the DOS (from a very deep energy state toward the TE) in a single hopping event.

In the equilibrium conditions, when electrons in the band tail states are provided by thermal excitation from the Fermi energy, a description of the electrical conductivity can easily be derived using (9.5)–(9.7) [9.39]. The maximal contribution to the integral in (9.5) comes from the electrons with energies in the vicinity of the TE ε_t, in an energy range with a width, W, described by (9.44). Neglecting the temperature dependence of the pre-exponential factor, σ_0, one arrives at the temperature dependence of the conductivity:

$$\sigma \approx \sigma_0 \exp \left(-\frac{2r(\varepsilon_t)}{B_c^{-1/3} \alpha} - \frac{\varepsilon_F - \varepsilon_t}{k_B T} \right) , \qquad (9.49)$$

where coefficient $B_c \approx 2.7$ is inserted in order to take into account the need for a charge carrier to move over macroscopic percolation distances in order to provide low-frequency charge transport.

A very similar theory is valid for charge transport in noncrystalline materials under stationary excitation of electrons (for example by light) [9.42]. In such a case, one first needs to develop a theory for the steady state of the system under stationary excitation. This theory takes into account various recombination processes for charge carriers and provides their stationary concentration along with the position of the quasi-Fermi energy. After solving this recombination problem, one can follow the track of the theory of charge transport in quasi-thermal equilibrium [9.39] and obtain the conductivity in a form similar to (9.49), where ε_F is the position of the quasi-Fermi level. We skip the corresponding (rather sophisticated) formulae here. Interested readers can find a comprehensive description of this sort of theory for electrical conductivity in the literature (see, for instance, *Shklovskii* et al. [9.42]).

Instead, in the next section we will consider a very interesting problem related to the nonequilibrium energy relaxation of charge carriers in the band tail states. It is well known that at low temperatures, $T \leq 50$ K, the photoconductivities of various inorganic noncrystalline materials, such as amorphous and microcrystalline semiconductors, do not depend on temperature [9.44–46]. At low temperatures, the TE ε_t lies very deep in the band tail and most electrons hop downward in energy, as described by (9.38) and (9.39). In such a regime, the photoconductivity is a temperature-independent quantity determined by the loss of energy during the hopping of electrons via the band-tail states [9.47]. During this hopping relaxation, neither the diffusion coefficient D nor the mobility of the carriers μ depend on temperature, and the conventional form of Einstein's relationship $\mu = eD/k_\mathrm{B}T$ cannot be valid. The question then arises as to what the relation between μ and D is for hopping relaxation. We answer this question in the following section.

Einstein's Relationship for Hopping Electrons

Let us start by considering a system of nonequilibrium electrons in the band tail states at $T = 0$. The only process that can happen with an electron is its hop downward in energy (upward hops are not possible at $T = 0$) to the nearest localized state in the tail. Such a process is described by (9.37)–(9.39). If the spatial distribution of localized tail states is isotropic, the probability of finding the nearest neighbor is also isotropic in the absence of the external electric field. In this case, the process of the hopping relaxation of electrons resembles diffusion in space. However, the median length of a hop (the distance r to the nearest available neighbor), as well as the median time, $\tau = \nu_\downarrow^{-1}(r)$, of a hop [see (9.38)] increases during the course of relaxation, since the hopping process brings electrons deeper into the tail. Nevertheless, one can ascribe a diffusion coefficient to such a process [9.42]:

$$D(r) = \frac{1}{6}\nu_\downarrow(r)r^2 \,. \tag{9.50}$$

Here $\nu_\downarrow(r)r^2$ replaces the product of the "mean free path" r and the "velocity" $r \cdot \nu_\downarrow(r)$, and the coefficient $1/6$ accounts for the spatial symmetry of the problem. According to (9.37)–(9.39) and (9.50), this diffusion coefficient decreases exponentially with increasing r and hence with the number of successive electron hops in the relaxation process.

In order to calculate the mobility of electrons during hopping relaxation under the influence of the electric field, one should take into account the spatial asymmetry of the hopping process due to the field [9.47, 48]. Let us consider an electron in a localized state at energy ε. If an external electric field with a strength F is applied along direction x, the concentration of tail states available to this hopping electron at $T = 0$ (in other words those that have energies deeper in the tail than ε) is [9.47]

$$N(\varepsilon, x) = N(\varepsilon)\left(1 + \frac{eFx}{\varepsilon_0}\right), \tag{9.51}$$

where

$$N(\varepsilon) = \int_\varepsilon^\infty g(\varepsilon)\,\mathrm{d}\varepsilon = N_0 \exp\left(-\frac{\varepsilon}{\varepsilon_0}\right). \tag{9.52}$$

It was assumed in the derivation of (9.51) that $eFx \ll \varepsilon_0$.

Due to the exponential dependence of the hopping rate on the hopping length r, the electron predominantly hops to the nearest tail state among the available states if $r \gg \alpha$, which we assume to be valid. Let us calculate the average projection $\langle x \rangle$ on the field direction of the vector $\boldsymbol{r}$ from the initial states at energy ε to the nearest available neighbor among sites with a concentration $N(\varepsilon, x)$ determined by (9.51). Introducing spherical coordinates with the angle θ between r and the x-axis, we obtain [9.48]

$$\langle x \rangle = \int_0^{2\pi} \mathrm{d}\phi \int_0^{\pi} \mathrm{d}\theta \sin\theta$$

$$\times \int_0^\infty [\mathrm{d}r \cdot r^3 \cos(\theta) \cdot N(\varepsilon, r\cos\theta)]$$

$$\times \exp\left[-\int_0^{2\pi}\mathrm{d}\phi\int_0^\pi \mathrm{d}\theta\sin\theta\right.$$

$$\left.\times \int_0^r \mathrm{d}r' r'^2 N(\varepsilon, r'\cos\theta)\right]. \tag{9.53}$$

Substituting (9.51) for $N(\varepsilon, r\cos\theta)$, calculating the integrals in (9.53) and omitting the second-order terms

$$\left(\frac{eN^{-1/3}(\varepsilon)F}{\varepsilon_0}\right)^2 \ll 1, \tag{9.54}$$

we obtain

$$\langle x \rangle = \frac{eFN^{-2/3}(\varepsilon)}{3\varepsilon_0} \frac{\Gamma(5/3)}{(4\pi/3)^{2/3}}, \tag{9.55}$$

where Γ is the gamma-function and $N(\varepsilon)$ is determined by (9.52). Equation (9.55) gives the average displacement in the field direction of an electron that hops downward from a state at energy ε to the nearest available neighbor in the band tail. The average length $\langle r \rangle$ of such a hop is

$$\langle r \rangle = \int_0^\infty \mathrm{d}r 4\pi r^3 N(\varepsilon) \exp\left[-\frac{4\pi}{3} N(\varepsilon) r^3\right]$$

$$= \left(\frac{4\pi N(\varepsilon)}{3}\right)^{-1/3} \Gamma\left(\frac{4}{3}\right). \quad (9.56)$$

One can ascribe to the hopping process a mobility

$$\mu = \frac{v}{F} = \frac{\langle x \rangle \nu(\langle r \rangle)}{F}$$
$$= \frac{eN^{-2/3}(\varepsilon) \nu(\langle r \rangle)}{3\varepsilon_0} \frac{\Gamma(5/3)}{(4\pi/3)^{2/3}} \quad (9.57)$$

and a diffusion coefficient

$$D = \frac{1}{6} \langle r \rangle^2 \nu(\langle r \rangle)$$
$$= \frac{1}{6} N^{-2/3}(\varepsilon) \nu(\langle r \rangle) \frac{\Gamma^2(4/3)}{(4\pi/3)^{2/3}}. \quad (9.58)$$

Expressions (9.57) and (9.58) lead to a relationship between μ and D of the form

$$\mu = \frac{2\Gamma(5/3)}{\Gamma^2(4/3)} \frac{e}{\varepsilon_0} D \approx 2.3 \frac{e}{\varepsilon_0} D. \quad (9.59)$$

This formula replaces the Einstein's relationship $\mu = eD/k_B T$ for electron hopping relaxation in the exponential band tail. Several points should be noted about this result. First of all, one should clearly realize that (9.59) is valid for nonequilibrium energy-loss relaxation in which only downward (in energy) transitions between localized states can occur. This regime is valid only at low temperatures when the TE ε_t is very deep in the band tail. As the temperature increases, the upward hops become more and more efficient for electron relaxation. Under these circumstances, the relation between μ and D evolves gradually with rising temperature from its temperature-independent form at $T = 0$ to the conventional Einstein's relationship, $\mu = eD/k_B T$ [9.50, 51]. Secondly, one should realize that (9.59) was derived in the linear regime with respect to the applied field under the assumption that $eFx \ll \varepsilon_0$. According to (9.55), the quantity $\langle x \rangle$ is proportional to $N^{-2/3}(\varepsilon) = N_0^{-2/3} \exp[2\varepsilon/(3\varepsilon_0)]$, in other words it increases exponentially during the course of the relaxation toward larger localization energies ε. This means that for deep localized states in the band tail, the condition $eFx \ll \varepsilon_0$ breaks down. The boundary energy for application of the linear theory depends on the strength of the electric field, F. As F decreases, this boundary energy drops deeper into the tail. However, for any F, there is always a boundary energy in the tail below which the condition $eFx \ll \varepsilon_0$ cannot be fulfilled and where nonlinear effects play the decisive role in the hopping conduction of charge carriers. In the next section we show how one can describe these nonlinear effects with respect to the applied electric field.

Nonlinear Effects in Hopping Conduction

Transport phenomena in inorganic noncrystalline materials, such as amorphous semiconductors, under the influence of high electric fields are the foci for intensive experimental and theoretical study. This is due to observations of strong nonlinearities in the dependencies of the dark conductivity [9.11, 52, 53], the photoconductivity [9.49] and the charge carrier drift mobility [9.54–56] on the field for high electric fields. These effects are most pronounced at low temperatures, when charge transport is determined by electron hopping via localized band tail states (Fig. 9.12).

Whereas the field-dependent hopping conductivity at low temperatures has always been a challenge to describe theoretically, theories for the temperature dependence of the hopping conductivity in low electric fields have been successfully developed for all of the transport regimes discussed: for the dark conductivity [9.39], for the drift mobility [9.41], and for the photoconductivity [9.42]. In all of these theories, hopping transitions of electrons between localized states in the exponential band tails play a decisive role, as described above in (9.37)–(9.59).

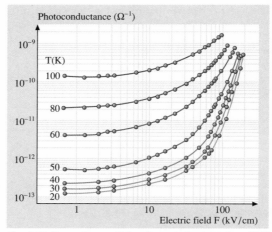

Fig. 9.12 Dependence of the photoconductivity in a-Si:H on the electric field at different temperatures [9.49]

Shklovskii [9.57] was the first to recognize that a strong electric field plays a similar role to that of temperature in hopping conduction. In order to obtain the field dependence of the conductivity $\sigma(F)$ at high fields, *Shklovskii* [9.57] replaced the temperature T in the well-known dependence $\sigma(T)$ for low fields by a function $T_{\text{eff}}(F)$ of the form

$$T_{\text{eff}} = \frac{eF\alpha}{2k_{\text{B}}}, \tag{9.60}$$

where e is the elementary charge, k_{B} is the Boltzmann constant, and α is the localization length of electrons in the band tail states. A very similar result was obtained later by *Grünewald* and *Movaghar* [9.58] in their study of the hopping energy relaxation of electrons through band tails at very low temperatures and high electric fields. The same idea was also used by *Shklovskii* et al. [9.42], who suggested that, at $T = 0$, one can calculate the field dependence of the stationary photoconductivity in amorphous semiconductors by replacing the laboratory temperature T in the formulae of the low-field finite-temperature theory by an effective temperature $T_{\text{eff}}(F)$ given by (9.60).

It is easy to understand why the electric field plays a role similar to that of temperature in the energy relaxation of electrons. Indeed, in the presence of the field, the number of sites available at $T = 0$ is significantly enhanced in the field direction, as shown in Fig. 9.13. Hence electrons can relax faster at higher fields. From the figure it is apparent that an electron can increase its energy with respect to the mobility edge by an amount $\varepsilon = eFx$ in a hopping event over a distance x in the direction prescribed by the electric field. The process is reminiscent of thermal activation. The analogy becomes tighter when we express the transition rate for this hop as

$$\nu = \nu_0 \exp\left(-\frac{2x}{\alpha}\right) = \nu_0 \exp\left(-\frac{2\varepsilon}{eF\alpha}\right)$$
$$= \nu_0 \exp\left(-\frac{\varepsilon}{k_{\text{B}} T_{\text{eff}}(F)}\right), \tag{9.61}$$

where $T_{\text{eff}}(F)$ is provided by (9.60).

This electric field-induced activation at $T = 0$ produces a Boltzmann tail to the energy distribution function of electrons in localized states as shown by numerical calculations [9.59, 60]. In Fig. 9.12, the field-dependent photoconductivity in a-Si:H is shown for several temperatures [9.49]. If we compare the photoconductivity at the lowest measured temperature,

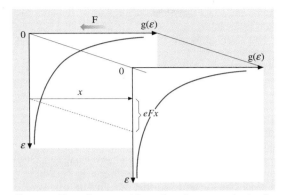

Fig. 9.13 Tunneling transition of a charge carrier in the band tail that is affected by a strong electric field. Upon traveling the distance x, the carrier acquires the energy eFx, where F is the strength of the electric field, and e is the elementary charge

$T = 20\,\text{K}$ in Fig. 9.12, with the low-field photoconductivity at $T = T_{\text{eff}} = \frac{eF\alpha}{2k_{\text{B}}}$ as measured by *Hoheisel* et al. [9.44] and by *Stradins* and *Fritzsche* [9.45], we come to the conclusion that the data agree quantitatively if one assumes that the localization length $\alpha = 1.05\,\text{nm}$ [9.42], which is very close to the value $\alpha \approx 1.0\,\text{nm}$ found for a-Si:H from independent estimates [9.11]. This comparison shows that the concept of the effective temperature based on (9.60) provides a powerful tool for estimating transport coefficient nonlinearity with respect to the electric field using the low-field results for the temperature dependencies of such coefficients.

However, experiments are usually carried out not at $T = 0$ but at finite temperatures, and so the question of how to describe transport phenomena in the presence of both factors, finite T and high F, arises. By studying the steady state energy distribution of electrons in numerical calculations and computer simulations [9.59, 60], as well as straightforward computer simulations of the steady-state hopping conductivity and the transient energy relaxation of electrons [9.61], the following result was found. The whole set of transport coefficients can be represented by a function with a single parameter $T_{\text{eff}}(F, T)$

$$T_{\text{eff}}(F, T) = \left[T^\beta + \left(\gamma \frac{eF\alpha}{k_{\text{B}}}\right)^\beta\right]^{1/\beta}, \tag{9.62}$$

where $\beta \approx 2$ and γ is between 0.5 and 0.9 depending on which transport coefficient is considered [9.61]. We are aware of no analytical theory that can support this numerical result.

To wrap up this section we would like to make the following remark. It is commonly claimed in the scientific literature that transport coefficients in the hopping regime should have a purely exponential dependence on the applied electric field. The idea behind such statements seems rather transparent. Electric field diminishes potential barriers between localized states by an amount $\Delta\varepsilon = eFx$, where x is the projection of the hopping radius on the field direction. The field should therefore diminish the activation energies in (9.24) and (9.25) by this amount, leading to the term $\exp(eFx/k_BT)$ in the expressions for the charge carrier mobility, diffusivity and conductivity. One should, however, take into account that hopping transport in all real materials is essentially described by the variable-range hopping process. In such a process, as discussed above, the interplay between spatial and energy-dependent terms in the exponents of the transition probabilities determine the conduction path. Therefore it is not enough to solely take into account the influence of the strong electric field on the activation energies of single hopping transitions. One should consider the modification of the whole transport path due to the effect of the strong field. It is this VRH nature of the hopping process that leads to a more complicated field dependence for the transport coefficients expressed by (9.60)–(9.62).

We have now completed our description of electron transport in inorganic disordered materials with exponential DOS in the band tails. In the next section we tackle the problem of charge transport in organic disordered materials.

9.3.4 Description of Charge Carrier Energy Relaxation and Hopping Conduction in Organic Noncrystalline Materials

Electron transport and energy relaxation in disordered organic solids, such as molecularly doped polymers, conjugated polymers and organic glasses, has been the subject of intensive experimental and theoretical study for more than 20 years. Although there is a wide array of different disordered organic solids, the charge transport process is similar in most of these materials. Even at the beginning of the 1980s it was well understood that the main transport mechanism in disordered organic media is the hopping of charge carriers via spatially randomly distributed localized states. Binary systems like doped polymeric matrices provide canonical examples of disordered organic materials that exhibit the hopping transport mechanism. Examples include polyvinylcarbazole (PVK) or bis-polycarbonate (Lexan) doped with either strong electron acceptors such as trinitrofluorenone acting as an electron transporting agent, or strong electron donors such as derivatives of tryphenylamine of triphenylmethane for hole transport [9.62,63]. To avoid the need to specify whether transport is carried by electrons or holes each time, we will use a general notation of "charge carrier" below. The results are valid for both types of carrier – electrons or holes. Charge carriers in disordered organic materials are believed to be strongly localized [9.18,62–64]. The localization centers are molecules or molecular subunits, henceforth called sites. These sites are located in statistically different environments. As a consequence, the site energies, which are to great extent determined by electronic polarization, fluctuate from site to site. The fluctuations are typically on the order of 0.1 eV [9.65]. This is about one order of magnitude larger than the corresponding transfer integrals [9.65]. Therefore carrier wavefunctions can be considered to be strongly localized [9.65].

As discussed above, the crucial problem when developing a theoretical picture for hopping transport is the structure of the energy spectrum of localized states, DOS. It is believed that, unlike inorganic noncrystalline materials where the DOS is believed exponential, the energy dependence of the DOS in organic disordered solids is Gaussian (see *Bässler* [9.18] and references therein),

$$g(\varepsilon) = \frac{N_0}{\varepsilon_0\sqrt{2\pi}}\exp\left(-\frac{\varepsilon^2}{2\varepsilon_0^2}\right), \quad (9.63)$$

where N_0 is the total concentration of states and ε_0 is the energy scale of the DOS. The strongest evidence in favor of such an energy spectrum in disordered organic materials is the ability to reproduce the observed experimentally temperature dependence of the carrier mobility and that of hopping conductivity assuming the Gaussian DOS in computer simulations [9.18, 66]. It has been observed in numerous experimental studies [9.67–73] that the temperature dependence of the drift mobility of charge carriers in disordered organic solids takes the form

$$\mu \propto \exp\left[-\left(\frac{T_0}{T}\right)^2\right] \quad (9.64)$$

with a characteristic temperature T_0, as shown in Fig. 9.14a. Computer simulations and theoretical calculations [9.65, 66, 74, 75] with the Gaussian DOS described by (9.63) give a dependence of the form

$$\mu \propto \exp\left[-\left(C\frac{\varepsilon_0}{k_B T}\right)^2\right], \quad (9.65)$$

where C is a numerical coefficient. Computer simulations [9.65, 66] give a value $C \approx 0.69$ for this coefficient, and analytical calculations [9.74, 75] predict a similar value of $C \approx 0.64$. Equation (9.65) is often used to determine the parameter ε_0 of the DOS from experimental measurements of the $\ln(\mu)$ versus $(1/T)^2$ dependences (see, for example, *Ochse* et al. [9.71]).

One may wonder whether the theoretical description of hopping conduction and carrier energy relaxation in a system with a Gaussian DOS (9.63) should differ significantly from the theory described above for disordered systems with a purely exponential DOS (9.37). The answer to this question is yes. The reason becomes clear if one considers the behavior of a single charge carrier in an empty system. In an empty system with an exponential DOS, a charge carrier always (on average) falls downward in energy if $k_B T < \varepsilon_0$ [see (9.45)–(9.48)], and its mobility continuously decreases with time; however, in a system with a Gaussian DOS, a particular energy level ε_∞ determines the equilibrium energy position of a charge carrier. When it is located at some site with high energy in the Gaussian DOS, the charge carrier first hops via localized states so that its average energy $\varepsilon_d(t)$ decreases until it achieves the energy level ε_∞ after some typical time period τ_{rel}. At times $t < \tau_{rel}$ the behavior of the carrier qualitatively resembles that seen for the purely exponential DOS. The downward hops are then replaced by relaxation hops that send the carrier upward to the transport energy, and the carrier mobility at $t < \tau_{rel}$ decreases with time. However, in contrast with the case for the exponential DOS, in a Gaussian DOS the carrier mobility becomes time-independent after a time τ_{rel}, when the average carrier energy reaches the level ε_∞. At $t > \tau_{rel}$, the dispersive transport regime with time-dependent carrier mobility is replaced by a quasi-equilibrium so-called "Gaussian transport" regime, in which the spatial spreading of the carrier packet with time can be described by the traditional diffusion picture with a time-independent diffusion coefficient.

The peculiarity of the hopping energy relaxation of charge carriers in a system with a Gaussian DOS described above makes it easier to describe charge transport at times $t > \tau_{rel}$ than in the case of the exponential DOS. In the latter case, only the presence of a significant number of carriers in a quasi Fermi level can make kinetic coefficients such as mobility, diffusivity and conductivity time-independent and hence conventionally measurable and discussible quantities. In the case of the Gaussian DOS, these kinetic coefficients are not time-dependent at times $t > \tau_{rel}$. Moreover, in diluted systems one can calculate these coefficients by considering the behavior of a single charge carrier. This makes theoretical considerations of electrical conductivity in organic disordered solids with a Gaussian DOS much easier than when considering inorganic noncrystalline materials with an exponential DOS. Let us now calculate ε_∞, τ_{rel} and μ in disordered organic solids with a Gaussian DOS.

Computer simulations [9.66] and analytical calculations [9.74, 75] show that the mean energy of the independently hopping carriers, initially distributed randomly over all states in the Gaussian DOS, decreases

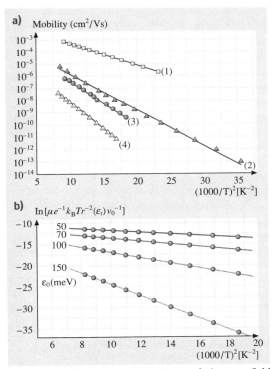

Fig. 9.14a,b Temperature dependence of the zero-field mobility in organic semiconductors. Experimental data (**a**): (1) di-*p*-tolylphenylamine containing (DEASP)-traps [9.69]; (2) (BD)-doped polycarbonate [9.70]; (3) (NTDI)-doped poly(styrene) [9.68]; (4) (BD)-doped TTA/polycarbonate [9.72]. Theoretical results (**b**) were obtained via (9.73)

with time until it approaches the thermal equilibrium value

$$\varepsilon_\infty = \frac{\int_{-\infty}^{\infty} \varepsilon \exp\left(-\frac{\varepsilon}{k_B T}\right) g(\varepsilon) d\varepsilon}{\int_{-\infty}^{\infty} \exp\left(-\frac{\varepsilon}{k_B T}\right) g(\varepsilon) d\varepsilon} = -\frac{\varepsilon_0^2}{k_B T}. \quad (9.66)$$

The time τ_{rel} required to reach this equilibrium is of key importance in the analysis of experimental data [9.65], since at $t < \tau_{\text{rel}}$ the carrier mobility decreases with time (dispersive transport) until it reaches its equilibrium, time-independent value at $t \approx \tau_{\text{rel}}$. It has been established by computer simulations that τ_{rel} strongly depends on temperature [9.18]:

$$\tau_{\text{rel}} \propto \exp\left[\left(B \frac{\varepsilon_0}{k_B T}\right)^2\right] \quad (9.67)$$

with $B \approx 1.07$. Given that the same hopping processes determine both μ and τ_{rel}, researchers were puzzled for many years by the fact that they had different coefficients B and C (in other words they have different temperature dependencies) [9.65]. Below we show how to calculate both quantities – μ and τ_{rel} – easily, and we explain their temperature dependencies (obtained experimentally and by computer simulations as expressed by (9.64), (9.65) and (9.67)).

Our theoretical approach is based on the concept of transport energy (TE), introduced in Sect. 9.3.3, where it was calculated for the exponential DOS given by (9.37). Literally repeating these calculations with the Gaussian DOS, given by (9.63), we obtain the equation [9.76, 77]

$$\exp\left(\frac{x^2}{2}\right)\left[\int_{-\infty}^{\frac{x}{\sqrt{2}}} \exp(-t^2) dt\right]^{4/3} = \left[9(2\pi)^{1/2} N_0 \alpha^3\right]^{-1/3} \frac{k_B T}{\varepsilon_0}. \quad (9.68)$$

If we denote the solution of (9.68) as $X_t(N_0\alpha^3, k_B T/\varepsilon_0)$, then the transport energy in the Gaussian DOS is equal to

$$\varepsilon_t = \varepsilon_0 \cdot X_t\left(N_0\alpha^3, k_B T/\varepsilon_0\right). \quad (9.69)$$

Charge carriers perform thermally activated transitions from states with energies below the TE, ε_t, to the states with energies close to that of the TE [9.76]. Charge carriers hop downward in energy from states with energies above the TE to the spatially nearest sites with rates determined by (9.38) and (9.39).

Now that we have clarified the relaxation kinetics of charge carriers in the Gaussian DOS, it is easy to calculate the relaxation time τ_{rel} and the drift mobility μ. We consider the case $\varepsilon_\infty < \varepsilon_t < 0$, which corresponds to all reasonable values of material parameters $N_0\alpha^3$ and $k_B T/\varepsilon_0$ [9.76]. The energy relaxation of most carriers with energies ε in the interval $\varepsilon_\infty < \varepsilon < \varepsilon_t$ occurs via a multiple trapping-like process, well described in the literature (see, for example, *Orenstein* and *Kastner* [9.38] or *Marschall* [9.78]). Below ε_t the average energy of the carriers $\varepsilon(t)$ moves logarithmically downward with time t. States above $\varepsilon(t)$ achieve thermal equilibrium with states at ε_t at time t, while states below $\varepsilon(t)$ have no chance at time t to exchange carriers with states in the vicinity of ε_t. Hence the occupation of those deep states does not correspond to the equilibrium one, being determined solely by the DOS of the deep states. The system reaches thermal equilibrium when the time-dependent average energy $\varepsilon(t)$ achieves the equilibrium level ε_∞, determined by (9.66). This happens at $t = \tau_{\text{rel}}$. Since the relaxation of carriers occurs via thermal activation to the level ε_t, the relaxation time τ_{rel} is determined by the time required for activated transitions from the equilibrium level ε_∞ to the transport energy ε_t. Hence, according to (9.40) and (9.47), τ_{rel} is determined by the expression

$$\tau_{\text{rel}} = \nu_0^{-1} \exp\left[\frac{2r(\varepsilon_t)}{\alpha} + \frac{\varepsilon_t - \varepsilon_\infty}{k_B T}\right]. \quad (9.70)$$

From (9.68)–(9.70) it is obvious that the activation energy of the relaxation time depends on the parameters $N_0\alpha^3$ and $k_B T/\varepsilon_0$. Hence, generally speaking, this dependence cannot be represented by (9.67) and, if at all, the coefficient B should depend on the magnitude of the parameter $N_0\alpha^3$. However, numerically solving (9.68)–(9.70) using the value $N_0\alpha^3 = 0.001$, which was also used in computer simulations by *Bässler* [9.18, 65], confirms the validity of (9.67) with $B \approx 1.0$. This result is in agreement with the value $B \approx 1.07$ obtained from computer simulations [9.18, 65]. A way to describe the temperature dependence of the relaxation time τ_{rel} by (9.67) is provided by the strong temperature dependence of ε_∞ in the exponent in (9.70), while the temperature dependencies of the quantities ε_t and $r(\varepsilon_t)$ in (9.70) are weaker and they almost cancel each other out. However, if $N_0\alpha^3 = 0.02$, the relaxation time is described by (9.67) with $B \approx 0.9$. This shows that (9.67) can only be considered to be a good approximation.

Now we turn to the calculation of the carrier drift mobility μ. We assume that the transition time t_{tr} necessary for a carrier to travel through a sample is longer than τ_{rel}, and hence the charge transport takes place under equilibrium conditions. As described above, every second jump brings the carrier upward in energy to the vicinity of ε_t, and is then followed by a jump to the spatially nearest site with deeper energy, determined solely by the DOS. Therefore, in order to calculate the drift mobility μ, we must average the hopping transition times over energy states below ε_t, since only these states are essential to charge transport in thermal equilibrium [9.77, 80]. Hops downward in energy from the level ε_t occur exponentially faster than upward hops towards ε_t. This means that one can neglect the former in the calculation of the average time $\langle t \rangle$. The carrier drift mobility can be evaluated as

$$\mu \approx \frac{e}{k_B T} \frac{r^2(\varepsilon_t)}{\langle t \rangle}, \qquad (9.71)$$

where $r(\varepsilon_t)$ is determined via (9.39), (9.63), (9.68) and (9.69). The average hopping time takes the form [9.80]

$$\langle t \rangle = \left[\int_{-\infty}^{\varepsilon_t} g(\varepsilon) d\varepsilon \right]^{-1} \times \int_{-\infty}^{\varepsilon_t} v_0^{-1} g(\varepsilon)$$
$$\times \exp\left(\frac{2r(\varepsilon_t) B_c^{1/3}}{a} + \frac{\varepsilon_t - \varepsilon}{k_B T} \right) d\varepsilon, \qquad (9.72)$$

where $B_c \approx 2.7$ is the percolation parameter. This numerical coefficient is introduced into (9.72) in order to warrant the existence of an infinite percolation path over the states with energies below ε_t. Using (9.63), (9.68), (9.69), (9.71) and (9.72), one obtains the following relation for the exponential terms in the expression for the carrier drift mobility:

$$\ln\left[\mu / \left(\frac{er^2(\varepsilon_t) v_0}{k_B T} \right) \right]$$
$$= -2 \left[\frac{4\sqrt{\pi}}{3 B_c} N_0 \alpha^3 \int_{-\infty}^{X_t/\sqrt{\pi}} \exp(-t^2) dt \right]^{-1/3}$$
$$- \frac{X_t \varepsilon_0}{k_B T} - \frac{1}{2} \left(\frac{\varepsilon_0}{k_B T} \right)^2. \qquad (9.73)$$

It is (9.73) that determines the dependence of the carrier drift mobility on the parameters $N_0 \alpha^3$ and $k_B T / \varepsilon_0$.

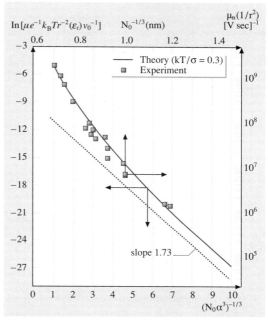

Fig. 9.15 Concentration dependence of the drift mobility evaluated from (9.73) (*solid line*), and the dependence observed experimentally (*circles*) for TNF/PE and TNF/PVK [9.79]

In Fig. 9.14b, the dependence of the drift mobility on the temperature at $N_0 \alpha^3 = 0.01$ is depicted for several values of ε_0. The sensitivity of the mobility to temperature is clear from this picture. Comparison of these dependencies with experimental measurements of $\ln(\mu)$ versus $(1/T)^2$ [some are shown in Fig. 9.14a] provides information on the energy scale, ε_0, of the DOS (see, for example, *Bässler* [9.18] and *Ochse* et al. [9.71]).

In Fig. 9.15, the dependence of the drift mobility on $N_0 \alpha^3$ is shown for $k_B T / \varepsilon_0 = 0.3$. Experimental data from *Gill* [9.81] are also shown in the figure. It is clear that the slope of the mobility exponent as a function of $(N_0 \alpha^3)^{-1/3}$ given by the theory described above agrees with the experimental data. At a very low concentration of localized states, N_0, when the probability of carrier tunneling in space dominates the transition rate in (9.24), charge carriers hop preferentially to the nearest spatial sites. In this regime of nearest-neighbor hopping, the concentration dependence of the drift mobility is described by (9.28), as illustrated by the dashed line in Fig. 9.15.

So far we have discussed the drift mobility of charge carriers under the assumption that the concentration of charge carriers is much less than that of the localized states in the energy range relevant to hopping transport. In such a case one can assume that the carriers perform independent hopping motion and so the conductivity can be calculated as the product

$$\sigma = en\mu, \tag{9.74}$$

where n is the concentration of charge carriers in the material and μ is their drift mobility. If, however, the concentration n is so large that the Fermi energy at thermal equilibrium or the quasi-Fermi energy at stationary excitation is located significantly higher (energetically) than the equilibrium energy ε_∞, a more sophisticated theory based on the percolation approach is required [9.82]. The result obtained is similar to that given by (9.49).

9.4 Concluding Remarks

Beautiful effects have been observed experimentally by studying the charge transport in disordered organic and inorganic materials. Among these, the transport coefficients in the hopping regime show enormously strong dependencies on material parameters. The dependence of the charge carrier mobility on the concentration of localized states N_0 (Fig. 9.15) spreads over many orders of magnitude, as does its dependence on the temperature T (Fig. 9.14) and on the (high) electric field strength F (Fig. 9.12). Such strong variations in physical quantities are typical, say, in astrophysics, but they are not usual in solid state physics. This makes the study of the charge transport in disordered materials absolutely fascinating. The strong dependencies of kinetic coefficients (like drift mobility, diffusivity and conductivity) in disordered materials on various material parameters makes these systems very attractive for various device applications. Since they are relatively inexpensive to manufacture too, it is then easy to understand why disordered organic and inorganic materials are of enormous interest for various technical applications.

These materials also provide a purely academic challenge with respect to their transport phenomena. While traditional kinetic theories developed for crystalline materials are largely dependent on the systems having translational symmetry, there is no such symmetry in disordered materials. However, we have shown in this chapter that it is still possible to develop a reliable theoretical approach to transport phenomena in disordered materials. Particularly interesting is the hopping transport regime. In this regime, charge carriers perform incoherent tunneling jumps between localized states distributed in space and energy. The enormously strong (exponential) dependence of the transition rates on the distances between the sites and their energies call for a completely new set of ideas compared to those for crystalline solids. Conventional transport theories based on the averaging of transition rates lead to absurd results if applied to hopping transport in disordered materials. One can use ideas from percolation theory instead to adequately describe charge transport. One of the most important ideas in this field is so-called variable-range hopping (VRH) conduction. Although the rate of transitions between two localized states is a product of exponential terms that are separately dependent on the concentration of localized states N_0, the temperature of the system T, and also on the field strength F (for high field strengths), it is generally wrong to assume that the carrier drift mobility, diffusivity or conductivity can also be represented as the product of three functions that are separately dependent on N_0, T and F. Instead one should search for a percolation path that takes into account the exponential dependences of the hopping rates on all of these parameters simultaneously. Such a procedure, based on strong interplay between the important parameters in the exponents of the transition rates, leads to very interesting and (in some cases) unexpected results, some of which were described in this chapter. For example, it was shown that the effect of a strong electric field on transport coefficients can be accounted for by renormalizing the temperature. Most of the ideas discussed in this chapter were discussed in the early works of Mott and his coauthors (see, for example, *Mott and Davis* [9.32]). Unfortunately, these ideas are not yet known to the majority of researchers working in the field of disordered materials. Moreover, it is often believed that transport phenomena in different disordered materials need to be described using different ideas. Mott based his ideas, in particular the VRH, mostly on inorganic glassy semiconductors. Most of the researchers that are studying amorphous inorganic semiconductors (like a-Si:H) are aware of these ideas. However, new researchers that are working on more modern disordered materials, such as organic disordered solids and dye-sensitized materials, are often not aware of these very useful and powerful ideas developed by Mott and his followers that can be used to describe charge transport in inorganic disordered systems. In this chapter we have shown that the most pronounced charge transport effects in inorganic and organic disordered materials can be successfully described in a general manner using these ideas.

Although we have presented some useful ideas for describing charge transport in disordered systems above, it is clear that the theoretical side of this field is still embyonic. There are still no reliable theories for charge transport via extended states in disordered materials. Nor are there any reliable theoretical descriptions for the spatial structure of the localized states (DOS) in organic and inorganic noncrystalline materials. All of the theoretical concepts presented in this chapter were developed using very simple models of localization centers with a given energy spectrum that are randomly distributed in space. No correlations between the spatial positions of the sites and the energies of the electronic states at these sites were considered here. Some theoretical attempts to account for such correlations can be found in the literature, although the correlations have not been calculated ab initio: instead they are inserted into a framework of model assumptions. This shows how far the field of charge transport in disordered materials is from a desirable state. Since these materials are already widely used in various technical applications, such as field transistor manufacture, light-emitting diodes and solar cells, and since the sphere of such applications is increasing, the authors are optimistic about the future of research in this field. The study of fundamental charge transport properties in disordered materials should develop, leading us to a better understanding of the fundamental charge transport mechanisms in such systems.

References

9.1   A. Bunde, K. Funke, M. D. Ingram: Solid State Ionics **105**, 1 (1998)
9.2   S. D. Baranovskii, H. Cordes: J. Chem. Phys. **111**, 7546 (1999)
9.3   C. Brabec, V. Dyakonov, J. Parisi, N. S. Sariciftci: *Organic Photovoltaics: Concepts and Realization* (Springer, Berlin, Heidelberg 2003)
9.4   M. H. Brodsky: *Amorphous Semiconductors* (Springer, Berlin, Heidelberg 1979)
9.5   G. Hadziioannou, P. F. van Hutten: *Semiconducting Polymers* (Wiley, New York 2000)
9.6   J. D. Joannopoulos, G. Locowsky: *The Physics of Hydrogenated Amorphous Silicon I* (Springer, Berlin, Heidelberg 1984)
9.7   J. D. Joannopoulos, G. Locowsky: *The Physics of Hydrogenated Amorphous Silicon II* (Springer, Berlin, Heidelberg 1984)
9.8   A. Madan, M. P. Shaw: *The Physics and Applications of Amorphous Semiconductors* (Academic, New York 1988)
9.9   M. Pope, C. E. Swenberg: *Electronic Processes in Organic Crystals and Polymers* (Oxford Univ. Press, Oxford 1999)
9.10 J. Singh, K. Shimakawa: *Advances in Amorphous Semiconductors* (Gordon and Breach/Taylor & Francis, London 2003)
9.11 R. A. Street: *Hydrogenated Amorphous Silicon*, Cambridge Solid State Science Series (Cambridge Univ. Press, Cambridge 1991)
9.12 K. Tanaka, E. Maruyama, T. Shimada, H. Okamoto: *Amorphous Silicon* (Wiley, New York 1999)
9.13 J. S. Dugdale: *The Electrical Properties of Disordered Metals*, Cambridge Solid State Science Series (Cambridge Univ. Press, Cambridge 1995)
9.14 B. I. Shklovskii, A. L. Efros: *Electronic Properties of Doped Semiconductors* (Springer, Berlin, Heidelberg 1984)
9.15 I. P. Zvyagin: *Kinetic Phenomena in Disordered Semiconductors* (Moscow University Press, Moscow 1984) (in Russian)
9.16 H. Böttger, V. V. Bryksin: *Hopping Conduction in Solids* (Wiley, New York 1985)
9.17 H. Overhof, P. Thomas: *Electronic Transport in Hydrogenated Amorphous Semiconductors* (Springer, Berlin, Heidelberg 1989)
9.18 H. Bässler: Phys. Status Solidi B **175**, 15 (1993)
9.19 P. W. Anderson: Phys. Rev. **109**, 1492 (1958)
9.20 A. L. Efros, M. E. Raikh: Effects of Composition Disorder on the Electronic Properties of Semiconducting Mixed Crystals. In: *Optical Properties of Mixed Crystals*, ed. by R. J. Elliott, I. P. Ipatova (Elsevier, New York 1988)
9.21 D. Chattopadhyay, B. R. Nag: Phys. Rev. B **12**, 5676 (1975)
9.22 J. W. Harrison, J. R. Hauser: Phys. Rev. B **13**, 5347 (1976)
9.23 I. S. Shlimak, A. L. Efros, I. V. Yanchev: Sov. Phys. Semicond. **11**, 149 (1977)
9.24 S. D. Baranovskii, A. L. Efros: Sov. Phys. Semicond. **12**, 1328 (1978)
9.25 P. K. Basu, K. Bhattacharyya: J. Appl. Phys. **59**, 992 (1986)
9.26 S. Fahy, E. P. O'Reily: Appl. Phys. Lett. **83**, 3731 (2003)
9.27 V. Venkataraman, C. W. Liu, J. C. Sturm: Appl. Phys. Lett. **63**, 2795 (1993)
9.28 C. Michel, P. J. Klar, S. D. Baranovskii, P. Thomas: Phys. Rev. B **69**, 165211–1 (2004)
9.29 T. Holstein: Philos. Mag. B **37**, 49 (1978)

9.30 H. Scher, T. Holstein: Philos. Mag. **44**, 343 (1981)
9.31 A. Miller, E. Abrahams: Phys. Rev. **120**, 745 (1960)
9.32 N. F. Mott, E. A. Davis: *Electronic Processes in Non-Crystalline Materials* (Clarendon, Oxford 1971)
9.33 A. L. Efros, B. I. Shklovskii: J. Phys. C **8**, L49 (1975)
9.34 M. Pollak: Disc. Faraday Soc. **50**, 13 (1970)
9.35 S. D. Baranovskii, A. L. Efros, B. L. Gelmont, B. I. Shklovskii: J. Phys. C **12**, 1023 (1979)
9.36 I. Shlimak, M. Kaveh, R. Ussyshkin, V. Ginodman, S. D. Baranovskii, H. Vaupel, P. Thomas, R. W. van der Heijden: Phys. Rev. Lett. **75**, 4764 (1995)
9.37 S. D. Baranovskii, P. Thomas, G. J. Adriaenssens: J. Non-Cryst. Solids **190**, 283 (1995)
9.38 J. Orenstein, M. A. Kastner: Solid State Commun. **40**, 85 (1981)
9.39 M. Grünewald, P. Thomas: Phys. Status Solidi B **94**, 125 (1979)
9.40 F. R. Shapiro, D. Adler: J. Non-Cryst. Solids **74**, 189 (1985)
9.41 D. Monroe: Phys. Rev. Lett. **54**, 146 (1985)
9.42 B. I. Shklovskii, E. I. Levin, H. Fritzsche, S. D. Baranovskii: Hopping photoconductivity in amorphous semiconductors: dependence on temperature, electric field and frequency. In: *Advances in Disordered Semiconductors*, Vol. 3, ed. by H. Fritzsche (World Scientific, Singapore 1990) p. 3161
9.43 S. D. Baranovskii, F. Hensel, K. Ruckes, P. Thomas, G. J. Adriaenssens: J. Non-Cryst. Solids **190**, 117 (1995)
9.44 M. Hoheisel, R. Carius, W. Fuhs: J. Non-Cryst. Solids **63**, 313 (1984)
9.45 P. Stradins, H. Fritzsche: Philos. Mag. **69**, 121 (1994)
9.46 J.-H. Zhou, S. D. Baranovskii, S. Yamasaki, K. Ikuta, K. Tanaka, M. Kondo, A. Matsuda, P. Thomas: Phys. Status Solidi B **205**, 147 (1998)
9.47 B. I. Shklovskii, H. Fritzsche, S. D. Baranovskii: Phys. Rev. Lett. **62**, 2989 (1989)
9.48 S. D. Baranovskii, T. Faber, F. Hensel, P. Thomas, G. J. Adriaenssense: J. Non-Cryst. Solids **198–200**, 214 (1996)
9.49 R. Stachowitz, W. Fuhs, K. Jahn: Philos. Mag. B **62**, 5 (1990)
9.50 S. D. Baranovskii, T. Faber, F. Hensel, P. Thomas: Phys. Status Solidi B **205**, 87 (1998)
9.51 S. D. Baranovskii, T. Faber, F. Hensel, P. Thomas: J. Non-Cryst. Solids **227–230**, 158 (1998)
9.52 A. Nagy, M. Hundhausen, L. Ley, G. Brunst, E. Holzenkämpfer: J. Non-Cryst. Solids **164–166**, 529 (1993)
9.53 C. E. Nebel, R. A. Street, N. M. Johanson, C. C. Tsai: Phys. Rev. B **46**, 6803 (1992)
9.54 H. Antoniadis, E. A. Schiff: Phys. Rev. B **43**, 13957 (1991)
9.55 K. Murayama, H. Oheda, S. Yamasaki, A. Matsuda: Solid State Commun. **81**, 887 (1992)
9.56 C. E. Nebel, R. A. Street, N. M. Johanson, J. Kocka: Phys. Rev. B **46**, 6789 (1992)
9.57 B. I. Shklovskii: Sov. Phys. Semicond. **6**, 1964 (1973)
9.58 M. Grünewald, B. Movaghar: J. Phys. Condens. Mat. **1**, 2521 (1989)
9.59 S. D. Baranovskii, B. Cleve, R. Hess, P. Thomas: J. Non-Cryst. Solids **164–166**, 437 (1993)
9.60 S. Marianer, B. I. Shklovskii: Phys. Rev. B **46**, 13100 (1992)
9.61 B. Cleve, B. Hartenstein, S. D. Baranovskii, M. Scheidler, P. Thomas, H. Baessler: Phys. Rev. B **51**, 16705 (1995)
9.62 M. Abkowitz, M. Stolka, M. Morgan: J. Appl. Phys. **52**, 3453 (1981)
9.63 W. D. Gill: J. Appl. Phys. **43**, 5033 (1972)
9.64 S. J. Santos Lemus, J. Hirsch: Philos. Mag. B **53**, 25 (1986)
9.65 H. Bässler: Advances in Disordered Semiconductors. In: *Hopping and Related Phenomena*, Vol. 2, ed. by M. Pollak, H. Fritzsche (World Scientific, Singapore 1990) p. 491
9.66 G. Schönherr, H. Bässler, M. Silver: Philos. Mag. B **44**, 369 (1981)
9.67 P. M. Borsenberger, H. Bässler: J. Chem. Phys. **95**, 5327 (1991)
9.68 P. M. Borsenberger, W. T. Gruenbaum, E. H. Magin, S. A. Visser: Phys. Status Solidi A **166**, 835 (1998)
9.69 P. M. Borsenberger, W. T. Gruenbaum, E. H. Magin, S. A. Visser, D. E. Schildkraut: J. Polym. Sci. Polym. Phys. **37**, 349 (1999)
9.70 A. Nemeth-Buhin, C. Juhasz: Hole transport in 1,1-bis(4-diethylaminophenyl)-4,4-diphenyl-1,3-butadiene. In: *Hopping and Related Phenomena*, ed. by O. Millo, Z. Ovadyahu (Racah Institute of Physics, The Hebrew University Jerusalem, Jerusalem 1995) pp. 410–415
9.71 A. Ochse, A. Kettner, J. Kopitzke, J.-H. Wendorff, H. Bässler: Chem. Phys. **1**, 1757 (1999)
9.72 J. Veres, C. Juhasz: Philos. Mag. B **75**, 377 (1997)
9.73 U. Wolf, H. Bässler, P. M. Borsenberger, W. T. Gruenbaum: Chem. Phys. **222**, 259 (1997)
9.74 M. Grünewald, B. Pohlmann, B. Movaghar, D. Würtz: Philos. Mag. B **49**, 341 (1984)
9.75 B. Movaghar, M. Grünewald, B. Ries, H. Bässler, D. Würtz: Phys. Rev. B **33**, 5545 (1986)
9.76 S. D. Baranovskii, T. Faber, F. Hensel, P. Thomas: J. Phys. C **9**, 2699 (1997)
9.77 S. D. Baranovskii, H. Cordes, F. Hensel, G. Leising: Phys. Rev. B **62**, 7934 (2000)
9.78 J. M. Marshall: Rep. Prog. Phys. **46**, 1235 (1983)
9.79 W. D. Gill: J. Appl. Phys. **43**, 5033 (1972)
9.80 O. Rubel, S. D. Baranovskii, P. Thomas, S. Yamasaki: Phys. Rev. B **69**, 014206–1 (2004)
9.81 W. D. Gill: Electron mobilities in disordred and crystalline tritrofluorenone. In: *Proc. Fifth Int. Conf. of Amorphous and Liquid Semiconductors*, ed. by J. Stuke, W. Brenig (Taylor and Francis, London 1974) p. 901
9.82 S. D. Baranovskii, I. P. Zvyagin, H. Cordes, S. Yamasaki, P. Thomas: Phys. Status Solidi B **230**, 281 (2002)

10. Dielectric Response

Nearly all materials are dielectrics, and the measurement of their dielectric response is a very common technique for their characterisation. This chapter is intended to guide scientists and engineers through the subject to the point where they can interpret their data in terms of the microscopic and atomistic dynamics responsible for the dielectric response, and hence derive useful information appropriate to their particular needs. The focus is on the physical concepts underlying the observed behaviour and is developed from material understandable by an undergraduate student. Emphasis is placed on the information content in the data, and the limits to be placed on its interpretation are clearly identified.

Generic forms of behaviour are identified using examples typical of different classes of material, rather than an exhaustive review of the literature. Limited-range charge transport is included as a special item. The theoretical concepts are developed from a basic level up to the ideas current in the field, and the points where these are controversial have been noted so that the readers can choose for themselves how far to rely on them.

10.1	Definition of Dielectric Response	188
	10.1.1 Relationship to Capacitance	188
	10.1.2 Frequency-Dependent Susceptibility	188
	10.1.3 Relationship to Refractive Index	189
10.2	Frequency-Dependent Linear Responses	190
	10.2.1 Resonance Response	190
	10.2.2 Relaxation Response	192
10.3	Information Contained in the Relaxation Response	196
	10.3.1 The Dielectric Increment for a Linear Response χ_0	196
	10.3.2 The Characteristic Relaxation Time (Frequency)	199
	10.3.3 The Relaxation Peak Shape	205
10.4	Charge Transport	208
10.5	A Few Final Comments	211
References		211

Nearly all materials are dielectrics, that is they do not exhibit a direct-current (DC) conductivity on the macroscopic scale, but instead act as an electrical capacitance i.e. they store charge. The measurement of the dielectric response is noninvasive and has been used for material characterisation throughout most of the 20th century, and consequently a number of books already exist that cover the technique from various points of view. Those that have stood the test of time are *Debye* [10.1], *Smyth* [10.2], *McCrum* et al. [10.3], *Daniels* [10.4], *Bordewijk* and *Bottcher* [10.5], and *Jonscher* [10.6]. These texts cover the subject in terms of the basic physics [10.1, 5], the material properties [10.2–4], and the electrical features [10.6]. An introduction to the wide range of dielectric response measurements that are undertaken can be obtained by referring to the proceedings publication of the International Discussion Meeting on Relaxations in Complex Systems [10.7]. In view of the enormous range of properties and materials covered by the topic it is not feasible or desirable to attempt to review the whole field in a chapter such as this. Instead the topic is approached from the viewpoint of a researcher who, having measured the dielectric *spectrum* (i.e. frequency-dependent complex permittivity) of a material sample, wishes to know what information can be taken from the measurements. Along the way the limits on the information content and the problems (and controversies) associated with the microscopic and molecular-scale interpretation will be identified. Emphasis will be placed on the physical concepts involved, but inevitably there will be some mathematical expressions whose features I aim to place in as simple a physical context as possible.

10.1 Definition of Dielectric Response

10.1.1 Relationship to Capacitance

Stated simply the dielectric response of a material is its response to an electric field within it. Let us start in the simplest way with some basic macroscopic definitions. Imagine that we have a flat slab of our material of thickness d with electrodes of area A on each opposing surface, and we apply a potential difference between opposing sides (Fig. 10.1). Since the material is a dielectric our system is a capacitor and electric charges $\pm Q$ will be stored on the surfaces between which the potential difference V is applied with Q given by

$$Q = CV, \tag{10.1}$$

where C is the capacitance of the system. For the sample geometry considered C is given by the expression

$$C = (\varepsilon_0 \varepsilon_r A)/d. \tag{10.2}$$

Here the factor ε_0 is a fundamental constant termed the permittivity of free space ($\varepsilon_0 = 8.854 \times 10^{-12}\,\mathrm{F\,m^{-1}}$) and ε_r is the permittivity of the material relative to that of free space and hence is termed the relative permittivity. Of course it is a pure number, which is a material property that contains information about the way the material responds to the application of the potential difference. Since however, the replacement of our sample by a vacuum yields a capacitance given by

$$C = (\varepsilon_0 A)/d, \tag{10.3}$$

i.e. $\varepsilon_r = 1$, the information about the response of the material to the electric field is contained in a factor χ,

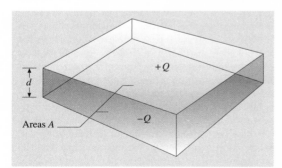

Fig. 10.1 A schematic representation of a dielectric in the form of a parallel-sided slab of surface area A and thickness d that has acquired surface charges $+Q$ and $-Q$ as a result of a potential difference applied across it. After [10.8]

which is called the susceptibility of the material and is given by

$$\chi = \varepsilon_0(\varepsilon_r - 1). \tag{10.4}$$

Noting that the electric field for the parallel electrode geometry described is given by $E = V/d$ and that the electrostatic flux density (electric displacement field) D within the material is $D = Q/A$, allows (10.1) and (10.2) to be transformed into the familiar constitutive relationship,

$$\boldsymbol{D} = \varepsilon_0 \varepsilon_r \boldsymbol{E} = \varepsilon_0 \boldsymbol{E} + \chi \boldsymbol{E} \tag{10.5}$$

which is valid irrespective of the geometry. Here the response of the material has been explicitly separated out in the form of an additive term that is called the electric polarisation $\boldsymbol{P}$, which has dimensions of charge/area or more familiarly electric dipole moment density (electric dipole moment/volume), and is given by

$$\boldsymbol{P} = \chi \boldsymbol{E}. \tag{10.6}$$

An electric dipole occurs when electric charges of opposite polarity are separated in space and the magnitude of the dipole moment is given by $\mu = \delta r$, where δ is the magnitude of the charge and r is the separation vector. The connection between $\boldsymbol{P}$ as defined above and the dipole moment density can be easily seen once it is recognised that the separation of the charges $\pm Q$ by the inter-electrode separation d constitutes a macroscopic dipole moment of magnitude Qd contained within the volume Ad. The definition of the polarisation as the dipole moment density makes it clear that the dielectric response to an electric field relates to the generation of a net dipole per unit volume in the material. In most cases the dielectric has a zero net dipole moment in the absence of an applied electric field and the action of the electric field is to produce one. Some materials such as electrets [10.9] and ferroelectrics lack a centre of symmetry, however, and possess a nonzero polarisation in the absence of an electric field. Therefore the definition of the polarisation P in (10.6) that is applicable to all cases is that P is the *change in net dipole moment density* produced by the applied field.

10.1.2 Frequency-Dependent Susceptibility

The outline given above has been written as if the electric field within the material was a static field ($f = 0$)

but (10.6) is equally valid if the field oscillates with a circular frequency ω ($\omega = 2\pi f$, where f is the frequency in Hertz). In this case (10.6) becomes

$$P(\omega) = \chi(\omega) E(\omega) , \qquad (10.7)$$

where $E(\omega) = E_0 \exp(i\omega t)$.

The fundamental reason for the dependence of $P(\omega)$ upon the frequency of the alternating-current (AC) field, as in (10.7), can be envisaged by constructing a general picture of the way that a material responds to an electric field. Let us imagine that we have our material in thermal equilibrium in the absence of an electric field and we switch on a constant field at a specified time. The presence of the electric field causes the generation of a net dipole moment density (or change in one already existing). This alteration in the internal arrangement of positive and negative charges will not be instantaneous. Instead it will develop according to some equation of motion appropriate to the type of charges and dipole moments that are present. Consequently some time will be required before the system can come into equilibrium with the applied field. Formally this time will be infinity (equivalent to an AC frequency of zero), but to all intents and purposes we can regard the system as coming into equilibrium fairly rapidly after some relevant time scale, τ, with the polarisation approaching the static value $P = P(0)$ for $t \gg \tau$. If now we think of the electric field as reversing sign before equilibrium is reached, as is the case for an AC field at a time $t = 1/4f$ after it is switched on, it is clear that the polarisation will not have reached its equilibrium value before the field is reversed and hence that $P(\omega) \lesssim P(0)$, and $\chi(\omega) \lesssim \chi(0)$. The frequency dependence of the dielectric susceptibility $\chi(\omega)$ is therefore determined by the equation of motion governing the evolution of the ensemble of electric dipole moments.

In general $\chi(\omega)$ will be a complex function with a real component $\chi'(\omega)$ defining the component of $P(\omega)$ that is in phase with the applied AC field $E(\omega) = \mathrm{Re}[E_0 \exp(\omega t)] = E_0 \cos(\omega t)$, and $\chi''(\omega)$ defining the component that is 90° out of phase. The conventional form is given by

$$\chi(\omega) = \chi'(\omega) - i \chi''(\omega) , \quad \left[i = \sqrt{-1} \right] . \qquad (10.8)$$

It is easy to see that $\chi'(\omega)$ determines the net separation of charge with the dielectric in the form of a macroscopic capacitor, but the nature of $\chi''(\omega)$ is not so obvious. The answer lies in considering the *rate of change of polarisation*, $\mathrm{d}[P(\omega)]/\mathrm{d}t$. This has the dimensions of a current density (current/area), is sometimes termed the polarisation current density, and is given by,

$$\begin{aligned}\mathrm{d}[P(\omega)]/\mathrm{d}t &= \left[\chi'(\omega) - i\chi''(\omega)\right] \mathrm{d}[E(\omega)]/\mathrm{d}t \\ &= \left[\chi'(\omega) - i\chi''(\omega)\right] i\omega[\cos(\omega t) \\ &\quad + i \sin(\omega t)] E_0 .\end{aligned} \qquad (10.9)$$

Thus $\chi''(\omega)$ determines the real component of the polarisation current density that is in phase with the electric field, i.e. $J_{\mathrm{pol}}(\omega)$ given by

$$J_{\mathrm{pol}}(\omega) = \chi''(\omega)\omega E_0 \cos(\omega t) = \sigma_{\mathrm{AC}}(\omega) E_0 \cos(\omega t) . \qquad (10.10)$$

Here $\chi''(\omega)\omega = \sigma_{\mathrm{AC}}(\omega)$ is the contribution to the AC conductivity due to the polarisation response to the electric field. If we remember Joule's Law for the power dissipated thermally by an electric current, i.e. power lost $= IV$, then we can see that $\chi''(\omega)\omega(E_0)^2$ is the power dissipated per unit volume resulting from the generation of a net polarisation by the electric field, i.e. the power dissipation density. The imaginary susceptibility $\chi''(\omega)$ is often termed the power dissipation component. It arises because the electric field has to carry out work on the dielectric in order to produce a net dipole moment density. Some of this energy is stored in the charge separations and is recoverable in an equivalent way to the elastic energy stored in a spring. The rest of the energy is used to overcome the friction opposing the establishment of the net dipole density. This energy is transferred to the dielectric in an unrecoverable way, i.e. it is dissipated within the dielectric. It can be seen that $\chi''(\omega)$ is dependent upon the form of the equations of motion governing the evolution of the net dipole moment density under the action of an electric field.

10.1.3 Relationship to Refractive Index

Equation (10.7) can be regarded as relating to the polarisation response purely to an oscillating electric field, but of course all electromagnetic waves contain such a field. In general the topic of dielectric response includes the response of the material to the electric field component of an electromagnetic field, i.e. the electromagnetic spectrum of a material is a form of dielectric response. This form of response is generally characterised by a complex frequency-dependent refractive index $n^*(f)$, with

$$n^*(f) = n(f) - i\kappa(f) , \qquad (10.11)$$

where n is the real refractive index expressing the velocity of light in the material, v, as $v = c/n$, and κ is the absorption coefficient, which defines the reduction of intensity of light of frequency f as it passes through the medium due to absorption of the photons by the medium, i.e. it relates to energy dissipated from the electromagnetic wave. The absorption coefficient can be determined from the Beer–Lambert law,

$$I = I_0 \exp(-4\pi\kappa f z/c). \tag{10.12}$$

For nonmagnetic materials an equivalence can be established between n^* and the relative permittivity (see [10.10] for example). If we think of two slabs of dielectric material placed in contact with an electromagnetic wave passing through them, the component of the field $\boldsymbol{D}$ perpendicular to the boundary is unchanged as the wave passes from one medium to another. Essentially the same electrostatic flux passes through the same perpendicular area. Equation (10.5) gives the ratio of the electric fields in the two dielectric media to the inverse of the ratio of their relative permittivities. A comparison with Snell's law [10.10, 11] then yields

$$\varepsilon_r = [n^*(f)]^2. \tag{10.13}$$

Measurement of the dielectric response of a material involves the determination of the polarisation and its frequency dependence in some form or other. This can be carried out in a large number of ways such as the absorption spectra of electromagnetic radiation as has been described above as well as the application of an AC electric field across a sample of defined dimensions. Those readers interested in the technical details of the measurement systems are referred to the general literature, which has an enormous number of works on these experimental techniques (see for example [10.12] for bridge techniques and [10.13] for recent microwave techniques). This chapter will have a different focus. It will in essence be a discussion of the microscopic origins of the polarisation $\boldsymbol{P}(\omega)$ and the physical reasons for its variation in frequency so that data obtained from such measurements can be used to gain information relevant to the nature of the material to be studied.

10.2 Frequency-Dependent Linear Responses

In this context a linear response is one in which $\boldsymbol{P}(\omega)$ is only dependent upon the first power of the electric field, i.e. $\chi(\omega)$ is independent of the electric field. The general form of frequency dependence expected for $\varepsilon(\omega) = [\varepsilon_0 + \chi(\omega)]$ is shown schematically in (Fig. 10.2), where it can be seen that two basic types of response can be distinguished: a resonance response at high frequencies in the quantum region, and what is termed a relaxation response at lower frequencies.

10.2.1 Resonance Response

Although our main topic in this chapter will be the relaxation behaviour I will start with the resonance response as this has been dealt with extensively [10.14] from the spectroscopic viewpoint. Here we shall approach it from the perspective of its identity as a dielectric response with the intention of identifying basic features that also occur in relaxation responses. The equation of motion is familiar: this form of response relates to a *net* electric dipole moment, ϕ, that undergoes damped simple harmonic oscillation at a natural oscillation frequency $\Omega = 2\pi\nu$ in the absence of an electric field,

$$d^2\phi/dt^2 + \gamma\, d\phi/dt + \Omega^2\phi = 0. \tag{10.14}$$

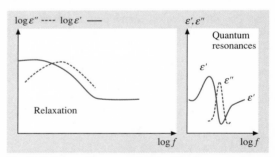

Fig. 10.2 Schematic representation of the frequency dependence of relaxation and resonance responses

The interpretation of (10.14) is not as obvious as it seems. We have to remember that we are always dealing with a sample of material that contains an enormous number of molecules or molecular moieties. Two types of situation may lead to this form of equation of motion. In one case a fluctuation in the positions of *groups* of positive and negative charges may form an oscillating net dipole moment. Plasma oscillations, found in metals [10.15], are an example of this type of behaviour. In the other case, individual molecules may possess identical dipoles oscillating independently at a frequency Ω.

Equation (10.14) then describes the behaviour of a *net* dipole moment produced by a fluctuation in the system of molecules independent of an applied electric field. Such fluctuations are deviations from the average equilibrium state produced by random impulses that act only for an infinitesimal time at $t = 0$, and (10.14) describes the subsequent evolution of the fluctuation. In both cases the damping (friction) term $\gamma \, d\phi/dt$, where γ is the damping factor, expresses the way in which the fluctuation dies away in the absence of an applied electric field and the net dipole moment density of the system returns to zero.

The situation applying in the case of most spectroscopic responses relates to dipole density fluctuations in which individual molecules contribute oscillating dipoles, and it is this form of behaviour that will be used to illustrate the resonance form of response. The first thing we have to do is to ask what kinds of dipole moments oscillate in our material at frequencies in the range $\nu \gtrsim 5 \times 10^{10}$ Hz where most resonance responses are found. The answer to this question lies in the electronic structure of the atoms, ions and molecules that make up our material. The uncertainty principle states the energy of a system cannot be specified at any precise instant of time. This means that the electronic structure of atoms and molecules is allowed to continuously fluctuate between quantum states as long as the average energy over a period of time remains constant. The averaging time is determined by the uncertainty relationship between energy and time. The molecule is thus continuously moving back and forth between ground state and its excited states. These quantum fluctuations (between ground and excited states) displace the negative charge of the electron cloud with respect to the positive charge of the nucleus, and produce oscillating dipoles, termed transition dipoles (Fig. 10.3a). Their oscillation frequencies have specific values depending upon the energy difference between excited and ground quantum states of the species concerned, typically in the range $\nu \gtrsim 4 \times 10^{14}$ Hz [10.14, 16]. These oscillating dipolar fluctuations give no net contribution to the dipole moment of the molecule or atom. The uncertainty principle similarly allows the nuclei of molecules, and the molecules themselves to vibrate and fluctuate between different vibration states. In many cases these nuclear fluctuations produce dipole moments [10.10, 16, 17] (Fig. 10.3b), which typically oscillate in the frequency range 4×10^{14} Hz $\gtrsim \nu \gtrsim 5 \times 10^{10}$ Hz. As with the electronic transition dipoles, these vibration fluctuations do not contribute to the dipole moment of the molecule or system of molecules.

When an AC electric field, such as that provided by an electromagnetic wave, is applied to the material, it couples to the transition dipoles to give a force that drives the system. As a result the population of dipoles is altered such that there is now a *net* average density of transition dipole moments. This oscillates with the frequency of the *field*, i.e. the system is polarised as defined by (10.7). Since transition dipoles do not contribute to a permanent dipole moment of a molecule, the net dipole moment density produced by the electric field is referred to as an induced dipole moment density. The response of the system to an electric field oscillating at a frequency ω is essentially obtained by adding the driving force to (10.14) and determining the solution for ϕ that oscillates with the same frequency, i.e. $\phi(t) \propto e^{i\omega t}$. The components of the frequency-dependent relative susceptibility, $\chi_r(\omega) = \chi(\omega)/\varepsilon_0$, then take the form

$$\chi_r'(\omega) = \frac{\chi_0 \Omega^2 (\Omega^2 - \omega^2)}{(\Omega^2 - \omega^2)^2 + \omega^2 \gamma^2} \,, \tag{10.15}$$

$$\chi_r''(\omega) = \frac{\chi_0 \Omega^2 \omega \gamma}{(\Omega^2 - \omega^2)^2 + \omega^2 \gamma^2} \,. \tag{10.16}$$

These equations exhibit the typical signature of a resonance response such as shown in Fig. 10.2. The imaginary component $\chi_r''(\omega)$ goes through a peak at $\omega = \Omega$ (the resonance frequency). The real component $\chi_r'(\omega)$ exhibits a rise in positive value as the driving

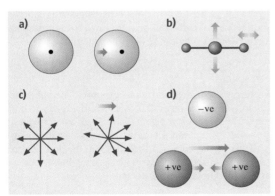

Fig. 10.3a–d Schematic representation of the various forms of dipole fluctuations: (**a**) atomic dipole, (**b**) dipoles formed by bending and stretching motion of a tri-atomic molecule of the form A–B–A, (**c**) net dipole fluctuation produced in a system of permanent dipoles, (**d**) reorienting dipole formed by transfer of an ion between two different centres of vibration. In all cases the *light arrow* shows the net dipole

frequency ω approaches the natural frequency Ω from below, passes through zero when ω and Ω are equal, and rises towards zero from a negative value when $\omega \gg \Omega$. In some cases the exact frequency dependence of the relative susceptibility can be slightly different from that given above. Typically the peak in $\chi''_r(\omega)$ is broadened due to the possibility that either the resonance frequencies of different transitions of the same molecule can be close together and their responses can overlap, or that local electrical interactions between molecules cause the transition energies of individual molecules to be slightly shifted in energy.

As long as the transition energies are sufficiently far apart to be resolved experimental data of the resonance type will yield three pieces of information, which can be related to the electronic (or vibration) structure of the molecules. These are: (a) the natural oscillation frequency Ω, (b) the damping constant γ, and (c) the amplitude factor χ_0. The natural frequency ν is equal to the energy difference of the electronic states between which the fluctuation occurs, ξ, divided by Planck's constant i.e. $\nu = \xi/h$, and so provides information about the different quantum states in relation to one another. The amplitude factor χ_0 is proportional to the square of the transition dipole and therefore yields information on the relative rearrangement of positive and negative charges within the molecule by the transition fluctuation. Damping in these types of systems arises from the sharing of the transition energy between many energy states of the molecule and its vibrations. It removes energy from the specific oscillating dipoles for which the field produces a net dipole moment density. It may act through a delay in returning energy to the electromagnetic wave, i.e. incoherent reradiation, or by transferring it to other energy states where it cannot be reradiated. The damping therefore expresses the way that the energy transferred from the electromagnetic field to the molecule is absorbed and dissipated in the system. The damping factor γ often will have a complicated form. There is of course one other piece of information that is implied by data that fit (10.15, 16) and that is that the equation of motion for the natural oscillating dipole moments is given by (10.14). In some cases, however, the damping factor may be frequency dependent as a result of changes in the interaction between different energy states of the system that occur on the same time scale as the relaxation time, $1/\gamma$. The equation of motion will now have a different and more complicated form than that of (10.14).

The above outline of spectroscopic responses is of necessity very sketchy as it is not the main theme of this chapter, and is dealt with in detail in many standard textbooks (e.g. *Heitler* [10.14]). There are however, a number of general features that can be used as a guide to what happens in the *linear* relaxation response. In the first place the dipoles involved are a property of natural fluctuations of the system, in this case quantum fluctuations in molecules. They are not produced by the electric field. In the absence of an electric field the fluctuating dipoles do not contribute to the net dipole moment of the system, in this case individual molecules. The action of the electric field in linear responses is solely to alter the population of the fluctuations such that a net dipole moment density is produced. This is achieved in the resonance cases considered above by the production of a net density of molecules in an excited state proportional to $\chi_0 E_0$. The irreversible transfer of energy from the electric field to the system relates to the sharing of this energy between the oscillating dipoles coupled to the electric field, and many equi-energetic states of the system that do not couple directly to the electric field. The energy shared in this way is dissipated among the many connected states. Dissipation is an essential consequence of natural fluctuations in an ensemble [10.18] and expresses the requirement that the fluctuation die away to zero at long times. The function $\phi(t)$ must therefore approach zero as t tends to infinity. In the absence of an electric field dipole density fluctuations utilise energy gained transiently from the ensemble and return that energy via the dissipation mechanism. When however an electric field is present, the relative number of fluctuations with dipole moments in different directions is altered and the dissipation term irreversibly transfers energy from the electric field to the ensemble.

10.2.2 Relaxation Response

We turn now to the relaxation response. The simplest way to view this behaviour is as an overdamped oscillation of the net dipole moment density, i.e. one for which $\gamma^2 > 4\Omega^2$. There are a number of ways of addressing this situation and below I shall develop the description starting from the simplest model whose behaviour is rarely found in condensed matter.

The Debye Response

In this case we can neglect the force constant term in (10.14), i.e. the term $\Omega^2 \phi$. This leads to an equation of motion with the form

$$\mathrm{d}\phi/\mathrm{d}t + \gamma\phi = 0 \,. \tag{10.17}$$

The solution to this equation is the very familiar exponential form, $\phi(t) \propto \mathrm{e}^{-t/\tau}$. Equation (10.17) can be interpreted by taking on board the lessons from the resonance response. As before we have to view it as describing the behaviour of a natural fluctuation in our system that produces a net dipole moment density as the result of a random impulse at $t=0$. Now however, there is no evidence for dipole oscillation, so we are not looking at the quantum fluctuations of electronic charge clouds and nuclei positions of molecules. In this case the response originates with the permanent dipoles that many molecules possess due to the asymmetry of their atomic construction. We should also remember that, though atoms do not possess a permanent dipole moment, ion pairs in a material will act as dipoles. Such systems contain a large number (ensemble) of permanent dipoles and this ensemble will obey the laws of thermodynamics. Therefore, with the exception of such materials as electrets and ferroelectrics the orientation of the permanent dipoles will be random in the absence of an electric field, i.e. the *average* net dipole moment of the system will be zero. Thermodynamic ensembles are however described by distributions that allow for fluctuations about the defined average values, thus for example canonical ensembles allow for fluctuations in energy about a defined average energy content, and grand canonical ensembles allow for fluctuations in the number of effective units (e.g. net dipole moments) as well. In the case of dipole responses we are looking at fluctuations that involve the orientations of the permanent dipoles and hence create a net dipole density (Fig. 10.3c,d). Such fluctuations are natural to the ensemble, but are transient, i.e. as in Sect. 10.2.1 $\phi(t) \to 0$ as $t \to \infty$. Equation (10.17) describes the way in which such a local fluctuation in the dipole moment density decays (regresses) to zero, i.e. the ensemble relaxes. An applied electric field couples with the permanent dipoles to produce a torque that attempts to line the dipole with the electric field vector where its energy is lowest. Consequently the linear response of the system to the application of an electric field is an increase in the population of the permanent dipole fluctuations with a component oriented in the field direction as compared to those which have components oriented in the reverse direction. This relative change in the populations of the natural fluctuations of the system gives a net dipole moment density that is driven at the frequency of the electric field as in (10.7) [10.19].

As in the resonance case the polarisation can be obtained by adding the AC driving force oscillating at frequency ω to (10.17) and determining the solution for ϕ oscillating with the same frequency. The corresponding relative susceptibility components have the form

$$\chi_r'(\omega) = \frac{\chi_0}{1+\omega^2/\gamma^2}, \quad (10.18)$$

$$\chi_r''(\omega) = \frac{\chi_0 \omega/\gamma}{1+\omega^2/\gamma^2}. \quad (10.19)$$

These functions show a peak in the imaginary susceptibility component, $\chi_r''(\omega)$, at a frequency $\Omega = \gamma$, which is sometimes called the loss peak frequency since $\chi_r''(\omega)$ is associated with the dissipation of energy, or equivalently the loss of energy from the driving electric field. The real component of the susceptibility, $\chi_r'(\omega)$, changes monotonically from zero at high frequencies to a limiting low-frequency value of χ_0. This is termed the dielectric dispersion.

Equations (10.17–10.19) define what has come to be known as the Debye response after *P. Debye* who first addressed the nature of relaxation dielectric responses [10.1]. It is characterised by two pieces of information: the magnitude of the dispersion χ_0 and the damping factor γ, more usually defined via the relaxation time $\tau = 1/\gamma$ of the dipole density fluctuations. The dispersion magnitude χ_0 is a measure of the net change in dipole density fluctuations that can be produced by a unit field (i.e. $E_0 = 1\,\text{V m}^{-1}$), and is proportional to the square of the individual permanent dipole moments. As with the resonance response an exact fit between the relaxation response data and (10.18), (10.19) implies a specific form for the equation of motion of the dipole density fluctuations of the permanent dipole ensemble, i.e. that of (10.17).

Frequency-Dependent Dielectric Response in Condensed Matter

In practice the Debye response is rarely observed outside of the gas phase. Instead the experimental data can usually be characterised through fractional power laws in the frequency dependence of $\chi_r''(\omega)$ [10.6, 8, 20] in the regions away from the peak (see Fig. 10.4), i.e. for $\omega \gg \gamma$, and $\omega \ll \gamma$, giving

$$\chi_r'(\omega) \propto \chi_r''(\omega) \propto \omega^{n-1}, \quad \omega \gg \gamma, \quad (10.20)$$

$$\chi_r'(0) - \chi_r'(\omega) \propto \chi_r''(\omega) \propto \omega^m, \quad \omega \ll \gamma. \quad (10.21)$$

Here n, and m are fractional exponents, i.e. $0 < n, m < 1$. This general form was first defined empirically as the Havriliak–Negami function [10.21, 22]. A number of special cases have been identified [10.5, 8]. Thus for example the Cole–Cole function is given by $n+m = 1$. When $m = 1$, and $0 < n < 1$, the Cole–Davidson form is produced, which obeys (10.20) and

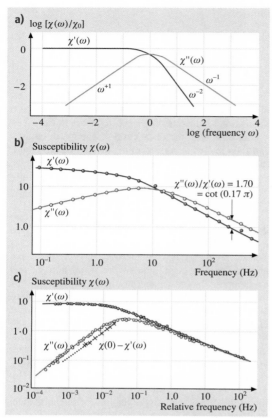

Fig. 10.4 (a) Schematic drawing of the Debye response. **(b)**, **(c)** Examples of measured data fitted to the response function $\chi(\omega)/\chi(0) \propto [(1+i\omega\tau)^{n-1} {}_2F_1(1-n; 1-m; 2-n; (1+i\omega\tau)^{-1})]$ resulting from the response function $\phi(t)$ of (10.22). The function ${}_2F_1(,;;)$ is the Gaussian hypergeometric function [10.25]. *Plot* **(b)** data from irradiated tri-glycine sulphate (TGS); *Plot* **(c)** data from polyvinylacetate. After [10.26]

follows $\chi''_r(\omega) \propto \omega^m$ ($m = 1$) for $\omega \ll \gamma$ but without $\chi''_r(\omega)$ being proportional to $\chi'_r(0) - \chi'_r(\omega)$ as in (10.21).

The physical theory of *Dissado* and *Hill* [10.23, 24] yields analytical expressions for $\chi'_r(\omega)$ and $\chi''_r(\omega)$ that contain both the Debye and Cole–Davidson functions as *exact* limiting cases, and the Cole–Cole and Havriliak–Negami functions as approximations with only minor differences in the curvature in the region of the peak in $\chi''_r(\omega)$. The frequency-dependent susceptibility given by this theory has come to be known as the Dissado–Hill function and is defined in terms of a hypergeometric function [10.25], which is an infinite series with similarities to that of the exponential function. The reader is referred to [10.23, 24] for details. Unlike the empirical functions the Dissado–Hill dielectric response function has a clearly defined equation of motion for $\phi(t)$ [10.24],

$$\frac{d^2\phi}{dt^2} + \frac{[2+n+(t/\tau)]}{t}\frac{d\phi}{dt} + \frac{[n+(t/\tau)(1+m)]}{t^2}\phi = 0 . \quad (10.22)$$

This equation has an analytical solution for $\phi(t)$ in terms of a confluent hypergeometric function [10.25], which has the limiting behaviour $\phi(t) \propto t^{-n}$ at times $t < \tau = 1/\gamma$ and $\phi(t) \propto t^{-(1+m)}$ when $t > \tau = 1/\gamma$. When $m = 1$ the solution is $\phi(t) \propto t^{-n} \exp(-t/\tau)$ giving the Cole–Davidson susceptibility function, and the Debye response function $\phi(t) \propto \exp(-t/\tau)$ is obtained when $n = 0$ and $m = 1$. The general analytical function for $\phi(t)$ was first derived in [10.27] and later re-derived and its equation of motion discussed in [10.23, 24] to which the reader is referred for details. The form of (10.22) illustrates clearly the development of the relaxation response from short-time high-frequency damped harmonic oscillations, characterised by (10.14), as a consequence of time-dependent damping functions and oscillation frequencies. More specifically the damping function approaches a constant value $1/\tau$ (equivalent to γ) at long times, while the oscillation frequency approaches zero. In contrast the Debye model assumes that the dipole motions are overdamped motions with no time-dependent transition from the damped oscillations occurring at short times. Of course there are many ways that time-dependent damping functions and oscillation frequencies may be introduced, most of which involve making similar approximations to that of Debye as regards the ensemble forces that control these factors, but only the specific forms of (10.22) give the power-law frequency dependencies observed in the susceptibility (10.20), (10.21). These forms are produced because the dipole density fluctuations retain some memory of the restoring and damping forces that act on it over very long periods of time, i.e. these forces are not random impulses. The reader is referred to [10.24] for more detail.

The Response Function

A different approach may be taken to deriving the frequency-dependent susceptibility of linear responses. Returning to (10.14), (10.17), and (10.20) it can be seen that the solutions for $\phi(t)$ in the absence of a driving field give the time evolution of the dipole density fluctuation caused by a dipole generating impulse at $t = 0$, i.e. the natural motions of the system. The resulting expression for $\phi(t)$ is called the response function [10.19], and the complex susceptibility is obtained through (10.23).

$$\chi_r(\omega) = \chi_r'(\omega) - i\,\chi_r''(\omega)$$
$$= \operatorname*{Lt}_{\varepsilon \to \infty} \left[\int_0^\infty \phi(t) e^{-i\omega t - \varepsilon t} \, dt \right] . \quad (10.23)$$

Here ε is an infinitesimal quantity that is taken to zero after the integral has been carried out. This equation is valid for all linear dielectric responses. The requirement that $\phi(t) \to 0$ as $t \to \infty$ for natural fluctuations ensures that $\chi_r'(\omega)$ and $\chi_r''(\omega)$ remain finite. The equations of motion, (10.14), (10.17), and (10.22) have solutions for $\phi(t)$ that can be transformed in this way to yield analytical expressions for the frequency-dependent susceptibility. An important consequence of (10.23) is that the frequency dependencies of $\chi_r'(\omega)$ and $\chi_r''(\omega)$ contain the same information since they are Laplace transforms of the same time-dependent function. This also means that a frequency-dependent conductivity $\sigma(\omega) \equiv [\omega \chi_r''(\omega)]$ contains no more information than the equivalent dielectric dispersion $\chi_r'(\omega)$. In fact the linear susceptibility components $\chi_r'(\omega)$ and $\chi_r''(\omega)$ are related to one another through the Kramers–Kronig relationships [10.5, 6]

$$\chi_r'(\omega) = \frac{2}{\pi} \int_0^\infty \frac{x \chi_r''(x)}{x^2 - \omega^2} \, dx , \quad (10.24)$$

$$\chi_r''(\omega) = -\frac{2\omega}{\pi} \int_0^\infty \frac{\chi_r'(x)}{x^2 - \omega^2} \, dx . \quad (10.25)$$

The imaginary contribution from the pole at $x = \omega$ is excluded from these integrals. Although formally values of $\chi_r'(\omega)$ and $\chi_r''(\omega)$ are required over the frequency range from zero to infinity, the reciprocal relationships are adequately reproduced as long as the major part of the dispersion from any specific relaxation process is used.

Equation (10.23) means that linear dielectric responses are characterised through the time dependence of $\phi(t)$. This led *Williams* and *Watts* [10.28] to approach the description of the frequency dependence commonly observed by proposing that the exponential behaviour of

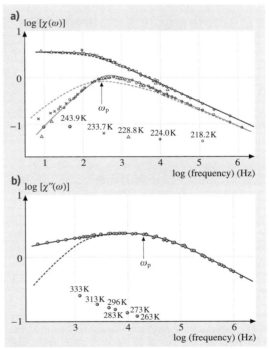

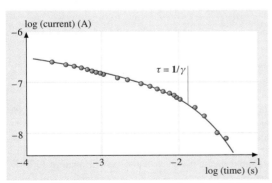

Fig. 10.5 The fit between the response function $\phi(t)/\phi(0) \propto (t/\tau)^{-n} e^{-t/\tau} \, {}_1F_1(1-m; 2-n; t/\tau)$ from (10.22) (where ${}_1F_1(;;)$ is the confluent hypergeometric function [10.25]), and experimental data for amorphous gallium arsenide. After [10.24]

Fig. 10.6a,b Two examples of the fit between the susceptibility functions resulting from the response function $\phi(t)$ of (10.22) (*continuous line*) and (10.26) (*broken line*) and experimental master curves: (**a**) Nematic form of N–(2–hydroxy–4–methoxybenzylidene)–pn–butylaniline (OHMBBA), and (**b**) Polyallylbenzene. The shift of the representative point required to construct the master curve is marked on the plot. After [10.24]

$\phi(t)$ resulting from the Debye model, (10.17), should be generalised to the form

$$\phi(t) \propto (t/\tau)^{-n} \exp\left[-(t/\tau)^{1-n}\right]. \tag{10.26}$$

This expression is sometimes called the expanded exponential function or the Kohlrausch–Williams–Watt function, as it was later found that it was first proposed in [10.29] for mechanical responses. It is not known to possess a simple equation of motion such as (10.14), (10.17), and (10.22) but its relaxation function $\Phi(t)$, defined by $\Phi(t) = \int_t^\infty \phi(t)\,\mathrm{d}t$ [10.19] obeys a relaxation equation of the form of (10.17) with a time-dependent damping factor $\gamma(t) \propto t^{-n}$. The corresponding frequency-dependent susceptibility has the same power-law form as (10.20) for $\omega \gg \gamma = 1/\tau$, but exhibits a slowly varying decrease of slope as the frequency γ is approached from below that, with suitable choices for the value n, can approximate a power law for $\chi_r''(\omega)$ such as is defined in (10.21). The relationship between experimental data for $\phi(t)$ and that derived in the Dissado–Hill cluster model, i. e. the solution to (10.22), is shown in Fig. 10.5. It can be seen that the data and the function for $\phi(t)$ approaches zero as t tends to infinity, with the time power law $t^{-(1+m)}$, but accurate experimental data for times several decades beyond τ is required if this behaviour is to be distinguished from that of (10.26). A better means of distinguishing the two results can often be had by recourse to their appropriate frequency-dependent susceptibilities, see Fig. 10.6.

10.3 Information Contained in the Relaxation Response

As described in Sect. 10.2.2 relaxation responses contain three pieces of information. The strength of the coupling of dipole density fluctuations to the electric field characterised by χ_0, a characteristic relaxation frequency $\gamma = 1/\tau$, where τ is the characteristic relaxation time, and the relaxation dynamics characterised by the frequency dependence of $\chi_r'(\omega)$ and $\chi_r''(\omega)$. This latter feature is open to different interpretations, as will be discussed later.

10.3.1 The Dielectric Increment for a Linear Response χ_0

The dielectric increment is proportional to the square of the permanent dipole moments that give rise to the dipole density fluctuation. It is a feature of the dielectric response that does not usually receive the most attention, mainly because a quantitative relationship to the molecular physics of the relaxation process is often difficult to achieve. Nevertheless it has been used to determine the dipole moments of polar molecules using measurement in the gas phase or if necessary dilute solutions in a nonpolar solvent. In these cases the permanent molecular dipoles, μ, can be assumed to be independent of one another and to be able to adopt all orientations with equal probability in the absence of an electric field, i. e. all dipole moment orientations are at the same energy. This section starts by outlining the derivation of χ_0 for this situation even though this is not the topic area of this book and chapter. The aim is to demonstrate the procedure and bring out the assumptions involved so that the more complicated nature of dipole density fluctuations in condensed-state materials can be better appreciated.

Independent Free Dipoles
In an electric field a dipole that is at an angle θ to the field direction is at the energy $-\mu E \cos\theta$. Those molecules aligned with the electric field are therefore at the lowest energy. The thermal motions of the molecules will however tend to randomise the dipole orientations and the probability of finding a dipole with an orientation angle θ becomes $\exp(\mu E \cos\theta/k_B T)$. The average value of $\mu \cos\theta$ is given by

$$M = \langle \mu \cos\theta \rangle$$
$$= \frac{\int_0^\pi \mu \cos\theta \exp(\mu E \cos\theta/k_B T) \sin\theta\,\mathrm{d}\theta}{\int_0^\pi \exp(\mu E \cos\theta/k_B T) \sin\theta\,\mathrm{d}\theta} \tag{10.27}$$

and the contribution to the static polarisation is given by $N\langle \mu \cos\theta \rangle$, where N is the number of permanent dipoles per unit volume. The term independent of the electric field is zero because all orientations are equally probable in the absence of the field. Equation (10.27) results in a nonlinear function in the electric field E for M, which is called the Langevin function, $L(\mu E/k_B T)$, with,

$$L(\mu E/k_B T) = \coth(\mu E/k_B T) - (k_B T/\mu E). \tag{10.28}$$

This function saturates at unity for very high values of $\mu E/k_B T$, reflecting the total alignment of all the permanent dipoles in the electric field. At low fields defined by $\mu E/k_B T \ll 1$, $L(\mu E/k_B T)$ is a linear function of E and gives the linear dielectric increment as

$$\chi_0 = \mu^2/3k_B T . \tag{10.29}$$

Dipoles in Condensed Matter

In condensed-phase systems, particularly solids, the approximations that lead to the Langevin function and (10.29) no longer apply, and hence these expressions no longer hold. In the first place the dipoles are constrained by the local structure and in general will not be able to assume all orientations with equal probability in the absence of an electric field. In the second place we cannot assume that the dipoles are independent of one another. This dependence may arise in more than one way. For example there may be electrostatic interactions between the dipoles, such as would be responsible for the formation of ferroelectric and anti-ferroelectric states. However when the dipoles concerned are of a low concentration such as those that originate with impurities, lattice defects, interstitial ions etc., these dipole–dipole interactions may be weak. The dipoles concerned may also be arranged in such a way that even though they can adopt one or more alternative orientations their dipole–dipole interactions essentially cancel, such as might be expected in dipole glasses [10.30]. The common way to deal with this situation is to assume that a dipole representative of the average dipole in the ensemble experiences the average electric field of all the other dipoles. This is called the *mean-field* approach [10.31]. Since the mean field will be a function of the average dipole moment due to the applied electric field it is usually possible to construct an equation that can be solved to yield M and hence χ_0. Another way in which the dipoles can interact arises because permanent dipoles are part of the lattice structure of the material. Those permanent dipoles that lead to a polarisation in the presence of an electric field must have two or more local orientations available to them, i. e. they must be able to adopt a different orientation that in the presence of an electric field has a lower energy. Any such change will inevitably alter the local atomic and molecular interactions around the dipole that has moved. This effect will travel through the structure and influence other permanent dipoles through changes in atomic and molecular positions in its environment [10.32]. The strength of such interactions will vary depending of the type of dipole and the way that it is connected to the structure. For example reorientable dipoles formed by small interstitial (or substitution) ions may not interact very strongly with the surrounding lattice, whereas polar groups attached to a polymer chain will in many cases interact very strongly when they adopt a different orientation. Similarly the reorientation of polar molecules in liquids may be expected to distort their surrounding solvent cage and create a disturbance that will be transmitted to other polar molecules. The special feature of this form of interaction is that it is transmitted along specific directions depending upon the lattice structure and hence is nonisotropic.

Order–Disorder Ferroelectrics

These are materials in which the permanent dipoles possess two or a limited number of possible orientations. At high temperatures the dipoles are randomly distributed between the alternative orientations in the absence of an electric field. As the temperature is lowered the electrostatic field of the dipoles acts on any one dipole to make one of the orientations more preferable than the others. This causes the permanent dipole system to adopt a specific orientation at the Curie temperature T_c. The mean-field approach results in an expression for χ_0 that diverges at T_c, i. e.

$$\chi_0 \propto \mu^2/|(T - T_c)| . \tag{10.30}$$

This expression is so common to us that it is easy to overlook the physical meaning that it contains, which is much better expressed in the renormalisation group approach [10.33]. Essentially the interactions between the dipoles cause their orientation and dynamics to become correlated to some extent. As T_c is approached from above, the dipole fluctuations in the system are correlated over increasingly long distances and involve increasingly larger groups of individual dipole moments μ. The dielectric increment increases in proportion to a power of the correlation length $\xi \propto |(T - T_c)^{-\delta}|$ and a more exact form for χ_0 is

$$\chi_0 \propto \mu^2/|(T - T_c)|^\alpha . \tag{10.31}$$

At temperatures below T_c the material will possess domains in which all the dipoles are aligned together. Dipole fluctuations in this state have the opposite orientation to that of the polarity of the domain dipole that is they are *changes in net dipole moment density*, see Sect. 10.1.1. These dipole fluctuations also produce an electrostatic field that causes them to be correlated. As the temperature reduces their correlation length reduces and hence so does χ_0. These materials show that the responding dipole in condensed-phase materials will not always be that of individual molecular or ionic components of the system, they may in fact be groups of individual dipoles that respond as a single unit. In fact the susceptibility increment may be written in terms of an effective dipole moment that is also a power of the correlation length, i.e. $\mu_{\text{eff}} \propto (\xi)^{\alpha/2\delta}$.

Kirkwood Correlation Factor

The particular feature of the ferroelectric order–disorder transition is that the correlation between dipoles in the dipole fluctuations is temperature dependent. In general this will not be the case; instead we can expect the correlation to be dependent upon the lattice structure of the material, and hence independent of temperature except for discrete step changes when the material undergoes a phase transition. One way of allowing for these correlations is to introduce a factor g, termed the Kirkwood factor [10.10, 34], into the expression for χ_0 such that it becomes

$$\chi_0 \propto Ng\mu^2 \, . \tag{10.32}$$

Here a value of $g = 1$ defines a system in which the dipoles are uncorrelated, i.e. they are independent of one another. Values of $g > 1$ indicate positive correlation i.e. the dipoles align in the same direction as in the ferroelectric case where g diverges as discussed. Values of $g < 1$ indicate dipoles that are anticorrelated i.e. pairs of dipoles tend to align in opposite directions. Both kinds of behaviour are known to occur.

General Features of the Temperature Dependence of the Dielectric Increment

Section 10.3.1 gives examples of behaviour in which an electric field imposes order, in terms of the polarity of the net dipole density fluctuations, upon an ensemble where the temperature acts as a disordering factor. In the case of independent free dipoles it is the applied field that attempts to align dipoles and for order–disorder ferroelectrics it is the electrostatic field of the other dipoles. It might therefore be expected that the temperature is always a factor that attempts to oppose the field and that the dielectric increment will decrease at high temperatures, but this is not always the case. For example, in situations where the dipole can adopt one of two orientations one of which is favoured by the local lattice structure, the equilibrium population will be heavily weighted towards the favoured orientation. The equilibrium value of M in such a region can be expressed as,

$$M = \mu \tanh[(\Delta_\chi/k_\text{B} T)] \, , \tag{10.33}$$

where $2\Delta_\chi$ is the energy difference between the alternative orientations produced by the local structure. Here μ is the reorientable component of the dipole moment, see Fig. 10.7. An applied electric field that will favour the one orientation over the other will change the energy difference between alternative orientations from $2\Delta_\chi$ to $2(\Delta_\chi - \mu E)$. Differentiation with respect to E then gives the susceptibility increment as $\chi_0 \propto (\mu^2/k_\text{B} T) \cosh^{-2}(\Delta_\chi/k_\text{B} T)$, with the approximately activated form

$$\chi_0 \propto (\mu^2/k_\text{B} T) \exp(-2\Delta_\chi/k_\text{B} T) \tag{10.34}$$

holding when $\Delta_\chi/k_\text{B} T \gg 1$. Although (10.33) implies that regions exist where dipoles are aligned by the structure this does not necessarily mean that the material has a net dipole in the absence of an electric field. Such regions may be local and with dipole vectors randomly arranged by the structure. This behaviour has been found in ferroelectric ceramics [10.35, 36] in both the ferroelectric and paraelectric phases outside the transition region where T approaches T_c. Expression (10.34) indicates that in the appropriate temperature range the effect of the thermal fluctuations is effectively to increase the density of dipoles that can respond to the electric field, and that this overcomes any randomising behaviour. In contrast, at high temperatures ($\Delta_\chi/k_\text{B} T \ll 1$) the distribution of dipoles between the alternative orienta-

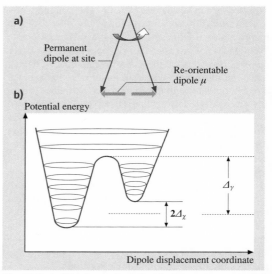

Fig. 10.7a,b Dipole reorientation between two potential wells: (**a**) shows the reorientable component of the dipole, (**b**) shows the potential-energy surface

tions becomes almost random and χ_0 approaches the free-dipole result (10.29).

In many experimental situations the value of the dielectric increment is essentially independent of temperature. It is difficult to see how this can occur in an ensemble where the dipole density fluctuations are produced by fluctuations in thermal energy about the average value, which couple to the electric field via changes in the heat content as in (10.27). However it may be possible to conceive of this behaviour as due to fluctuations in the configuration entropy of the molecular system, of which the dipole is a part, that take place without any change in the heat content. The effect of the field would be to change the configuration entropy S rather than the heat content H. As a result the susceptibility would be independent of temperature. This picture implies that we must think of the dipoles in this case not as local elements embedded within the material matrix moving in a fixed local potential, but as an integral part of the matrix whose dynamics is described by fluctuations in the Gibbs free energy $G = H - TS$ of the whole ensemble. In this case correlations between dipoles would be expected to occur mainly via the indirect route through their interaction with their local environment, rather than their direct electrostatic interactions.

Equation (10.31) describes the behaviour of a system undergoing an order–disorder transition *among the permanent dipole orientations*. A similar behaviour will be found for the relaxation response of a first-order ferroelectric or dipole alignment transition [10.37]. In general phase transitions will not give rise to a divergence in χ_0, which occurs because the phase transition in these cases is defined through the dipole, i.e. the dipole orientation is the order parameter. In other types of phase transition the dipoles are not the primary cause and what can be expected is an abrupt change of χ_0 as the dipoles find themselves embedded into a different lattice structure with different local potentials and orientation positions, different ensemble energies, and different correlations with one another and the material matrix.

The Information Content of the Dielectric Increment

As is clear from the above discussion it is not easy to make definite quantitative statements about the dipole system based on the dielectric increment. The basic reason for this situation is that the measurements are made on a macroscopic sample that contains an ensemble of an enormous number of dipoles, up to $\approx 10^{28}$ m^{-3}. The description of such systems can be carried through if the elements, here electric dipoles, are independent and their orientation is defined by a static local potential; as discussed above this will not be the case in general. More typically the dipoles will be correlated with the matrix in which they are embedded and/or one another. This means that the dipoles that are involved in the dipole density fluctuations are not site dipoles but groups of molecules/ions including dipoles, i.e. the responding features have a size intermediate between that of the molecule/unit cell and that of the sample. Determination of the temperature dependence will give some clues as to how to regard the dipole system through the definition of an effective dipole. The way the effective dipole changes with temperature will allow some interpretation of the kind of system that is present. Variation with other control parameters will produce more information, and systematic variation of the structure, for example replacement of side groups in polymers by longer or different side groups, or substitution of impurity ions by similar ions of different oxidation state or ionic radius, will help to identify the local dipole moment contributing the dielectric increment. However, even if the form of the site dipole is known, the quantitative evaluation of a factor such as the Kirkwood factor g (10.32) is not trivial. In condensed matter, as can be seen from Fig. 10.7, only a component of the site dipole is likely to be involved in reorientation. Unless the local structure is very well known it will not be possible to determine the actual value of the reorientable component in order to obtain a quantitative estimate of g. What can be achieved is a fingerprint of the dipole fluctuations in the material that can be used to characterise it. However at best this will be a partial picture of the dipole fluctuations in the material and information gained from the relaxation time and the relaxation dynamics should be used to enhance it further. In this way a holistic view of the dipole fluctuation can be attempted. It is important to realise that the picture obtained from these three features must be complementary. It is not acceptable to regard them as three independent features, as in fact they just yield different facets of the same process.

10.3.2 The Characteristic Relaxation Time (Frequency)

Equation (10.14), (10.17), (10.22), and (10.23) define a characteristic relaxation rate γ or relaxation time $\tau = 1/\gamma$ for the dipole density fluctuations. In the case of the Debye response, whose susceptibility functions are given by (10.18) and (10.19), γ is the frequency at which the imaginary (dielectric loss) component $\chi''_\mathrm{r}(\omega)$

exhibits a peak. It has therefore become customary to determine the dependence of the relaxation time (rate) upon the control parameters (e.g. temperature, pressure, etc.) that are varied via that of the loss peak frequency. As long as the frequency dependence of $\chi_r''(\omega)$ (i.e. the loss peak shape) remains unchanged this procedure is valid because essentially the dielectric response investigated can be treated as a single composite process, even though it has a wider frequency dependence than that predicted for the independent free dipoles. Both the theoretical equation of motion (10.22) and response function (10.26) do in fact describe the response as a single composite process with a characteristic relaxation rate (time). However the frequency of the peak in $\chi_r''(\omega)$ (loss peak frequency) predicted from (10.22) is not γ but γ multiplied by a numerical factor depending upon the power-law exponents, n and m. In many cases the loss peak is very broad ($n \to 1, m \to 0$) and it is difficult to locate the peak precisely and to be sure that the point located is at the same position with respect to the functional dependence of $\chi_r''(\omega)$ upon ω. Under these circumstances a better procedure is to construct a master curve, which is done by plotting $\chi_r''(\omega)$ as a function of ω in log–log coordinates. Translation of the data along the $\log(\omega)$ and $\log[\chi_r''(\omega)]$ axes will bring the data into coincidence if the susceptibility frequency dependence is unchanged. The translation required to achieve coincidence gives the dependence of the susceptibility increment [$\log(\chi'')$-axis] and characteristic relaxation frequency [$\log(\omega)$-axis] on the controlled variable. For example it gives the ratios $\gamma(T_1)/\gamma(T_2)$ [or $\chi_0(T_1)/\chi_0(T_2)$] for the temperature change T_1 to T_2. This technique also has the advantage of illustrating clearly whether or not the frequency dependence is independent of the variation in temperature (or other parameter), i.e. whether or not the different sets of data can be brought into coincidence. It can also be used to determine any relationship between χ_0 and γ. This is done by selecting a reference point (e.g. the point $\chi_0 = A, \omega = B$) and marking the position on the master curve of this point from each data set after it has been translated to achieve coincidence. A trace is formed giving the dependence of $\log[(\chi_0)^{-1}]$ as a function of $\log(\gamma^{-1})$.

The relaxation rate is the dielectric response feature that shows most dependence upon the variation in the control parameters and so is the feature that is most often studied. In the following sections I will outline some of the most common types of behaviour and discuss their implications for the physics of the relaxation process.

Site Dipole Reorientation

The simplest form of relaxation rate is that described by Debye for independent molecular dipoles suspended in a viscous continuum. As described in Sect. 10.3.1 these dipoles are regarded as free to adopt any orientation in the absence of an electric field. Relaxation of a dipole density fluctuation involves the rotation of the molecular dipoles in the fluctuation to a state in which the net dipole density is zero. In such a situation the rotation of each individual dipole occurs at the same speed determined by the viscosity, η, of the medium, and the relaxation time ($\tau = 1/\gamma$) of the dipole density fluctuation is governed by that speed. For a molecular dipole whose effective length is a the relaxation time has the form [10.10],

$$\tau \propto \eta a^3 / k_B T . \tag{10.35}$$

The more viscous the medium, or the bigger the molecular dipole, the slower it rotates and the slower will be the relaxation of a fluctuation, giving a net dipole moment to the ensemble.

Of course the conditions for this behaviour to be exactly applicable cannot be met except in a gaseous medium. Condensed-phase materials are not continua. Even liquids possess a local structure and molecular dipoles will either be part of that structure if they are contributed by the medium, or will be surrounded by a solvent shell if they are dissolved in the medium. In solids the molecular (or ionic) dipole is of necessity part of the structural matrix, and even though this must be irregular enough to permit rotational displacement to at least one other orientation the matrix can be expected to maintain some structural correlation to distances well away from the site of a reorientable dipole. These are the conditions that must be included in any description of the relaxation frequency (time).

The first point of departure from the picture of a free dipole in a continuum is that the dipole will possess only a limited number of orientations that it can adopt. Consequently there will be a potential barrier between these alternative equilibrium orientations. The rate of transition between alternative orientations and hence the relaxation frequency will be determined by the rate at which a dipole or, to be more specific, the atoms or ions that form the local dipole can pass over the potential barrier to switch orientation, as shown for example in Fig. 10.7. In this case the relaxation frequency will possess an activated (Arrhenius) form where the activation energy Δ_γ is the mean potential barrier height between the alternative orientations, i.e.

$$\gamma = 1/\tau = A \exp(-\Delta_\gamma / k_B T) . \tag{10.36}$$

The expression to be used for the pre-exponential factor A depends on the way in which the atoms/ions comprising the local dipole pass through the transition region at the top of the barrier [10.38, 39]. In the schematic drawing of Fig. 10.7 the dipole is described as having an atom (ion) at its head that performs quantum oscillations in one of two potential wells. As long as it can be assumed that there is a thermal equilibrium between all the vibration states and that the dipole head passes into the alternative well in a single transit of the barrier region, then $A = \nu$, where ν is the frequency of the quantum vibrations at the bottom of the wells. This result continues to hold even if thermal equilibrium is established only for the states at the bottom of the well as long as the effective friction acting on the dipole head in the barrier region is weak. The type of potential surface with these properties is one that remains essentially rigid during the actual transit of the barrier region, which takes place in a time typically of $\approx 10^{-14}$ s. The activated factor in (10.36) expresses the thermal probability of finding a dipole in a quantum state at the top of the barrier. The other extreme situation occurs when the friction ς_d in the barrier region is high, for which $A \propto (1/\varsigma_d)$. This occurs when the transit of the barrier region is slow enough to allow interactions with neighbouring vibrating atoms to overdamp the motion. In this case the potential surface distorts during the transit of the barrier. Such situations can be expected when the barrier is broad and ill-defined, and correspond to local structures that are flexible, such as may be expected in viscoelastic materials. A general expression $A = \lambda_r(\nu/\nu_b)$ has been developed by *Grote* and *Hynes* [10.39] where λ_r is a function that describes the change from low to high friction and ν_b is the quantum frequency in the barrier region.

An interesting consequence of this type of potential surface is that, regardless of the magnitude of the barrier energy Δ_γ, a temperature should exist below which reorientation over the barrier would take so long that any dipole fluctuation would essentially remain unrelaxed, i.e. the dipole system becomes frozen. However when the moving atom in the permanent dipole is a hydrogen atom this is not the case; relaxation can occur by the tunnelling of the hydrogen atom through the barrier [10.40]. This has been demonstrated by experiments on deuterated oxidised polyethylene molecules at millikelvin temperatures [10.41, 42]. In this case the relaxation frequency is determined by the tunnelling probability of the deuterium/hydrogen atom through the barrier, which is dependent upon the atomic mass, the barrier width and height, but *not the temperature*, i.e. the relaxation frequency becomes temperature independent at temperatures below ≈ 100 mK.

Relaxation on a Free Energy Surface

The situation described in the previous section is one in which the dipole moves on a potential surface provided by the surrounding structural matrix. The only dynamic effect of the matrix is via elastic and inelastic interaction between the quantum vibrations of the dipole and the matrix. In many cases however, the atoms (ions) comprising the dipole will cause the displacement of the centres of motion of the surrounding atoms during its transit between alternative orientations. In this case the expression for the relaxation frequency has to refer to the group of atoms affected as a unit, and an appropriate form is that derived by *Eyring* [10.43] for chemical reactions

$$\gamma = (k_B T/h) \exp(-G^\# / k_B T). \quad (10.37)$$

Here $G^\#$ is the change in Gibbs free energy on passing from the ground state to the transition state in the process of reorientation. The barrier is now a free energy rather than a potential barrier and reflects the need for the involvement of displacements in a number of atoms, ions or molecules in order to achieve the dipole reorientation. If we refer again to Fig. 10.7 the difference is that the normal coordinate of the quantum vibrations in the barrier region is a mixture of several different normal coordinates of the surrounding matrix as well as that of the dipole in the well. In general $G^\#$ will be composed of an activation entropy contribution $S^\#$ as well as an activation enthalpy contribution $H^\#$ with $G^\# = H^\# - TS^\#$, and both will be properties of the group of atoms/molecules involved and their structural relationship. The expression for γ therefore takes a form similar to that of (10.36):

$$\begin{aligned}\gamma &= \left[(k_B T/h) \exp(S^\#/k)\right] \exp\left(-H^\#/k_B T\right) \\ &= A_{\text{eff}} \exp\left(-H^\#/k_B T\right). \end{aligned} \quad (10.38)$$

The activation entropy $S^\#$ will reflect the configuration rearrangement required for the dipole to reorient. Thus for example, when reorientation requires the surrounding matrix to adopt a more irregular (disordered) arrangement so as to remove a steric hindrance to reorientation the transition region entropy will be greater than that of the dipole in the bottom of the well and $S^\#$ will be positive. Alternatively the transition region may require specific local arrangements in order that the dipole can avoid such hindrances. In this case the entropy of the transition state will be less than that of the dipole in the bottom of the well and $S^{\#}$ will be negative. A nonzero value of $S^{\#}$ will therefore lead to a relaxation frequency with an effective pre-exponential factor, A_{eff}, that is either greater or less than $k_B T/h$.

An interesting variation on this behaviour has been suggested by *Hill* and *Dissado* [10.44] who showed that several experimental relaxation frequencies could be described via what they termed an activated tunnelling expression. Here they allowed the possibility that the transfer between alternative orientations could take place by thermal activation to an energy state for which the tunnelling of hydrogen atoms through the barrier was feasible. An expression for the relaxation frequency was obtained by determining the optimum transfer rate, with all processes considered from that of an unactivated quantum tunnelling to activation over the top of the barrier. If the energy difference between the alternative orientations is zero the resulting relaxation frequency had the form

$$\gamma \propto \exp(BT) \tag{10.39}$$

over a considerable temperature range, where B is a temperature-independent factor dependent upon the tunnelling distance, the barrier height and shape, and the mass of the tunnelling particle. If the *barrier* in this work is taken to be a free-energy barrier rather than a potential barrier, the analysis can be seen to be equivalent to a situation whereby the dipole relaxes by finding a route between the alternative orientations that allows for lower values of $H^{\#}$ at the expense of ordering the surrounding molecules/atoms, i.e. a negative $S^{\#}$. Routes with higher values of $H^{\#}$ require values of $S^{\#}$ that are less negative. In this case the optimum relaxation frequency will take a form such as (10.39) and only at very high temperatures will γ become purely activated with $S^{\#} \approx 0$. The picture of dielectric relaxation provided by this interpretation is consistent with the *defect diffusion* mechanism [10.45], in which defects diffusing in the structural matrix, such as a kink in a polymer chain or a dislocation in a crystal, lower or remove barriers when they reach a dipole, allowing it to take an alternative orientation of equal energy. When the defect moves on, the dipole is locked into the new position until another defect arrives.

The Glass Transition

A glass is essentially a material that has become macroscopically rigid without attaining its thermodynamically favoured crystalline state. The manner in which this occurs for a liquid or viscoelastic (rubbery) material that has been rapidly cooled has been the subject of intense investigation over a large number of years (see *Angell* [10.46]). A simple, some would say oversimple, view of the situation is to regard it as a concatenation of two effects. In the first place supercooling the liquid phase will at some temperature result in a situation where the liquid no longer possesses a heat content in excess of the crystal state at that temperature [10.47]. The liquid will be unstable with respect to a disordered solid possessing only local crystalline order. In the second place the lowering of the temperature during the supercooling will cause the lattice to contract thereby introducing and increasing barriers to local molecular/atomic motions. At some temperature the thermal fluctuations [described by the Boltzman factor in (10.36), (10.37)] that are responsible for raising the site energy to that of the barrier become so rare that the time required for rearrangement becomes enormously long. At this temperature, termed T_g, the structure is essentially locked into a macroscopically rigid state, termed a glass. The glass transition cannot be regarded as a thermodynamic transition of state (unlike a melting/crystallisation temperature for example) and there are different ways of defining and determining T_g (see [10.46]).

During the approach to the glass state the rate of reorientation of permanent dipoles will become slower and eventually reach zero as shown in Fig. 10.8. It is common to denote the response due to a set of dipoles that are frozen during glass formation as the α-response. A second response can be seen in the figure: termed the β-response. This is provided by dipoles that are able to reorient without requiring any substantial rearrange-

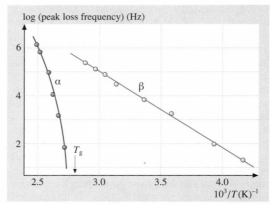

Fig. 10.8 The Arrhenius plot of the glass-forming polyvinylchloride system showing the slowing down of the α-response ($T > T_g$) as T_g is approached and the activated β-response. After [10.8]

ment of the surrounding structural matrix. It should not be expected however that this response is due just to a dipole reorientation with respect to the molecule it is attached to. In many cases the β-response involves the displacement of the molecule or part thereof as a whole [10.48]. In polymers this is a local *inter*-chain motion and either the free-energy expression (10.38) or the potential-energy expression (10.36) will apply, depending on whether the surrounding chains remain rigid during the relaxation or rearrange locally. These dipoles are active in the glass state and can be expected to have a relaxation rate of the form of (10.36), i. e. reorientation over a potential barrier. In the case of the α-response it is clear that the relaxations must involve displacements in a number of molecules/atoms other than just those comprising the permanent dipole, and hence it is instructive to discuss the behaviour in terms of the rate expression (10.37). What can be seen is that, as the temperature at which the system becomes rigid is approached, the gradient in the Arrhenius plot gets steeper, and $H^{\#}$ therefore becomes larger. The non-thermally-activated pre-exponential factor, A_{eff}, in (10.36) is greater than $k_\text{B} T/h$ and hence $S^{\#} > 0$. As the temperature approaches T_g there is an increase in A_{eff}, by many decades in frequency, which must be due to an increase in $S^{\#}$. These changes in $H^{\#}$ and $S^{\#}$ indicate that, as T_g is approached, dipole-orientation relaxation not only requires an increased amount of energy in order to enter the transition (barrier) region but also a larger amount of configuration disordering in the surrounding structure that makes up the molecular/atom group involved in relaxation. Although it is possible that such a situation may come about because reducing the temperature produces a local increase in density that increases steric hindrances for the same group of atoms and molecules, it is more likely that the number of molecules that are displaced in order to allow the dipole to pass through the transition region has increased. These considerations are consistent with a structure that is becoming either tangled or interlocked as the temperature decreases. Detailed expressions based on these concepts but involving macroscopic parameters have been attempted (see for example [10.46, 49–52]).

The glass formation discussed above has a structural basis and dipole–dipole interactions will play at most a minimal role. In some situations however, the glass is a disordered array of dipole orientations [10.30]. This sort of state is most likely to occur at very low temperatures in materials that possess dipoles occupying the sites of a regular lattice. At high temperatures the dipole orientations will be disordered but, as the temperature is reduced to low values, each individual dipole would be expected to adopt their lowest-energy orientation, resulting in a state of ordered dipole orientation. A *dipole glass* will result instead when the dipole–dipole interactions produce forces that generate barriers to the local reorientation and frustrate the ordering process at temperatures low enough that the barriers generated cannot be overcome in any conceivable time.

Ferroelectric Transition

The dielectric response of ferroelectrics at temperatures in the vicinity of their Curie (critical) temperature also exhibit relaxation frequencies that approach zero, just as their dielectric increment approaches infinity (Fig. 10.9a,b) as discussed in Sect. 10.3.1. In this case both the dielectric increment and the relaxation frequency are functions of a hidden variable that characterises the system, the correlation length ξ of the dipole fluctuations. Just as the dielectric increment increases with a power of the correlation length, the relaxation frequency will decrease. Put simply the more dipoles are correlated in the fluctuation the longer the time that is required for its relaxation. Scaling theory [10.33] describes the system by a hierarchy of self-similar correlations. The strongest correlations are between the dipole and its nearest neighbours. This gives a local geometrical arrangement of correlations. The next-strongest correlations are between the same geometrical arrangement of groups of nearest neighbours, and the next strongest is between the same geometry of groups of groups. Eventually the whole system up to the correlation length is constructed in this way. Because the geometrical arrangement is preserved at each stage the properties for each stage have to be proportional to a power of the size. This gives

$$\gamma \propto |(T - T_\text{c})|^\beta \propto \xi^{-\beta/\delta} \qquad (10.40)$$

and using (10.31) the relationship

$$\chi_0 \propto (\gamma)^{-\alpha/\beta} \qquad (10.41)$$

follows. But the theory can go further and predict the frequency dependence of $\chi_\text{r}''(\omega)$ and $\chi_\text{r}'(\omega)$ for $\omega > \gamma$. This follows because we can think of the response of the system to a field of frequency ω as being due to the correlation scale that can relax at the frequency ω, i. e. $\xi_\omega \propto \omega^{-\delta/\beta}$. The dielectric increment appropriate to this length scale can be obtained from (10.31) as $(\xi_\omega)^{\alpha/\delta}$ and hence,

$$\chi_\text{r}''(\omega) \propto \chi_\text{r}'(\omega) \propto (\xi_\omega)^{\alpha/\delta} \propto \omega^{-\alpha/\beta} = \omega^{n-1} \qquad (10.42)$$

with $n = 1 - \alpha/\beta$. These relationships describe what is called dynamic scaling. A detailed description of the derivation of the frequency dependence in (10.42) is given in [10.33] and the relationship of the power-law frequency dependence to self-similar relaxations is shown explicitly using scaled electrical circuits in [10.53]. Because of the small range of temperatures around T_c over which the power-law relationships (10.40, 10.41) are expected to hold it is difficult to determine the exponents α and β with any accuracy, however the dynamic scaling law can be easily demonstrated by using the master-curve technique and determining the locus of the representative point (see the beginning of Sect. 10.3.2). An example is shown in Fig. 10.9c where it can be seen that the representative-point locus of χ_0 as a function of γ is a power law with the same exponent as that of the frequency dependence of $\chi_r''(\omega)$ and $\chi_r'(\omega)$ for $\omega > \gamma$.

The Information Content of the Relaxation Frequency (Time)

The typical starting point in investigating the relaxation frequency (time) is to make an Arrhenius plot, i.e. $\log(\gamma)$ is plotted as a function of $1/T$. A straight line is taken to indicate an activated process of the form of (10.36), with the gradient yielding the activation energy Δ_γ. However in most cases data is only available from a restricted temperature range and so the variation of $k_B T/h$ with temperature will be small. In this case the free energy expression (10.38) would give an equally good straight line. Since there are many situations in which the pre-exponential factor in (10.36) will be less than a quantum vibration frequency 10^{12} Hz $\leq \nu \leq 3 \times 10^{13}$ Hz, it is difficult to discriminate between the potential-barrier situation and the free-energy barrier in which $S^\# < 0$. Only when $A \gg 3 \times 10^{13}$ Hz will identification with relaxation via a free-energy barrier be certain. As can be seen from the foregoing it is difficult to obtain any definite information from the pre-exponential factor in an activated relaxation, except for the latter case where an evaluation of $S^\#$ is possible. Where $S^\#$ can be determined it gives us a qualitative picture of a relaxing dipole centre which has to be structurally distorted in order for the dipole to adopt a different orientation. The corresponding value of $H^\#$ gives the amount of energy that has to be supplied to the group in order for the reorientation to occur even allowing for the distortion. What cannot be determined from this information is the size of the group that is involved, though the behaviour and magnitude of χ_0 may give an idea as to the dipole magnitude and density. To give an idea of the difficulty of a molecular interpretation let us take the case of cyclo-hexanol [10.54]. This molecule forms a plastic crystal, i.e. the molecular centres are located on a crystal lattice but their orientations are disordered. The dipole moment is associated with the only strong polar feature, the alcohol (−OH) group, which can take one of two orientations with respect to the molecule. It would therefore be expected that the measured activation energy of ≈ 0.5 eV would be the potential barrier to the transfer of the alcohol group between the two positions, with its contributions from the neigh-

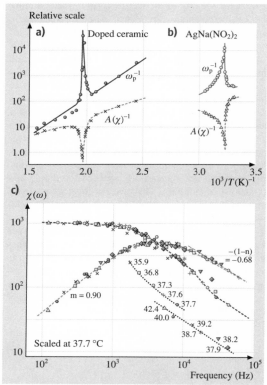

Fig. 10.9a–c An Arrhenius plot of the inverse of the susceptibility increment $A(\chi)(\equiv \chi_0)$ and loss peak frequency $\omega_p (\equiv \gamma)$ is shown for (**a**) a ceramic and (**b**) AgNa(NO$_2$)$_2$. The frequency-dependent susceptibility of AgNa(NO$_2$)$_2$ is shown in (**c**) together with the locus of the representative point as $T_c = 37.8\,°C$ is approached from above and below. Note the power-law relationship between the dielectric increment and the relaxation frequency, and the parallelism between this locus and the frequency dependence at $\omega > \gamma$, demonstrating an identical power-law behaviour. After [10.35]

bouring molecular cage as well as the molecular energy change. However, the molecular structure of the cyclohexanol can itself exist in two conformations, the chair and the boat, and can rearrange its orientation in a lattice by passing through the alternative conformation as an intermediate. The free-energy barrier to this interconversion is also ≈ 0.5 eV. So we cannot decide from the relaxation frequency whether the relaxation involves just $-$OH group transfer or transfer via a boat-to-chair transition or a mixture of both. In this case the pre-exponential frequency $A_{\text{eff}} \approx 6 \times 10^{16}$ Hz, so relaxation has to have a positive $S^{\#}$ and involve a number of atoms rather than a dipole reorienting on a rigid potential surface.

In the case of near-crystalline materials where the dipole is associated with defect centres we would expect the potential-barrier approach to be the best, but even here the fact that local reorientation is possible implies some sort of interaction between the surroundings and the moving dipole head. Calculations based on a rigid cage should (see for example [10.55]) however be possible, and comparison with experiment can be expected to determine how well this represents the situation and to what extent the transit of the barrier is affected by the barrier friction. Information provided by the dielectric increment should be of help here. The calculation ought to be able to yield an estimate of the reorientable component of the dipole, and if as seems likely the alternative orientations are at different energies, the temperature dependence of χ_0 should follow (10.34).

Although it is conceptually simple to think of dipoles relaxing upon a potential surface that remains unchanged during the relaxation, this is likely to be only an approximation to reality. The fact that alternative orientations exist indicates that in most cases the surrounding structure must be modified to some extent to accommodate the change; at the very least we can expect the dipole to polarise its surroundings differently according to its orientation. The expressions in Sect. 10.3.2 for the relaxation frequency of dipoles relaxing on a free-energy surface and dipoles in ferroelectrics reflect this fact in different ways. The ferroelectric behaviour described in Sect. 10.3.2 shows that when the *dipoles* become extensively correlated the relaxation frequency reduces as an inverse power of the correlation length and the dielectric increment increases as a power of the relaxation length. The self-similar scaling relates this behaviour to the frequency dependence of the susceptibility. The behaviour of the α-response of a glass-forming system involves dipole–structure interaction in a different way. The relaxation frequency approaches zero as T approaches T_g from above. Whatever the details of the process this behaviour has to indicate an increased difficulty for the dipole to reorient, which here is associated with structural ordering, densification, and atomic packing, rather than long-range correlations as in ferroelectrics. This response is also one for which the dielectric increment is often insensitive to temperature. If we put the two dielectric response features together we come to a picture in which the electric field effectively modifies the configuration entropy of the system in generating a net dipole density fluctuation. The net dipole density produced is essentially the same at different temperatures, so the change in configuration density generated by the electric field does not vary, but the relaxation time increases as the activation enthalpy $H^{\#}$ and entropy $S^{\#}$ increases. Put together with the fact that in structural glass formation small local regions are attempting to adopt a crystalline structure, this data indicates that there are *local* values of ground-state configuration entropy that reduce as T_g is approached, with a transition state involving a disordering of the local regions to free the dipole enough to let it adopt an alternative orientation in an equally ordered but different configuration. The dipole density fluctuations that couple to the electric field seem to involve reorganisations of the structure that can occur without a change in the value of the heat content H, i. e. they are essentially configuration entropy fluctuations rather than thermal fluctuations. In contrast to the ferroelectric situation the slowing down of the relaxation is not caused by longer-range correlations but by the increasingly larger numbers of molecular adjustments required to achieve a dipole reorientation.

The message of this section is that in most cases a detailed molecular description of the dipole motions is generally not accessible just from an analysis of the dielectric response. The reason is that, in general, dipole reorientations involve adjustments in the surrounding molecules/atoms that are not easy to define in molecular terms. However by putting together the behaviour of the dielectric increment and relaxation frequency it should be possible to obtain some general idea as to the extent of the connection of the reorientation to the molecular environment and the way in which it takes place.

10.3.3 The Relaxation Peak Shape

The explanation of the frequency dependence of the susceptibility is currently the most contentious of the features of the dielectric response. Many workers are content with just defining the shape by one or other of the empirical functions mentioned in Sect. 10.2.2, or through the power-law exponents of (10.20, 10.21). This gives a fingerprint of the dipole dynamics but no more. In particular it does not provide a description of the equation of motion of the dipole density fluctuation. Others determine what is termed a *distribution of relaxation times* for the loss peak in $\chi_r''(\omega)$. Essentially this approach is predicated on the assumption that the broadening of the loss peak compared to that of the Debye response (10.19) is the result of dipoles of the same type and dipole moment that each relax according to the Debye equation of motion (10.17) but possess different relaxation times with a distribution denoted by $g(\tau)$, which is defined via (10.43)

$$\chi_r''(\omega) = \int_0^\infty g(\tau) \frac{\chi_0 \omega \tau}{1 + \omega^2 \tau^2} \, d\tau \; . \qquad (10.43)$$

This construction is still no more than a fingerprint unless a physical reason for the distribution $g(\tau)$ can be found. Usually this is ascribed to a distribution of local activation energies associated with dipoles that each exist in their own potential surface independent of one another. The system usually quoted as an example is that of the β-response in the rigid glassy phase, which typically has a very broad loss peak. In this case it is assumed that each dipole that can reorient to contribute to the β-response is essentially trapped in a local potential surface that is held rigid in the glass state. Of course the potential surface is not truly rigid, molecular/atomic vibrations must take place, but it is assumed that their effect on the potential averages out during the relaxation and their only effect is to raise the energy state of the reorienting dipole head to the state at the top of the barrier. One problem associated with this explanation of the origin of $g(\tau)$ is that, if the *function $g(\tau)$* is independent of temperature the values of exponents n and m (10.20), (10.21) will be temperature dependent. This does not seem to be the case in general, with these exponents usually either constant or changing at the most slowly or discretely at a transition of state (see for example [10.8, 56]), but there is no real agreement on this point. Of course a temperature-dependent distribution function $g(\tau)$ may be assumed, but then the question arises as to why it *is* temperature dependent in a system that is presumed to be macroscopically rigid. Another facet of the problem associated with non-Debye loss peaks that does not seem to have received any consideration is the possibility that the magnitude of the reorientable dipole moment associated with each site of a given activation energy is also distributed. It is clear that this is highly likely even if the dipole moment that changes direction is the same everywhere, as illustrated in Fig. 10.7. Also, as described in Sects. 10.2 and 10.3, the local dipole may be correlated with other dipoles or its surroundings, and in this case we can expect the Debye rate equation not to hold. The ferroelectric result (Sect. 10.3.2) already shows that this is the case when the dipoles motions are correlated giving the system a scale relationship in its dynamics, and even correlation between the dipole and its surroundings, for which there is considerable evidence (Sects. 10.2 and 10.3) can be expected to alter the form of the equation of motion from (10.17), by for example anharmonic coupling between the various modes. Even if we assume that all the criteria for the application of (10.43) are met, the $g(\tau)$ that are required to fit the experimental form of response defined by (10.20), (10.21) [and its corresponding theoretical response function, (10.22)] have unique features that require a physical justification, i. e. there is a cusp or sharp peak at the value of τ corresponding to the characteristic frequency ($\tau_c = 1/\gamma$), and power-law wings to either side whose power exponents are $1 - n [g(\tau) \propto \tau^{1-n}]$ for $\tau < \tau_c$, and $-m [g(\tau) \propto \tau^{-m}]$ for $\tau > \tau_c$. In the Debye case the distribution becomes a delta function at the characteristic relaxation time. Essentially the *distribution of relaxation times* approach is convenient but it is not as easy to justify as would seem at first sight.

The Williams and Watt response function [10.28] started life as a heuristic suggestion but has received some later theoretical support [10.57–62]. The dynamic scaling behaviour appropriate to ferroelectrics gives a clue as to the way in which a frequency-dependent susceptibility of the form of (10.20) can come about, which results from both the equation of motion (10.22) and the response function (10.26). Essentially there has to be a self-similarity (or scaling) between the relaxation frequency of subcomponents of the system and their contribution to the dielectric increment (as illustrated in the circuit model of [10.8]). The theory proposed by *Palmer* et al. [10.60] refers this scaling to the removal of a hierarchy of constraints, thus for example we may imagine that close neighbours move quickest and remove the constraints imposed on larger groups of molecules and so on. This picture would be appropriate to a system such as a glass-forming material. The assumption however is that the motions are overdamped at all levels of the hierarchy, and hence no bridge is provided to the oscillatory motions known to occur at times close to quantum vibrations. A rather different stochastic approach has been taken by *Weron* and *Jurlewicz* [10.61, 62] who assumed that the system relaxation followed a path in which the fastest dipoles out of a distribution relaxed first and then the fastest out of the residual distribution and so on. The key feature is that the relaxing dipole is the extreme fastest from the distribution existing at the time. It was argued that the extreme-value statistical distribution function then led automatically to the response function of (10.26). The choice of appropriate extreme-value distribution was made on the grounds that the relaxation time was a positive definite variable. However this is not a sufficient criterion [10.63]. In order for (10.26) to apply the continuous distribution density of relaxation times (i.e. the distribution the system would have if it were of infinite size) has to be stable to scale changes (see for example [10.64]) and thus has to approach the extreme of long times as an inverse power law, otherwise a different extreme value statistic or *none at all* applies. The required form of distribution from which an extreme selection has to be made is one that applies to the size distribution of scaling systems [10.65] such as percolation clusters [10.66] for example. So even with this stochastic approach we are led back to a system for which the dynamics scale in some way.

The Dissado–Hill function [10.23, 24] for which the response function obeys (10.22) also has scaling features as its basis, however unlike the other approaches it starts with the vibration dynamics of the system. It is assumed that a dipole that can reorient couples local vibration modes to itself. These are no longer extended normal modes but modes centred on the dipole that reduce in frequency according to the molecular mass involved. Their frequencies lie in the region between optical modes and the relaxation frequency and have a scaling relationship one to another. In the theory of *Nigmatullin* and *Le Mehaute* [10.67, 68], the modes are impulses that are involved in the dipole relaxation process whose time of action is scaled, i.e. the longer the time of action the more correlated they are to the dipole motion. In general these modes are local versions of coupled optical and acoustic modes and it is not surprising that they extend to such low frequencies as those involved in relaxation, as acoustic modes essentially extend to zero frequency. Their coupling with the dipole leads to the high-frequency power law of (10.20), where n expresses the extent to which the dipole reorientation couples to the surroundings, i.e. $n = 0$ corresponds to no coupling and the dipole moves independently of its surroundings, and $n = 1$ corresponds to full coupling in which the dipole motion is just part of the local mode. In a sense the short-time development of the response function of (10.22) is that of the changes in the configuration entropy as various amounts of different local modes are progressively coupled into the dipole motion [10.23]. In this case there is no necessity for n to be temperature dependent. At the characteristic relaxation frequency, the *characteristic* dipole group relaxes and transfers energy to the heat bath. The low-frequency behaviour of (10.21) is the result of a distribution in the ensemble of locally coupled dipole motions. This occurs because the motions of local dipole centres may be weakly coupled to one another. As a result the relaxation of the centres proceeds in a scaled or self-similar manner. First the dipole in a local centre relaxes with respect to its own environment, this leaves each dipole centre unrelaxed with respect to one another. Next groups of dipole centres, with some arrangement depending upon the specific structure involved, relax as a group. Then groups of groups relax and so on. Each level of inter-group complexity essentially has a time scale associated with its relaxation that cannot be reached until the preceding level has been completed. This is rather similar to the constraint relaxation concept of *Palmer* et al. [10.60]. The power-law exponent m expresses the way that this hierarchy of relaxing groups is scaled, by defining the power-law tail of the distribution of inter-group relaxation times in the ensemble [10.69]. A value of $m = 1$ corresponds to a sequence of inter-group relaxations with a relaxation time that is proportional to the number of groups involved in the sequence [10.69]. This implies that the sequential events are uncorrelated, i.e. the long-time relaxation is a white-noise (random) process [10.23]. When combined together with $n = 0$ the Debye response is recovered. On the other hand a value of m approaching zero corresponds to relaxation times that are a very high power of the number of groups involved [10.69] and indicates a very strong connection between groups at all levels of the hierarchy. This will spread the response to very low frequencies, as observed. Essentially m is a measure of the extent to which energy is transferred to the heat bath (dissipated) at each level of the hierarchy compared to being stored in the inter-group interactions of the next level. Again scaling is at the basis of the theory, but now with two different ways in which it can be involved. This theory is not generally accepted. The controversial parts of the theory are firstly the coupling of the dipole motions with vibration modes, which modifies the oscillator behaviour towards an overdamped form, and secondly the hierarchy of relaxations whereby energy is transferred to the heat bath. However it should be noted that the susceptibility function that results has a general form that agrees well with experiment. In addition the concepts are reasonable given the complexity that is likely to occur in the internal motions once an ideal crystalline regularity is ruled out by the possibility of dipole reorientation. Thus for example this concept would apply to the dipoles involved in the β-response of the glass state as well as correlated motions of dipoles over long distances, since even in a macroscopically rigid material local vibrations take place. In fact the limited regions of local order in a glass phase can be expected to favour such local modes and increase the coupling of the dipole motions with them, as observed.

The Information Content of the Loss Peak Shape

It is clear from the foregoing discussion that for all theoretical models of the loss peak shape the characteristic or loss peak frequency is but the culmination of a process in which subsections of the dipole and environment (with or without dipoles) are mixed into the motion of the dipole centre. In these models the dipole is not an independent entity, but rather an entity that is connected to some extent over a region that may be small or large. This implies that the dipole is not a particle that relaxes on a rigid potential surface independently of its environment. Only the *distribution of relaxation times* approach preserves the latter concept. If the theoretical models are correct they reflect the fact that we are looking at entities that are not truly of molecular scale but are of a mesoscopic nature. The correlations noted to occur in χ_0 and the need to use free energy rather than potential surfaces in describing the relaxation frequency support this view. The local entities involved are however not rigid features like permanent dipoles, and for this reason we should expect there to be weak connections between them that can be expected to relate to the way in which the relaxation of the whole system takes place. That is, not all entities relax at the characteristic time. As one entity relaxes its neighbours have to come into equilibrium with its new orientation and the system approaches equilibrium more slowly [i.e. as the time power law $t^{-(1+m)}$] than the exponential behaviour of the Debye response function or the expanded exponential function. The information contained in the loss peak shape indicates the way in which a dipole density fluctuation evolves from its state when initially created to an ensemble of mesoscopic dipole centres. The broadening of the peak from that of a Debye peak indicates the involvement of faster and slower processes as part of the overall mechanism, whatever their detailed origin, and in particular processes that have a scale relationship to one another. This must apply even to a distribution of relaxation times because of the unique form required for that distribution. Equation (10.22) implies an equivalent description that refers the overall relaxation process to a conversion of the vibration oscillation at short times to an overdamped motion as the dipole density fluctuation dissipates its energy irreversibly. In this sense evaluation of the shape parameters n and m give a means of describing this conversion process. At the very least they give a sense of the scaling involved in spreading the relaxation process around the characteristic relaxation frequency or equivalently the characteristic relaxation time.

10.4 Charge Transport

All dielectrics possess a constant (DC) conductivity (σ_{DC}), although usually it is very weak. Since $\chi''(\omega) = \sigma(\omega)/\omega$ as demonstrated in Sect. 10.1 (10.10), it would be expected that a dielectric response at low frequencies ($f \lesssim 10^{-2}$ Hz) would take a form in which $\chi''(\omega) = \sigma_{DC}/\omega$ and χ' is independent of frequency. In many cases however the conduction process is blocked at the electrodes or internal interfaces. In this case the DC conduction charges the interface, which behaves as a capacitor, and the whole system behaves as a single dipole. As long as the interface does not possess relaxation dynamics of its own, the response that would be observed is that given by the Debye response of (10.18, 10.19), with $\tau = 1/\gamma = RC_i$, where R is the resistance of the body of the material and C_i is the capacitance of the interface. The measured dielectric increment $\chi_0 = dC_i/A$, where A is the electrode area and d is the sample thickness, and can be very large depending upon the ratio of the sample thickness to that of the interface. The situation where the interface has a frequency-dependent capacitance has been thoroughly discussed by *Jonscher* [10.6] who has shown that $\chi'_r(\omega)$ is modified from $\chi'_r(\omega) \propto \omega^{-2} (\omega > \gamma)$ to $\chi'_r(\omega) \propto \omega^{-q}$, while $\chi''_r(\omega) \propto (1/\omega)(\omega > \gamma)$ as in (10.19). The value of q lies in the range $1 < q < 2$ with its value depending upon the frequency dependence of the interface capacitance.

The bulk DC conductance arises from charged particles whose movements are not bound to a charge of the opposite polarity as in dipoles but are free to move independently of their countercharge, resulting in a net charge displacement in the same way that a liquid flows. However the transport of charged particles within the body of the sample can give rise to a very different form of response when their movement lies along defined paths such that the longer the displacement of the charge the lower the number of paths or equivalently the more difficult the transport becomes. This behaviour was called low-frequency dispersion by *Jonscher* [10.6, 20] and quasi-DC conduction (q-DC) in the theoretical model of *Dissado* and *Hill* [10.70] who wished to distinguish it from low-frequency dipole responses. At frequencies below some characteristic value ω_c this form of response takes the form,

$$\sigma(\omega) \propto \omega^{1-p}$$
$$\left[\text{i.e. } \chi_r''(\omega) \propto \chi_r'(\omega) \propto \omega^{-p} \right], \quad \omega < \omega_c \quad (10.44)$$

and at frequencies above ω_c,

$$\sigma(\omega) \propto \omega^n$$
$$\left[\text{i.e. } \chi_r''(\omega) \propto \chi_r'(\omega) \propto \omega^{n-1} \right], \quad \omega > \omega_c. \quad (10.45)$$

The power-law exponents p in (10.44) and n in (10.45) have positive fractional values near to unity. It is obviously difficult to identify a value of p close to unity from measurements of $\chi_r''(\omega)$ [or equivalently $\sigma(\omega)$] and in many cases it is assumed that the measured behaviour shows a static (DC) conductivity. It is then common to subtract its supposed value from the measured data for $\sigma(\omega)$ to obtain an expression for the dipole relaxation response supposedly responsible for the behaviour at $\omega > \omega_c$. The values obtained in this way for $\chi_r''(\omega)$ at frequencies $\omega < \omega_c$ will not be zero as $\sigma(\omega)$ is not in fact constant, instead they will reduce as the frequency is reduced. This procedure yields a spurious loss peak in $\chi_r''(\omega)$ if the response is actually due to the q-DC mechanism, for which the high-frequency behaviour is an essential component of the whole q-DC mechanism and can never be resolved as a separate peak in $\chi_r''(\omega)$. The way to be certain that the response is really of the q-DC form is to measure the frequency response for $\chi_r'(\omega)$ and show that it takes the same frequency dependence. A convenient check is to determine the ratio of $\chi_r''(\omega)$ to $\chi_r'(\omega)$ (i.e. $\tan\delta$) which will have a constant value [10.6, 20] given by

$$\chi_r''(\omega)/\chi_r'(\omega) = \tan\delta = \cot[(1-p)\pi/2]. \quad (10.46)$$

Here $\tan\delta$ is called the loss tangent and δ is the phase angle between the real and imaginary components of the susceptibility. This relationship holds for pairs of values of $\chi_r'(\omega)$ and $\chi_r''(\omega)$ at the same frequency even if the measurements are noisy and so make it difficult to determine accurately the value of p from the frequency dependence. Another situation where it is difficult to detect the q-DC behaviour occurs in heterogeneous materials when one component has a low DC conductivity. This will add to the AC component, (10.44) and obscure the q-DC behaviour. In this case the DC conductivity can be eliminated from the data, if it is available over a large enough frequency range, by applying the Kramers–Kronig transform of (10.25) to obtain the function $\chi_r''(\omega)$ without the DC component (σ_{DC}/ω). The validity of the procedure can be checked by applying the inverse transform (10.24) to the measured data for $\chi_r''(\omega)$. This should yield the measured $\chi_r'(\omega)$ since the DC conductivity does not contribute to the real component of the susceptibility.

The q-DC behaviour, (Fig. 10.10a), is most often found in materials that are heterogeneous on a mesoscopic scale such as ceramics [10.71], rocks [10.72], porous structures [10.73], and biological systems [10.74]. In these materials charged particles are transported via structured paths over some finite range. The transported charge and its countercharge give rise to an effective dipole with a large dipole moment. However the q-DC behaviour rarely appears as an isolated response. Because of the heterogeneous nature of the materials it is usually found to be electrically in series with other dielectric response elements such as interface capacitances, and electrically in parallel with a capacitive circuit element. The origin of the q-DC behaviour lies in a hidden scale relationship, with the dipole contribution to the susceptibility increment and its relaxation time both being a power of the length over which the transport takes place. The circuit models of *Dissado* and *Hill* et al. [10.8, 75] show how this behaviour can be produced when the system is represented by a geometrically self-similar arrangement of transport paths and blocking capacitive regions. Such geometrical regularity is not essential however [10.76]; a random arrangement of conductors (transport paths) in a dielectric (i.e. residual set of capacitances) will also result in the q-DC behaviour. It is clear that this construction yields percolation clusters below the size necessary to span the material, and these sub-percolation clusters will of necessity possess scaling relationships dependent on their size and the number of paths within them, and between clusters of different sizes. Such percolation systems also show q-DC behaviour when below their critical limit [10.77]. The theory proposed in [10.70] over much longer distances. The range allowed by the AC frequency increases as the frequency reduces, but the number of effective routes decreases as the distance becomes longer and hence the current becomes smaller. This leads to the weak dependence of the conductivity upon the AC frequency defined in (10.44). In the zero-frequency limit $\sigma(\omega) = 0$, which means that there are no paths across a system of infinite length, i.e. the system is formally below the percolation limit. Of course in reality the sample is of finite size and some routes may cross the sample, leading to a termination of the q-DC behaviour at low frequencies, either by a blocking capacitance at the electrodes or a weak DC current [10.78]. The theory leads to an analytical expression for the q-DC susceptibility in terms of a confluent hypergeometric function [10.25] (see [10.70] for the detailed expression), which is obtained from a response function that obeys an equation of motion similar to that of the permanent dipoles

$$\frac{d^2\phi}{dt^2} + \frac{(2+n+t/\tau)}{t}\frac{d\phi}{dt} + \frac{[n+(t/\tau)(1-p)]}{t^2}\phi = 0. \tag{10.47}$$

At times $t < \tau = 1/\omega_c$ the behaviour of the dipole density fluctuation is the same as that appropriate for permanent dipoles (10.22) with the same limiting solution, i.e. $\phi(t) \propto t^{-n}$ at $t < \tau$. At long times $t > \tau = 1/\omega_c$, however, it takes a different form in which the dipolar fluctuation relaxation has only a weak dependence on t, i.e. $\phi(t) \propto t^{p-1}$ at $t > \tau$. An example of the q-DC behaviour in frequency is given in Fig. 10.10b. The reader is referred to [10.74, 79, 80] for other examples.

In the limit of $p = 1 [\sigma(\omega) = \text{constant}; \omega < \omega_c]$, the last term of (10.47) is zero in the limit $t > \tau$ at which time the decay of ϕ is governed by the first two terms. The solution for $\phi(t)$ now takes the form

$$\phi(t) \propto t^{-n}, \quad t < \tau,$$
$$\phi(t) \propto \exp(-t/\tau), \quad t > \tau. \tag{10.48}$$

In this case we can see that the dipole density fluctuations are produced within the local clusters and that they relax by freeing the charges to move independently at $t = \tau$. At this time there is a density of completely free charged particles with charge and countercharge cancelling throughout the system giving a net dipole of zero as in a true DC-conduction process.

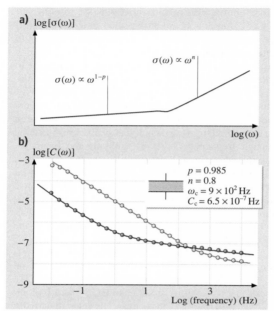

Fig. 10.10 (a) A schematic representation of the frequency dependence of $\sigma(\omega)$ for the q-DC process, (b) the equivalent q-DC susceptibility function $\chi(\omega)/\chi_c \propto (1+i\omega\tau)^{n-1}$ $_2F_1[1-n; 1+p; 2-n; (1+i\omega\tau)^{-1}]$ derived from the response function of (10.47) fitted to data from the leaf of *Nicotania* (Solanaceae), the *inset* gives the fitting parameters, with $\omega_c \equiv 1/\tau$, and $C_c \equiv \chi_c$. This data was chosen as it exhibits an isolated q-DC process. As before $_2F_1(,;;)$ is the Gaussian hypergeometric function; see [10.25]. After [10.79]

described the q-DC mechanism in similar terms to that used in percolation theory. At high frequencies they assumed that the charged particles had a short range of motion restricted to local clusters in which they were bound to their countercharges. The power-law exponent n reflected the extent of binding of the motions of the charges in the cluster and the local motions in the same way as for the dipole motions. Its value close to one indicates that the binding is strong and that the polarisation of the local clusters increases only to a small extent as the frequency is lowered, i.e. the net displacement of positive and negative charge is local and small. At the characteristic frequency the charged particles become free of their locality and are able to move

10.5 A Few Final Comments

The basic difficulty associated with the interpretation of dielectric responses is that they are of necessity macroscopic measurements made on samples that contain enormous numbers of atoms and molecules. In condensed-phase materials it is not possible to consider these systems as made up of local entities each moving independently of one another. All entities that contribute a permanent dipole are part of the condensed-phase structure, and even though they have a degree of freedom associated with the possibility of dipole reorientation, they will have motions that are correlated or connected to some extent to the molecules/atoms in their environment. This means that dipole reorientation is not that of a bare entity; instead it involves to some extent a local region. These regions will behave differently in different kinds of material and their definition and the way of describing their behaviour has not yet been established with any sort of rigour. Since we are dealing with a macroscopic measurement, there will of necessity be an ensemble of the local entities. This will result in a distribution of entities, but since these are part of the structure there will be some sort of connection between them unless the structure itself is disconnected dynamically. This means that fluctuations will take place among our entities, and perhaps even dissociation and amalgamation. These effects will also have an influence on the relaxation dynamics. In the foregoing I have tried to give some simple pictures as to what is happening and to do so in a holistic way by correlating information from different facets of the measurement. What is abundantly clear is that the dynamics of such systems are very complicated in detail, but I hope that I have done enough to convince you that there are some basic features of the relaxation process that are common to all systems of this type, even though a full understanding of their nature does not yet exist.

References

10.1   P. Debye: *Polar Molecules* (Dover, New York 1945)
10.2   C. P. Smyth: *Dielectric Behaviour and Structure* (McGraw–Hill, New York 1955)
10.3   N. G. McCrum, B. E. Read, G. Williams: *Anelastic and Dielectric Effects in Polymeric Solids* (Wiley, New York 1967)
10.4   V. V. Daniels: *Dielectric Relaxation* (Academic, New York 1967)
10.5   C. J. F. Bottcher, P. Bordewijk: *Theory of Electric Polarisation*, Vol. I, II (Elsevier, Amsterdam 1978)
10.6   A. K. Jonscher: *Dielectric Relaxation in Solids* (Chelsea Dielectric, London 1983)
10.7   K. L. Ngai, G. B. Wright (Eds.): Relaxations in Complex Systems. In: *Proc. The International Discussion Meeting on Relaxations in Complex Systems* (Elsevier, Amsterdam 1991)
10.8   R. M. Hill: Electronic Materials from Silicon to Organics. In: *Dielectric properties and materials*, ed. by L. S. Miller, J. B. Mullin (Plenum, New York 1991) pp. 253–285
10.9   T. Furukawa: IEEE Trans. E.I. **24**, 375 (1989)
10.10   C. J. F. Bottcher: *Theory of Electric Polarisation* (Elsevier, Amsterdam 1952) p. 206
10.11   M. Born, E. Wolf: *Principles of Optics* (Pergamon, Oxford 1965)
10.12   A. R. von Hippel (ed.): *Dielectric Materials and Applications* (Wiley, New York 1958)
10.13   R. N. Clarke, A. Gregory, D. Connell, M. Patrick, I. Youngs, G. Hill: Guide to the Characterisation of Dielectric Materials at RF and Microwave Frequencies. In: *NPL Good Practice Guide* (Pub. Inst. Measurement and Control, London 2003)
10.14   W. Heitler: *The Quantum Theory of Radiation*, 3rd edn. (Dover, London 1984)
10.15   D. Bohm, D. Pines: Phys. Rev. **82**, 625 (1951)
10.16   H. Eyring, J. Walter, G. E. Kimball: *Quantum Chemistry* (Wiley, New York 1960)
10.17   P. Wheatley: *The Determination of Molecular Structure* (Clarendon, Oxford 1959) Chap. XI
10.18   R. Kubo: Rep. Prog. Phys. **29**, 255–284 (1966)
10.19   R. Kubo: J. Phys. Soc. Jpn. **12**, 570 (1957)
10.20   A. K. Jonscher: J. Phys. D Appl. Phys. **32**, R57 (1999)
10.21   S. Havriliak, S. Negami: J. Polym. Sci. C **14**, 99 (1966)
10.22   S. Jr. Havriliak, S. J. Havriliak: *Dielectric and Mechanical Relaxation in Materials* (Hanser, New York 1997)
10.23   L. A. Dissado, R. M. Hill: Proc. R. Soc. London **390**(A), 131 (1983)
10.24   L. A. Dissado, R. R. Nigmatullin, R. M. Hill: Dynamical Processes in Condensed Matter. In: *Adv. Chem. Phys*, Vol. LXIII, ed. by R. Evans M. (Wiley, New York 1985) p. 253
10.25   M. Abramowitz, I. A. Stegun: *Handbook of Mathematical Functions* (Dover, New York 1965)
10.26   L. A. Dissado, R. M. Hill: Chem. Phys. **111**, 193 (1987)
10.27   L. A. Dissado, R. M. Hill: Nature (London) **279**, 685 (1979)
10.28   G. Williams, D. C. Watt: Trans. Farad. Soc. **66**, 80 (1970)
10.29   R. Kohlrausch: Pogg. Ann. Phys. **91**, 198 (1854)

10.30 A. K. Loidl, J. Knorr, R. Hessinger, I. Fehst, U. T. Hochli: J. Non-Cryst. Solids **269**, 131 (1991)
10.31 C. Kittel: *Introduction to Solid State Physics* (Wiley, New York 1966)
10.32 J. Joffrin, A. Levelut: J. Phys. (Paris) **36**, 811 (1975)
10.33 P. C. Hohenberg, B. I. Halperin: Rev. Mod. Phys. **49**, 435–479 (1977)
10.34 J. G. Kirkwood: J. Chem. Phys. **7**, 911 (1939)
10.35 L. A. Dissado, R. M. Hill: Phil. Mag. B **41**, 625–642 (1980)
10.36 L. A. Dissado, M. E. Brown, R. M. Hill: J. Phys. C **16**, 4041–4055 (1983)
10.37 L. A. Dissado, R. M. Hill: J. Phys. C **16**, 4023–4039 (1983)
10.38 H. A. Kramers: Physica **VII**(4), 284–304 (1940)
10.39 R. F. Grote, J. T. Hynes: J. Chem. Phys. **73**, 2715–2732 (1980)
10.40 W. A. Phillips: Proc. R. Soc. London A **319**, 535 (1970)
10.41 J. le G. Gilchrist: Cryogenics **19**, 281 (1979)
10.42 J. le G. Gilchrist: Private communication with R. M. Hill, reported in 10.44, (1978)
10.43 H. Eyring: J. Chem. Phys. **4**, 283 (1936)
10.44 R. M. Hill, L. A. Dissado: J. Phys. C **15**, 5171 (1982)
10.45 S. H. Glarum: J. Chem. Phys. **33**, 1371 (1960)
10.46 C. A. Angell: Encyclopedia of Materials. In: *Science and Technology*, Vol. 4, ed. by K. H. J. Buschow, R. W. Cahn, M. C. Fleming, B. Ilschner, E. J. Kramer, S. Mahajan (Elsevier, New York 2001) pp. 3565–3575
10.47 W. Kauzmann: Chem. Rev. **43**, 219 (1948)
10.48 G. P. Johari, M. Goldstein: J. Chem. Phys. **53**, 2372 (1970)
10.49 M. L. Williams, R. F. Landel, J. D. Ferry: J. Am. Chem. Soc. **77**, 3701 (1955)
10.50 M. Goldstein: J. Chem. Phys. **39**, 3369 (1963)
10.51 D. Turnbull, M. H. Cohen: J. Chem. Phys. **14**, 120 (1961)
10.52 R. R. Nigmatullin, S. I. Osokin, G. Smith: J. Phys. Cond. Matter **15**, 1 (2003)
10.53 R. M. Hill, L. A. Dissado, R. R. Nigmatullin: J. Phys. Cond. Matter **3**, 9773 (1991)
10.54 M. Shablakh, L. A. Dissado, R. M. Hill: J. Chem. Soc. Faraday Trans. 2 **79**, 369 (1983)
10.55 R. Pirc, B. Zeks, P. Goshar: Phys. Chem. Solids **27**, 1219 (1966)
10.56 K. Pathmanathan, L. A. Dissado, R. M. Hill: Mol. Cryst. Liq. Cryst. **135**, 65 (1986)
10.57 K. L. Ngai, A. K. Jonscher, C. T. White: Nature **277**, 185 (1979)
10.58 K. L. Ngai, A. K. Rajgopal, S. Tietler: J. Phys. C **17**, 6611 (1984)
10.59 K. L. Ngai, R. W. Rendell, A. K. Rajgopal, S. Tietler: Ann. Acad. Sci. NY **484**, 150 (1986)
10.60 R. G. Palmer, D. Stein, E. S. Abrahams, P. W. Anderson: Phys. Rev. Lett. **53**, 958 (1984)
10.61 K. Weron: J. Phys. Cond. Matter **4**, 10507 (1992)
10.62 K. Weron, A. Jurlewicz: J. Phys. A **26**, 395 (1993)
10.63 E. J. Gumbel: *Statistics of Extremes* (Columbia University Press, New York 1958)
10.64 J. T. Bendler: J. Stat. Phys. **36**, 625 (1984)
10.65 J. Klafter, M. F. Schlesinger: Proc. Natl. Acad. Sci. **83**, 848 (1986)
10.66 D. Stauffer: *Introduction to Percolation Theory* (Taylor Francis, London 1985)
10.67 R. R. Nigmatullin: Theor. Math. Phys. **90**, 354 (1992)
10.68 R. R. Nigmatullin, A. Le Mehaute: Int. J. Sci. Geores. **8**, 2 (2003)
10.69 L. A. Dissado, R. M. Hill: J. Appl. Phys. **66**, 2511 (1989)
10.70 L. A. Dissado, R. M. Hill: J. Chem. Soc. Faraday Trans. 1 **80**, 325 (1984)
10.71 T. Ramdeen, L. A. Dissado, R. M. Hill: J. Chem. Soc. Faraday Trans. 2 **80**, 325 (1984)
10.72 R. R. Nigmatullin, L. A. Dissado, N. N. Soutougin: J. Phys. D **25**, 113 (1992)
10.73 A. Puzenko, N. Kozlovich, A. Gutina, Yu. Feldman: Phys. Rev. B **60**, 14348 (1999)
10.74 L. A. Dissado: Phys. Med. Biol. **35**, 1487 (1990)
10.75 L. A. Dissado, R. M. Hill: Phys. Rev. B **37**, 3434 (1988)
10.76 D. P. Almond, C. R. Brown: Phys. Rev. Lett. **92**, 157601 (2004)
10.77 Yu. Feldman, N. Kozlovich, Yu. Alexandrov, R. Nigmatullin, Ya. Ryabov: Phys. Rev. E **54**, 20–28 (1996)
10.78 L. A. Dissado, R. M. Hill, C. Pickup, S. H. Zaidi: Appl. Phys. Commun. **5**, 13 (1985)
10.79 R. M. Hill, L. A. Dissado, K. Pathmanathan: J. Biol. Phys. **15**, 2 (1987)
10.80 M. Shablakh, L. A. Dissado, R. M. Hill: J. Biol. Phys. **12**, 63 (1984)

11. Ionic Conduction and Applications

Solid state ionic conductors are crucial to a number of major technological developments, notably in the domains of energy storage and conversion and in environmental monitoring (such as battery, fuel cell and sensor technologies). Solid state ionic membranes based on *fast ion conductors* potentially provide important advantages over liquid electrolytes, including the elimination of sealing problems and the ability to miniaturize electrochemical devices using thin films. This chapter reviews methods of optimizing ionic conduction in solids and controlling the ratio of ionic to electronic conductivity in mixed conductors. Materials are distinguished based on whether they are characterized by intrinsic vs. extrinsic disorder, amorphous vs. crystalline structure, bulk vs. interfacial control, cation vs. anion conduction and ionic vs. mixed ionic–electronic conduction. Data for representative conductors are tabulated.

A number of applications that rely on solid state electrolytes and/or mixed ionic–electronic conductors are considered, and the criteria used to choose such materials are reviewed. Emphasis is placed on fuel cells, sensors and batteries, where there is strong scientific and technological interest. The chapter concludes by considering how solid state ionic materials are likely to be used in the future, particularly in light of the trend for miniaturizing sensors and power sources.

11.1	**Conduction in Ionic Solids**	214
11.2	**Fast Ion Conduction**	216
	11.2.1 Structurally Disordered Crystalline Solids	216
	11.2.2 Amorphous Solids	219
	11.2.3 Heavily Doped Defective Solids	219
	11.2.4 Interfacial Ionic Conduction and Nanostructural Effects	220
11.3	**Mixed Ionic–Electronic Conduction**	221
	11.3.1 Defect Equilibria	221
	11.3.2 Electrolytic Domain Boundaries	222
11.4	**Applications**	223
	11.4.1 Sensors	223
	11.4.2 Solid Oxide Fuel Cells (SOFC)	224
	11.4.3 Membranes	225
	11.4.4 Batteries	225
	11.4.5 Electrochromic Windows	226
11.5	**Future Trends**	226
References		226

The ionic bonding of many refractory compounds allows for ionic diffusion and correspondingly, under the influence of an electric field, ionic conduction. This contribution, for many years, was ignored as being inconsequential. However, over the past three to four decades, an increasing number of solids that support anomalously high levels of ionic conductivity have been identified. Indeed, some solids exhibit levels of ionic conductivity comparable to those of liquids. Such materials are termed *fast ion conductors*. Like solid state electronics, progress in *solid state ionics* has been driven by major technological developments, notably in the domains of energy storage and conversion and environmental monitoring, based on ongoing developments in battery, fuel cell and sensor technologies. Some of the most important applications of solid state electronics and solid state ionics, and their categorization by type and magnitude of conductivity (such as dielectric, semiconducting, metallic and superconducting), are illustrated in Fig. 11.1 [11.1]. This figure also emphasizes that solids need not be strictly ionic or electronic, but may and often do exhibit mixed ionic–electronic conductivity. These *mixed conductors* play a critical role – particularly as electrodes – in solid state ionics, and are receiving comparable if not more attention than solid electrolytes at the present. Such solids are the result of a combination of the fields of *solid state ionics* and *solid state electrochemistry*, and they have grown in importance as our society has become more acutely concerned with efficient and environmentally clean methods for energy conversion, conservation and storage [11.2].

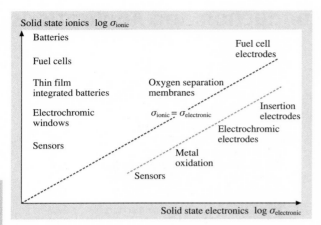

Fig. 11.1 Illustration of typical applications of ionic and electronic conductors as a function of the magnitude of electrical conductivity. Applications requiring mixed ionic electronic conductivity fall within the quadrant bounded by the two axes. After [11.1]

Solid state ionic membranes provide important potential advantages over liquids. The most important of these include: (a) elimination of sealing problems associated with chemically reactive liquid or molten electrolytes; (b) minimization of discharge under open circuit conditions, and; (c) the ability to miniaturize electrochemical devices through the use of thin films. In the following, we begin by discussing methods of optimizing ionic conduction in solids and controlling the ratio of ionic to electronic conductivity. We then consider a number of applications that rely on solid state electrolytes and/or mixed ionic–electronic conductors and the criteria that should be used when selecting materials. We conclude by considering how solid state ionic materials are likely to be used in the future, particularly in light of trends related to the miniaturization of sensors and power sources.

11.1 Conduction in Ionic Solids

The electrical conductivity, σ, the proportionality constant between the current density j and the electric field E, is given by

$$j/E = \sigma = \sum_i c_i Z_i q \mu_i \quad (11.1)$$

where c_i is the carrier density (number/cm^3), μ_i the mobility (cm^2/Vs), and $Z_i q$ the charge ($q = 1.6 \times 10^{-19}$ C) of the ith charge carrier. The huge (many orders of magnitude) differences in σ between metals, semiconductors and insulators generally result from differences in c rather then μ. On the other hand, the higher conductivities of electronic versus ionic conductors are generally due to the much higher mobilities of electronic versus ionic species [11.3].

Optimized ionic conduction is a well-known characteristic of molten salts and aqueous electrolytes wherein all ions move with little hindrance within their surroundings. This leads to ionic conductivities as high as 10^{-1}–10^1 S/cm in molten salts at temperatures of 400–900 °C [11.4]. Typical ionic solids, in contrast, possess limited numbers of mobile ions, hindered in their motion by virtue of being trapped in relatively stable potential wells. Ionic conduction in such solids easily falls below 10^{-10} S/cm for temperatures between room temperature and 200 °C. In the following sections, we examine the circumstances under which the magnitude of ionic conduction in solids approaches that found in liquid electrolytes.

The motion of ions is described by an activated jump process, for which the diffusion coefficient is given by [11.5]

$$D = D_0 \exp(-\Delta G/kT)$$
$$= \gamma(1-c)Za^2 \nu_0 \exp(\Delta S/k)\exp(-E_m/kT), \quad (11.2)$$

where a is the jump distance, ν_0 the attempt frequency, and E_m the migration energy. The factor $(1-c)Z$ defines the number of neighboring unoccupied sites, while γ includes geometric and correlation factors. Note that the fractional occupation c here should not be confused with c_i, the charge carrier concentration, nor should the number of nearest neighbors Z be confused with Z_i, the number of charges per carrier defined in (11.1). Since the ion mobility is defined by $\mu_i = Z_i q D_i / k_B T$, where D_i and k_B are the diffusivity and Boltzmann constant respectively, and the density of carriers of charge $Z_i q$ is Nc, where N is the density of ion sites in the sublattice of interest, the ionic conductivity becomes

$$\sigma_{\text{ion}} = \gamma \left[N(Z_i q)^2 / k_B T \right] c(1-c) Z a^2 \nu_0$$
$$\times \exp(\Delta S/k_B) \exp(-E_m/k_B T)$$
$$= (\sigma_0/T) \exp(-E/k_B T). \quad (11.3)$$

or

$$\sigma_{\text{ion}} = \gamma N(Z_i q)^2 c(1-c) Z a^2 \gamma_0 / k_B T$$
$$\times \exp(\Delta S/k_B) \exp(-E_m/k_B T) \quad (11.4)$$

This expression shows that σ_{ion} is nonzero only when the product $c(1-c)$ is nonzero. Since all normal sites are fully occupied ($c=1$) and all interstitial sites are empty ($c=0$) in a perfect classical crystal, this is expected to lead to highly insulating characteristics. The classical theory of ionic conduction in solids is thus described in terms of the creation and motion of atomic defects, notably vacancies and interstitials.

Three mechanisms for ionic defect formation in oxides should be considered. These are (1) thermally induced intrinsic ionic disorder (such as Schottky and Frenkel defect pairs), (2) redox-induced defects, and (3) impurity-induced defects. The first two categories of defects are predicted from statistical thermodynamics [11.6], and the latter form to satisfy electroneutrality. Examples of typical defect reactions in the three categories, representative of an ionically bonded binary metal oxide, are given in Table 11.1, in which the $K_i(T)$s represent the respective equilibrium constant and $a_{N_2O_3}$ the activity of the dopant oxide N_2O_3 added to the host oxide MO_2. Schottky and Frenkel disorder (1, 2) leave the stoichiometric balance intact. Reduction–oxidation behavior, as represented by (3), results in an imbalance in the ideal cation-to-anion ratio and thus leads to *nonstoichiometry*. Note that equilibration with the gas phase, by the exchange of oxygen between the crystal lattice and the gas phase, generally results in the simultaneous generation of both ionic and electronic carriers. For completeness, the equilibrium between electrons and holes is given in (4).

Altervalent impurities [for example N^{3+} substituted for the host cation M^{4+} – see (5)] also contribute to the generation of ionic carriers, commonly more than intrinsic levels do. This follows from the considerably reduced ionization energies required to dissociate impurity-defect pairs as compared to intrinsic defect generation. For example, E_A might correspond to the energy required to dissociate an acceptor–anion vacancy pair or E_D to the energy needed to dissociate a donor–anion interstitial pair. Such dissociative effects have been extensively reported in both halide and oxide literature [11.7]. A more detailed discussion is provided below in the context of achieving high oxygen ion conductivity in solid oxide electrolytes.

The oxygen ion conductivity σ_i is given by the sum of the oxygen vacancy and interstitial partial conductivities. In all oxygen ion electrolytes of interest, the interstitial does not appear to make significant contributions to the ionic conductivity, and so it is the product of the oxygen vacancy concentration $[V_O^{\cdot\cdot}]$, the charge $2q$, and the mobility (μ_v):

$$\sigma_i \approx [V_O^{\cdot\cdot}] \, 2q\mu_v \quad (11.5)$$

Optimized levels of σ_i obviously require a combination of high charge carrier density and mobility.

Classically, high charge carrier densities have been induced in solids by substituting lower valent cations for the host cations [11.2]. Implicit in the requirement for high carrier densities are:

1. High solid solubility of the substituent with the lower valency,
2. Low association energies between the oxygen vacancy and dopant,
3. No long-range ordering of defects.

Additives which induce minimal strain tend to exhibit higher levels of solubility. The fluorite structure is the most well-known of these structures, with stabilized zirconia the best-known example. In this case, Y^{3+} substitutes for approximately 10% of Zr in $Zr_{1-x}Y_xO_{2-x/2}$, leading to $\sigma_i \approx 10^{-1}$ S/cm at 1000 °C and an activation energy of ≈ 1 eV. Other examples include CeO_2 [11.8], other fluorite-related structures such as the pyrochlores $A_2B_2O_7$ [11.9], and perovskites such as $La_{1-x}Sr_xGa_{1-y}Mg_yO_{3-\delta}$ (LSGM) [11.10].

Since the dopant and vacancy are of opposite charge (for example, Y_{Zr}' and $V_O^{\cdot\cdot}$), they tend to associate. With cations being much less mobile than oxygen ions, this serves to trap the charge carrier. It is of interest to examine how the concentration of "free" mobile carriers

Table 11.1 Typical defect reactions

Defect reactions	Mass action relations	
$MO \Leftrightarrow V_M'' + V_O^{\cdot\cdot}$	$[V_M''][V_O^{\cdot\cdot}] = K_S(T)$	(1)
$O_O \Leftrightarrow V_O^{\cdot\cdot} + O_i''$	$[V_O^{\cdot\cdot}][O_i''] = K_F(T)$	(2)
$O_O \Leftrightarrow V_O^{\cdot\cdot} + 2e' + 1/2 O_2$	$[V_O^{\cdot\cdot}]n^2 = K_R(T) P_{O_2}^{-1/2}$	(3)
$0 \Leftrightarrow e' + h^\bullet$	$np = K_e(T)$	(4)
$N_2O_3(MO_2) \Leftrightarrow 2N_M' + 3O_O + V_O^{\cdot\cdot}$	$[N_M']^2 \cdot [V_O^{\cdot\cdot}]/a_{N_2O_3} = K_N(T)$	(5)

depends on the dopant concentration and the association energy. Consider the neutrality relation representing vacancy compensation of acceptor impurities by

$$N_V = \beta\, N_I ,\qquad (11.6)$$

where N_V and N_I are the vacancy and impurity densities while β reflects the relative charges of the two species and normally takes on values of 1 (for A_M'') and $\frac{1}{2}$ (for A_M'). The association reaction is given by

$$(I-V)^{x-y} \Leftrightarrow I^x + V^y ,\quad (\beta = x/y) ,\qquad (11.7)$$

where x and y are the relative charges of the impurity and vacancy, respectively. The corresponding mass action relation is then

$$N_I N_V / N_{Dim} = K_A^\circ \exp(-\Delta H_A / kT) \qquad (11.8)$$

where N_{Dim} is the concentration of dimers and N_I and N_V are the corresponding defects remaining outside the complexes. It is straightforward to show that for weak dissociation (low temperatures or high association energies) one obtains the following solutions:

$$\beta = 1 : \quad N_V = \left(N_I K_A^0\right)^{\frac{1}{2}} \exp(-\Delta H_A / 2kT) , \qquad (11.9)$$

$$\beta < 1 : \quad N_V = \left(\frac{1-\beta}{\beta}\right) K_A^0 \exp(-\Delta H_A / kT) . \qquad (11.10)$$

The solution for condition $\beta=1$ is the more familiar one. As in semiconductor physics [11.11], the number of free electrons or holes is proportional to the square root of the dopant density at reduced temperature, and it exhibits an Arrhenius dependence with activation energy that is equal to one half of the association or ionization energy. The solution for condition $\beta < 1$ is more unusual. Here one predicts that N_V is independent of dopant density! Also, the activation energy is predicted to be equal to the association energy.

At sufficiently high temperatures or low association energies, essentially all of the dimers are dissociated and

$$N_V = \beta N_I = \beta N_I \text{ (total)} . \qquad (11.11)$$

In general, therefore, two energies contribute to ionic conduction: a defect energy, E_D (which may either be related to the Frenkel or Schottky formation energy, or to a dissociation energy), and a migration energy E_m. The value of E in (11.3) therefore takes on different values in three characteristic temperature regimes. These include:
1. $E = E_m + E_A/2$: extrinsic associated regime at low T,
2. $E = E_m$: extrinsic fully dissociated regime at intermediate T,
3. $E = E_m + E_F/2$: intrinsic defect regime at elevated T (for instance, for Frenkel equilibrium).

For optimized ionic conduction to exist, two criteria must be satisfied simultaneously. First the term c in (11.3) must approach 1/2. This corresponds to nearly all of the ions on a given sublattice being mobile. Second, the crystal structure must be arranged so as to enable easy motion of ions from one equivalent site to the next. This is reflected in exceptionally low values for the migration energy E_m. In the next section we discuss the conditions under which these criteria are satisfied.

11.2 Fast Ion Conduction

A number of routes leading to exceptionally high ion carrier densities in solids have been identified over the last few decades. These are subdivided into two major categories below (structurally disordered solids and highly defective solids). An important new development in recent years is the focus on the role of interfaces in creating ionic disorder localized in the vicinity of the boundaries. For nanosized structures, these disordered regimes may represent a large fraction of the overall volume of the material. Whatever the source of the enhanced ionic conductivity, such solids are commonly designated as *fast ion conductors* (FIC).

11.2.1 Structurally Disordered Crystalline Solids

In contrast to the idealized picture of crystal structures, many solids exist in which a sublattice of sites is only partially occupied. Strock [11.12, 13] already came to this conclusion in the 1930s in relation to the Ag sublattice in the high-temperature form (α-phase) of AgI. More recent neutron diffraction studies [11.14] differ with regard to the number of equivalent Ag sites. The special feature of partial occupancy of sites is nevertheless sustained.

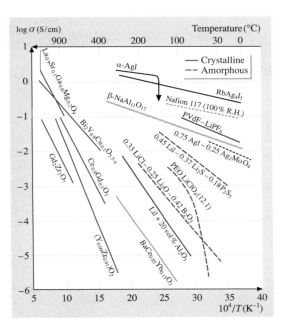

Fig. 11.2 The temperature dependences of representative FICs, including cation and anion conductors, crystalline and amorphous conductors, and inorganic and organic conductors. After [11.3]

Other notable systems characterized by sublattice disorder include Nasicon ($Na_3Zr_2PSi_2O_{12}$), sodium beta alumina ($1.2\,Na_2O - -0.11\,Al_2O_3$) and $LiAlSiO_4$,

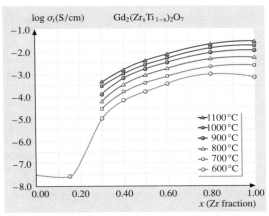

Fig. 11.3 Log ionic conductivity versus mole fraction Zr in $Gd_2(Zr_xTi_{1-x})_2O_7$ at a series of temperatures. After [11.16]

which exhibit fast ion transport in three, two and one dimensions, respectively. Hundreds of other structurally disordered conductors may be found listed in review articles on the subject [11.2, 15]. Figure 11.2 illustrates the $\log\sigma - 1/T$ relations for representative FICs, while Table 11.2 summarizes data on representative materials in tabular form.

Similarities between FICs and liquid electrolytes are often noted; the most important of these is that the disordered sublattice in the solid resembles the disordered nature of ions in a liquid. For this reason one often hears the term *lattice melting* used to describe phase transitions in solids which relate to the conversion of a conventional ionic conductor to a FIC (such as β to α transition in AgI at approximately 150 °C). Nevertheless, most investigators now believe that transport in FICs occurs via correlated jumps between well-defined sites rather than the liquid-like motion characteristic of aqueous or molten salt electrolytes.

The major structural characteristics of FICs include: (a) a highly ordered, immobile or framework sublattice providing continuous open channels for ion transport, and (b) a mobile carrier sublattice which supports a random distribution of carriers over an excess number of equipotential sites. FICs exhibit framework sublattices that minimize strain, electrostatic and polarization contributions to the migration energy while offering high carrier concentrations within the mobile carrier sublattice. Given the high concentration of carriers, correlation effects between carriers must be taken into account. Calculations by *Wang* et al. [11.17], for example, have demonstrated that cooperative motions of ions can lead to significantly lower calculated migration energies than those based on consideration of isolated jumps alone. Although no precise criterion now exists for categorizing FICs, they normally exhibit unusually high ionic conductivities $(\sigma > 10^{-2}\,\text{S/cm})$, well below their melting points, and generally low activation energies (commonly $E \approx 0.05 - 0.5\,\text{eV}$, but they can be as high as $\approx 1.0\,\text{eV}$).

Few intrinsically disordered oxygen ion conducting FICs are known. The pyrochlores, with general formula $A_2^{3+}B_2^{4+}O_7$, represent a particularly interesting system, given that the degree of disorder can be varied almost continuously from low to high values within select solid-solution systems. The pyrochlore crystal structure is a superstructure of the defect fluorite lattice with twice the lattice parameter and one out of eight oxygens missing. This can be viewed as resulting from a need to maintain charge neutrality after substituting trivalent ions for 50% of the quadravalent ions in the fluorite structure, as in $Gd_2Zr_2O_7$. Although

Table 11.2 Representative solid electrolytes and mixed conductors: mobile ions, electrical properties and applications

Material	Mobile ion	Properties	Remarks and applications
$Zr_{0.85}Y_{0.15}O_{2-x}$ [11.20]	O^{2-}	$\sigma_{O^{2-}}(1000\,°C) = 0.12$ S/cm $E = 0.8$ eV	Material of choice in auto exhaust sensors; prime solid electrolyte candidate for SOFCs
$Ce_{0.95}Y_{0.05}O_{2-x}$ [11.21]	O^{2-}	$\sigma_{O^{2-}}(1000\,°C) = 0.15$ S/cm $E = 0.76$ eV	Semiconducting at low PO_2; prime candidate solid electrolyte for intermediate temperature SOFCs
$Ce_{0.1}Gd_{0.9}O_{2-x}$ [11.22]		$\sigma_{O^{2-}}(800\,°C) = 0.10$ S/cm $E = 0.56$ eV	
$La_{0.8}Sr_{0.2}Ga_{0.8}Mg_{0.2}O_{3-x}$ [11.23]	O^{2-}	$\sigma_{O^{2-}}(750\,°C) = 0.35$ S/cm $E = 0.55$ eV	Readily converts to mixed conductor by addition of transition metals; candidate solid electrolyte for intermediate temperature SOFCs
$Gd_{1.8}Ca_{1.2}Ti_2O_{7-x}$ [11.24]	O^{2-}	$\sigma_{O^{2-}}(1000\,°C) = 0.05$ S/cm $E = 0.67$ eV	The related pyrochlore, $Gd_2Zr_2O_7$, is an intrinsic FIC
$Bi_2V_{0.9}Cu_{0.1}O_{5.5-x}$ [11.25]	O^{2-}	$\sigma_{O^{2-}}(700\,°C) = 0.15$ S/cm $E = 0.47$ eV	Mixed ionic–electronic conductor, $t_{O^{2-}} \approx 0.9$ at 900 K; of interest as a permeation membrane
$La_{0.8}Sr_{0.2}MnO_{3+x}$ [11.26]	O^{2-}	$\sigma_{O^{2-}}(1000\,°C) \approx 3\times 10^{-7}$ S/cm $E = 2.81$ eV	This material is largely an electronic conductor with $\sigma_e(1000\,°C) \approx 100$ S/cm; prime candidate as a cathode in SOFC.
$BaCe_{0.85}Yb_{0.15}O_{3-x}$ [11.27]	H^+	$\sigma_{H^+}(300\,°C) = 7\times 10^{-4}$ S/cm $E = 0.54$ eV	Of interest in SOFC based on protonic conduction
Nafion [11.28]	H^+	$\sigma_{H^+}(75\,°C) \approx 10^{-2}$ S/cm (at 20% relative humidity)	Organic; prime candidate for low-temperature solid state fuel cell based on protonic conduction
α–AgI [11.29]	Ag^+	$\sigma_{Ag^+}(200\,°C) = 1.6$ S/cm $E = 0.1$ eV	Phase transition at 146°C; first recognized fast ion conductor
$RbAg_4I_5$ [11.29]	Ag^+	$\sigma_{Ag^+}(30\,°C) = 0.3$ S/cm $E = 0.09$ eV	One of the most conductive ionic conductors at room temperature
α–CuI [11.30]	Cu^+	$\sigma_{Cu^+}(450\,°C) = 10^{-1}$ S/cm $E = 0.15$ eV	Phase transition at 407°C
Na β-alumina [11.31]	Na^+	$\sigma_{Na^+}(300\,°C) = 0.13$ S/cm $E \approx 0.3$ eV	Stoichiometry varies between $Al_2O_3/Na_2O = 5.3$–8.5
$60Li_2S$–$40SiS_2$ [11.32]	Li^+	$\sigma_{Li^+}(25\,°C) = 5\times 10^{-4}$ S/cm $E = 0.25$ eV	Amorphous
poly(vinylidene fluoride) (PVdF) – propylene carbonate (PC) – Li salt (LiX = $LiSO_3CF_3$ $LiPF_6$ or $LiN(SO_2CF_3)_2$) [11.33]	Li^+	$\sigma_{Li^+}(20\,°C) \approx 10^{-3}$ S/cm	Organic conductor; of interest for lithium batteries
β-PbF_2 [11.34]	F^-	$\sigma_{F^-}(100\,°C) = 10^{-4}$ S/cm $E = 0.48$ eV	Basis of a variety of gas sensors

oxygen vacancies occur at random throughout the anion sublattice in an ideal defect fluorite (such as YSZ), they are ordered onto particular sites in the pyrochlore structure. Thus, one properly views these as empty interstitial oxygen sites rather than oxygen vacancies. As a consequence, nearly ideal pyrochlore oxides, such as $Gd_2Ti_2O_7$, are ionic insulators [11.18]. Figure 11.3 illustrates the large increases in ionic conductivity induced by systematically substituting zirconium for titanium. *Moon* and *Tuller* [11.16] explain this on the basis of increased A and B cation antisite disorder as the radius of the B ion approaches that of the A ion. Thus, as the cation environments of the oxygen ions becoming more homogeneous, exchange between regular and interstitial sites also becomes more favorable, leading to increased Frenkel disorder. This interpretation has been confirmed by neutron diffraction studies on a closely related system [11.19].

Other important intrinsically disordered oxygen ion conductors are based on Bi_2O_3. At 730 °C [11.35], the low-temperature semiconducting modification transforms to the δ phase, which is accompanied by an oxide-ion conductivity jump of nearly three orders of magnitude. This is tied to the highly disordered fluorite-type structure, where a quarter of the oxygen sites are intrinsically empty, and to the high polarizability of the bismuth cation. *Takahashi* and *Iwahara* [11.36] succeeded in stabilizing the high-temperature δ phase to well below the transition temperature by doping with various oxides, including rare-earth oxides such as Y_2O_3. High oxide ion conductivity was also discovered above 570 °C in the Aurivillius-type γ phase of $Bi_4V_2O_{11}$ [11.37, 38], where one quarter of the oxygen sites coordinating V^{5+} are empty. Partial substitution of vanadium by lower valence cations, such as copper, nickel or cobalt, led to a new family of so-called BIMEVOX compounds [11.39], with a remarkably high oxygen ion conductivity at moderate temperatures. The copper-substituted compound has an oxide ion conductivity above 200 °C which is ≈ 2 orders of magnitude higher than other oxide ion conductors. The bismuth-based electrolytes unfortunately suffer from instability under reducing conditions, which is a limitation for some applications, such as in the solid oxide fuel cells discussed below.

11.2.2 Amorphous Solids

One of the oft-mentioned criteria for FIC in solids (see before) is the existence of a highly ordered framework which provides channels for the ready motion of ions in the complementary, disordered sublattice. Reports of FIC in inorganic glasses [11.40] raised serious doubts concerning the relevance of this feature. The amorphous state, viewed as being liquid-like, is known to lack long-range order, with short-range order typically extending to, at most, a few atom spacings. Although highly oriented channels may be helpful in FIC, they are not essential, as demonstrated by the existence of FIC in glasses.

Fast ionic conductivity is observed in many glasses containing smaller cations with mole fractions greater than about 0.20, such as silver, copper, lithium and sodium [11.41, 42]. These glasses typically contain one or more network formers (such as SiO_2, B_2O_3, P_2O_5 or GeS_2), network modifiers (such as Ag_2O, Li_2O, Cu_2O or Ag_2S), and dopant compounds, largely halides (such as AgI, CuI and LiCl). The network structure and therefore its physical and chemical properties can be substantially modified by addition of the modifier. Dopant salts, on the other hand, do not strongly interact with the network, but *dissolve* into the interstices of the glass structure. A number of phenomenological trends have been noted including: ion conduction increases (1) in the order K, Na, Li, Ag, (2) with increasing modifier concentration, (3) with halide additions in the order Cl, Br, I, (4) in sulfide versus oxide glasses, and (5) in correlation with decreasing density and glass transition temperature.

Some authors speculate that ionic transport in glasses is enhanced upon addition of the halide anions by lowering the association energy between the mobile charge carriers and the network and thereby increasing the free carrier density [11.43]. An alternate model attributes the increased conductivity to major changes induced in the glass structure by the additives, as reflected in changes in glass transition temperature T_g and the density ρ. In this latter model, a large fraction of the carriers are already assumed to be unassociated and free to move, but with increased ionic mobility driven by structural changes. Here [11.44–47], the predominant influence of the halide addition is believed to impact the strain component of the migration energy.

Another important class of amorphous fast ion conductors is those based on organic or polymer electrolytes, which (analogous to the inorganic systems) are composed of a backbone polymer and a salt complex in which the counter-ion is covalently bound to the backbone. A classic example is the one based on polypropylene oxide $(CH_2CH(CH_3)O)_n$ (PPO) complexed with $LiCF_3SO_3$ to form $PPO_n \cdot LiCF_3SO_3$. Upon forming the complex, the Li ion conductivity increases by as much as a factor of $\approx 10^5$ [11.48]. In contrast to the inorganic glasses, which exhibit an Arrhenius temperature dependence, however, these polymers follow a curved dependence best expressed by

$$\sigma = \sigma_0 \exp\left(-\frac{B}{T-T_0}\right), \qquad (11.12)$$

where T_0 is the glass transition temperature. This suggests a coupling between transport and network relaxation, a situation more closely coupled to transport in a liquid than in a solid, albeit a highly viscous liquid. Polymer electrolytes are now materials of choice for Li batteries and proton-based solid electrolytes given their attractive mechanical properties (ability to relax elastically upon stresses induced by volume changes related to charge/discharge of adjacent electrodes) and ease of processing [11.49].

11.2.3 Heavily Doped Defective Solids

Anomalously high concentrations of ionic carriers may also be induced in intrinsically insulating solids. In the following we briefly discuss two approaches for generating such *highly defective solids*.

We already know that ionic defect densities may be greatly enhanced above intrinsic levels by doping with altervalent impurities. However, the solubility limit of such impurities is often limited to only tens or hundreds of ppm. This corresponds to roughly 10^{17}–10^{18} defects/cm^3, a value 10^3–10^4 times smaller than in typical FICs. Compounds do exist, however, in which the solubility limit is extensive, reaching the 10–20% level even at reduced temperatures. Perhaps the most familiar example of such a system is stabilized zirconia, which due to its wide solid solubility with cations of lower valency such as Ca^{2+} and Y^{3+}, exhibits exceptionally high oxygen ion conductivity ($\sigma \approx 10^{-1}$ S/cm) at temperatures approaching 1000 °C.

As discussed above, high carrier densities must be coupled with high ion mobilities in order to attain high magnitudes of ionic conduction. The cubic fluorite structure, exhibited by stabilized zirconia (ZrO_2) and ceria (CeO_2), for example, supports high oxygen ion mobility due to the low four-fold coordination of cations around the oxygens, coupled with the interconnected nature of the face-shared polyhedra which surround the oxygen sites. Migration energies as low as ≈ 0.6 eV are reported for oxygen vacancy motion in ceria-based solid solutions [11.50]. High fluorine ion mobility is also observed in fluorite CaF_2 and related crystal systems.

More recently Ishihara demonstrated that very high oxygen conductivity can be achieved in the perovskite $LaGaO_3$ by acceptor doping on both the La and Ga sites [11.51]. The solid solution $(La_{1-x}Sr_x)(Ga_{1-y}Mg_y)O_3$ exhibits ionic conductivity levels above that of ZrO_2 and CeO_2, for example 3×10^{-1} S/cm at 850 °C. Perovskites also support some of the highest proton conductivities at elevated temperatures. The most popular of these are ABO_3-type compounds with A = Ba, Sr, and B = Ce or Zr. Upon acceptor doping, as in $SrCe_{0.95}Yb_{0.05}O_3$, oxygen vacancies are generated as in the gallate above. However, in the presence of moisture, water is adsorbed and protons are generated [11.52]:

$$H_2O + V_O^{\bullet\bullet} + O_O \Leftrightarrow 2OH^{\bullet} \,. \tag{11.13}$$

Given the high proton mobility, this is sufficient to induce large proton conductivity. Perovskite-related structures with the general formula $A_3B'B''O_9$ also exhibit high protonic conductivity [11.53]. Atomistic calculations simulating proton diffusion in numerous perovskite-type oxides are reported by the group of *Catlow* [11.54].

Ionic conductivities do not generally increase linearly with foreign atom additions. At the levels of defects being discussed here, defect–defect interactions become important, generally leading to defect ordering. This results in a maximum in ionic conductivity at some level of doping that depends on the particular system being investigated. *Nowick* and coworkers [11.55, 56] have demonstrated, in a series of studies, that the deviations from ideality are caused initially by composition-dependent activation energies rather than pre-exponentials (11.9), a feature also observed in a number of FIC glasses.

The formation of ionic defects that accompany excursions in composition away from stoichiometry due to redox reactions (Table 11.1, reaction 3) may also be large. CeO_2, for example, may be readily reduced to $CeO_{1.8}$ at 1000 °C [11.57], resulting in oxygen vacancy concentrations of 5×10^{21} cm^{-3} ($c = 0.9$). It should be noted, however, that comparable concentrations of electrons are also formed during such stoichiometry excursions.

11.2.4 Interfacial Ionic Conduction and Nanostructural Effects

Interfaces can significantly modify the ionic conductivities of polycrystalline or composite materials and thin films. Modified levels of ionic conductivity near interfaces may result from space-charge regions formed near interfaces to compensate for charged defects and impurities segregated to surfaces, grain and phase boundaries. Grain boundaries, for example, serve as source and sink for impurities and point defects and thus often take on a net negative or positive charge relative to the grains. To maintain overall charge neutrality, a space charge of opposite charge forms in the grains adjacent to the grain boundaries with a width related to the Debye length L_D given by

$$L_D = \left(\varepsilon_r \varepsilon_0 k / T q^2 n_b\right)^{1/2} \tag{11.14}$$

in which n_b is the majority charge carrier concentration within the grain, $\varepsilon_r \varepsilon_0$ the dielectric constant, k_B the Boltzmann constant, T the temperature and q is the electron charge. Depending on the sign of the charge at the interface, a depletion or accumulation of mobile ions in the vicinity of the boundary will form. Liang provided

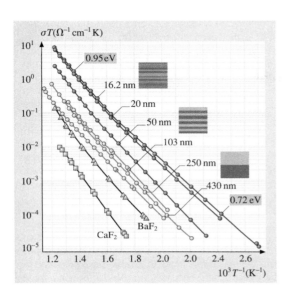

Fig. 11.4 Parallel ionic conductivity of CaF_2/BaF_2 nanometer-scale, artificially modulated heterolayers, with various periods and interfacial densities in the 430 to 16 nm range. After [11.59]

on the difference between the bulk and the local electrical potential (ϕ^∞ and ϕ). For positive values of ϕ, the concentrations of all negative defects are increased by the exponential factor, while those of the positive defects are decreased by the same factor and vice versa for negative values.

Films and/or polycrystalline materials with very small lateral dimensions can be expected to exhibit particularly strong space-charge effects on ionic conduction. This follows from the fact that the space-charge width approaches the dimensions of the film or grain. In this case, the space-charge regions overlap, and the defect densities no longer reach bulk values, even at the center of the particles [11.60]. In the limit of very small grains, local charge neutrality is not satisfied anywhere, and a full depletion (or accumulation) of charge carriers can occur with major consequences for ionic and electronic conductivity. Strong nanoscale effects on ionic and mixed ionic conductivity have been demonstrated for artificially modulated heterolayers of the solid ionic conductors CaF_2/BaF_2 [11.59] and nanocrystalline CeO_2 [11.61]. Figure 11.4 illustrates the orders of magnitude increase in fluorine ion conductivity possible with space-charge accumulation of mobile carriers in nanoscale CaF_2/BaF_2 multilayers.

one of the first demonstrations of enhancement in the $LiI:Al_2O_3$ system [11.58].

The defect concentration profile in the space-charge region can be expressed as [11.60]

$$c_i/c_i^\infty = \exp\left[-q_i(\phi - \phi^\infty)/k_B T\right]. \qquad (11.15)$$

The bulk concentration (c_i^∞) is a function of temperature, chemical potential and doping. The local concentration in the space-charge region (c_i) depends

11.3 Mixed Ionic–Electronic Conduction

11.3.1 Defect Equilibria

Deviations from stoichiometry in the direction of oxygen excess (MO_{1+x}) or deficiency (MO_{1-x}) form defect states that act identically in every way to impurity-related acceptor or donor states, respectively. In general, the electrical behavior of solids depends on defects formed in response to both impurities and deviations from stoichiometry. At or near stoichiometry, impurities predominate, while under strongly reducing or oxidizing conditions, defects associated with deviations from stoichiometry often take control. To characterize the electrical response of a metal oxide to temperature and atmosphere excursions, a series of simultaneous reactions of the form represented by (Table 11.1, reactions 1–5) must be considered. Furthermore, a representative electroneutrality equation for the case considered in Table 11.1 would be:

$$2[O_i''] + [N_M'] + n = 2[V_O^{\bullet\bullet}] + p. \qquad (11.16)$$

Note that: (1) intrinsic Frenkel disorder is assumed to predominate, so that (Table 11.1, reaction 1) may be ignored in subsequent discussions; (2) $a_{N_2O_3}$ is often assumed to be sufficiently low that all of N goes into solid solution.

A piecewise solution to such problems is commonly attempted by sequentially choosing conditions for which only one term on either side of (11.16) need be considered. The region corresponding to mixed ionic conductivity is where the predominant charge carrier is

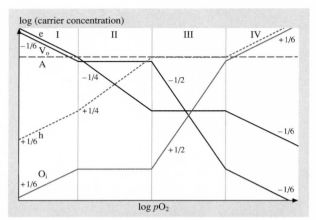

Fig. 11.5 Defect diagram for acceptor-doped oxide. After [11.62]

an ion. In acceptor-doped material (A' being a generic acceptor), this corresponds to the condition (Region II in Fig. 11.5) for which (11.16) may be simplified to read

$$[N'_M] = [A'_c] = 2[V^{\bullet\bullet}_O] . \quad (11.17)$$

Combining this with (Table 11.1, reaction 3) one obtains

$$n = \left[2K_R(T)/[N'_M]\right]^{1/2} P_{O_2}^{-1/4} \quad (11.18)$$

and from (Table 11.1, reactions 2 and 4),

$$p = K_e(T)\left(2K_R(T)/[N'_M]\right)^{-1/2} P_{O_2}^{1/4} , \quad (11.19)$$

$$[O''_i] = 2[A'_c]^{-1} K_F(T) . \quad (11.20)$$

Note that, in this defect regime, the ionic defects are pO_2-independent while the electronic species exhibit a $pO_2^{\pm 1/4}$ dependence. One obtains predictions for the corresponding dependencies of the partial conductivities of each of these charged species by multiplying carrier concentration by the respective charge and mobility. Experimentally, one normally observes the same pO_2 dependence of the partial conductivity as that predicted for the defect concentration, demonstrating that the mobility is pO_2-independent. One then uses the predicted pO_2 dependencies of the partial conductivities to deconvolute the ionic and electronic contributions to the electrical conductivity as discussed below.

The three other defect regimes most likely to occur, beginning at low P_{O_2} and moving on to increasing P_{O_2}, are depicted in Fig. 11.5 and include $n = 2[V^{\bullet\bullet}_O]$ (Region I), $p = [N'_M]$ (Region III) and $p = 2[O''_i]$ (Region IV).

In the case where μ_n, μ_p are sufficiently greater than $\mu(V^{\bullet\bullet}_O)$, then even in the defect regime where $V^{\bullet\bullet}_O$ is the predominant defect (so that $2[V^{\bullet\bullet}_O] = [N'_M]$), the total conductivity remains electronic. When the carrier mobility inequality is not nearly so pronounced, so that at the pO_2 at which electronic defects are at a minimum ($n = p$), conduction is predominantly ionic. Under these circumstances the oxide acts as a solid electrolyte, and in this regime of temperature and pO_2, one designates this as the *electrolytic domain*. Aside from the electrolytic domain, the neighboring zones on either side are designated as mixed zones within which both ionic and electronic conductivities are of comparable magnitude.

11.3.2 Electrolytic Domain Boundaries

In applications where solid electrolytes are to be utilized, it is essential to know a priori under which conditions the material is likely to exhibit largely electrolytic characteristics. Expressions for the *electrolytic domain boundaries* can be obtained by first writing down general expressions for the partial conductivities (11.17–11.19):

$$\sigma_i = \sigma_i^\circ \exp(-E_i/kT) , \quad (11.21)$$

$$\sigma_p = \sigma_p^\circ P_{O_2}^{+\frac{1}{4}} \exp(-E_p/kT) , \quad (11.22)$$

$$\sigma_n = \sigma_n^\circ P_{O_2}^{-\frac{1}{4}} \exp(-E_n/kT) . \quad (11.23)$$

One commonly defines the electrolytic domain boundary as that condition of T and pO_2 for which the ionic conductivity drops to 0.5. Under reducing conditions, this pO_2 is designated by P_n and under oxidizing conditions by P_p. Consequently, one equates σ_i and σ_n or σ_i and σ_p to solve for P_n and P_p, respectively. These are given by

$$\ln P_n = \frac{-4(E_n - E_i)}{k}\frac{1}{T} + 4\ln\left(\frac{\sigma_n^0}{\sigma_i^0}\right) , \quad (11.24)$$

$$\ln P_p = \frac{-4(E_i - E_p)}{k}\frac{1}{T} + 4\ln\left(\frac{\sigma_i^0}{\sigma_p^0}\right) . \quad (11.25)$$

Note that since the mobilities of vacancies in such oxides have been found to be much greater than those of interstitials [11.16], we ignore the latter's contributions. The domain boundaries for stabilized zirconia are shown plotted in Fig. 11.6 [11.63]. Note that, as commonly observed, the electrolytic domain shrinks with increasing temperature due to the fact that E_n and E_p are typically greater in magnitude than E_i.

11.4 Applications

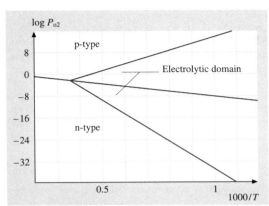

Fig. 11.6 Domain boundaries of stabilized zirconia as projected onto the log $P_{O_2} - 1000/T$ plane [11.63]

Fast ionic conducting ceramics and MIECs are finding extensive application in various solid state electrochemical devices. Some of these include fuel cells and electrolyzers [11.2, 64–66], high energy density Li batteries [11.67], electrochromic windows [11.68] and auto exhaust sensors [11.69].

11.4.1 Sensors

The monitoring of our environment has become essential for effective emissions control. Likewise, monitoring of chemical processes in real time enables closer quality control of products. Electrochemical sensors transform a chemical signal into an electrical signal, which is easy to measure, monitor and process [11.70]. Ionic and mixed conducting solids are basic materials for

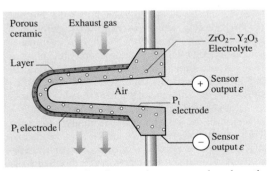

Fig. 11.7 Schematic of auto exhaust sensor based on the Nernst equation. After [11.64]

this development, because they can be easily miniaturized, for instance in thin film form, and they can often be operated at elevated temperatures or in an aggressive environment. The two major types of electrochemical sensors, potentiometric and amperometric, are summarized below.

Potentiometric Sensors

In a potentiometric gas sensor, the concentration or partial pressure of a species is determined by measuring the emf of a solid electrolyte concentration cell. The most successful commercial sensor is the oxygen sensor [11.64], which uses stabilized zirconia as the solid oxygen ion electrolyte (Fig. 11.7). The emf of the cell, $O_{2,\text{ref}}$, Pt / YSZ / Pt, O_2, can be written according to the Nernst equation:

$$E = t_i (kT/4q) \ln[P(O_2)/P(O_2)_{\text{ref}}] . \tag{11.26}$$

$P(O_2)_{\text{ref}}$ is the oxygen partial pressure of the reference gas, generally air, and t_i is the ionic transference number. For proper operation, t_i needs to be kept very close to unity. All other terms have their common meanings.

The zirconia auto exhaust sensor monitors the air-to-fuel ratio, which is maintained within close limits for optimum operating efficiency of the three-way exhaust catalyst that serves to reduce the amount of pollutants, including unburned hydrocarbons, CO and NO_x. Such sensors are designed to provide response times on the order of tens of milliseconds. A voltage near to zero corresponds to an oxygen-rich "lean" mixture, and a voltage near to 1 V to an oxygen-poor "rich" mixture. The fuel injector of the engine is controlled via a closed-loop system. The zirconia oxygen sensor sees a wide range of exhaust temperatures, up to values as high as $\approx 900\,°\text{C}$. Fortunately, the large step in voltage in going from lean to rich conditions can be easily detected at all temperatures. The molecular mechanisms operating at the electrolyte/platinum interface have been examined in detail [11.71].

Another example of a potentiometric sensor is the one reported by *Maier* et al., of the type [11.72]

$$\text{Au, } O_2, CO_2, Na_2CO_3/Na^+\text{-conductor}/$$
$$Na_2ZrO_3, ZrO_2, CO_2, O_2, \text{Au} \tag{11.27}$$

This is used to monitor pCO_2; it eliminates the need for a gas-tight reference electrode. Here the two-phase reference electrode Na_2ZrO_3, ZrO_2, which fixes the Na activity on the right side of the cell, is insensitive to

CO_2, while the Na_2CO_3 electrode on the left side can be sensed by the Na^+-conductor, typically NASICON or β-alumina. Yamazoe and Miura have reviewed the possible different types of potentiometric sensors by using single or multicomponent auxiliary phases [11.73].

Amperometric Sensors

By applying a voltage across an electrolyte, it is possible to electrochemically pump chemical species from one chamber to the other. Amperometric sensors rely on limiting current due to diffusion or interfacial phenomena at the electrode, which are linearly dependent on the partial pressure of the gas constituent [11.74, 75]. These become particularly important in so-called lean burn engines. Here the partial pressure of oxygen does not strongly vary with the air-to-fuel ratio, in contrast to engines operating at or near the stoichiometric air-to-fuel ratio. Under these circumstances, sensors are needed which have a stronger than logarithmic sensitivity to oxygen partial pressure variations. Sensors based on this principle are also being developed to detect other gases including NO_x.

11.4.2 Solid Oxide Fuel Cells (SOFC)

Solid oxide fuel cells (SOFC) provide many advantages over traditional energy conversion systems, including high energy conversion efficiency, fuel flexibility (due to internal reforming), low levels of NO_x and SO_x emissions, versatile plant size and long lifetimes [11.76]. Quiet, vibration-free operation also eliminates noise associated with conventional power-generation systems. However, operation at elevated temperature is necessary given relatively low ionic conductivities and slow electrode processes at temperatures below 800 °C. Recent progress with thinner electrolytes and advanced electrodes holds promise for reducing operating temperatures by as much as several hundred degrees.

The three major components of the elemental solid oxide fuel cell (SOFC) include the cathode, electrolyte and anode. While the solid electrolyte is selected so that it only conducts ions to ensure a Nernst open circuit potential that is as close to ideal as possible, the electrodes must support the reduction/oxidation reactions that occur at the electrolyte/electrode/gas interfaces. For example, when current is being drawn, the following reaction occurs at the cathode:

$$1/2 O_2 + 2e' + V_O^{\bullet\bullet} \Leftrightarrow O_O . \quad (11.28)$$

This reaction is accelerated if the cathode can provide both electrons, as in a typical current collector, as well as oxygen vacancies. An example of such a mixed conducting cathode is $La_{1-x}Sr_xCoO_3$ (LSCO), which has an electronic conductivity of $> 100 S/cm$ and an oxygen ion conductivity of $> 1 S/cm$ at temperatures above 800 °C [11.77]. Given its importance to performance, modeling of the electrode processes has also received a great deal of attention recently [11.78, 79]. Unfortunately, while exhibiting highly attractive mixed conducting properties, LSCO is unstable in contact with yttria-stabilized zirconia, the electrolyte of choice.

In an attempt to take advantage of LSCO's attractive features, there is growing interest in marrying this electrode with doped ceria electrolytes, such as $Ce_{1-x}M_xO_2$:M=Gd or Sm, for operation at reduced temperatures of 550–750 °C, given ceria's higher ionic conductivity (albeit higher mixed ionic electronic conductivity at the anode). Electronic conduction degrades solid electrolyte performance in several ways. Because electronic conduction serves as an alternate path for charged species through the electrolyte, it decreases the power that can be dissipated through the load. Further, the short circuiting factor also serves to allow permeation of gaseous species through the electrolyte, even under open circuit conditions (see the section on membranes below). These primary figures of merit are summarized in the context of a solid oxide fuel cell (SOFC) in Fig. 11.8. E represents the potential induced across the cell under open circuit conditions for a given P_{O_2} gradient, $\hat{t}_i$ is the ionic transference number, E_N is the Nernst potential, R_{INT}, R_C, R_{SE} and R_A are the internal cell, cathode, solid electrolyte and anode resistances, respectively, J_{O_2} is the oxygen permeation flux, and L the thickness across which the P_{O_2} gradient is imposed. All other terms have their normal meanings. Fortunately, the

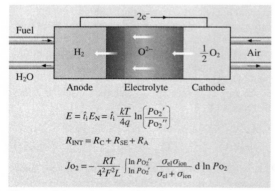

Fig. 11.8 Schematic of a solid oxide fuel cell and the primary figures of merit [11.66]

electronic conductivity in ceria electrolytes drops exponentially with decreasing temperature, and the overall power output exceeds that of zirconia-based systems at reduced temperatures, so it can be used with LSCO and other electrodes incompatible with YSZ.

The system $(La_{1-x}Sr_x)(Ga_{1-y}Mg_y)O_3$ (LSGM) was mentioned above as exhibiting one of the highest oxygen ion conductivities ($\approx 3 \times 10^{-1}$ S/cm at 850 °C) due to high levels of acceptor doping (Sr and Mg) on both the La and Ga sites. As a consequence, it is now being considered as one of several candidates for the electrolyte in solid oxide fuel cells. Experiments have shown that mixed ionic electronic conduction in a fuel cell electrode contributes to reduced overpotentials [11.66]. It has also been recognized that a single-phase *monolithic* fuel cell structure would benefit from the minimization of chemical and thermomechanical degradation [11.80]. Consequently, an electrode based on LSGM would satisfy all requirements. *Long* et al. proposed to add a transition metal in solution, which would introduce an additional 3d *conducting* band within the wide band-gap of the initially electronically insulating gallate [11.81]. As expected, as the Ni content in the system $La_{0.9}Sr_{0.1}Ga_{1-x}Ni_xO_3$ (LSGN) increased, the electronic conductivity increased, finally reaching ≈ 50 S/cm without decreasing the already high ionic conductivity. Improved electrode performance was indeed observed with the LSGM/LSGN interface [11.82].

11.4.3 Membranes

Oxygen-permeable ceramic membranes are used for the separation of oxygen from air or for industrial-scale oxygen separation in the conversion of natural gas to syngas (CO + H_2) for example [11.83]. They are made from mixed conducting oxides in which ambipolar diffusion of ionic and electronic charge carriers in an oxygen potential gradient assures a high oxygen permeation flux through the membrane (see Fig. 11.8 for an expression for the permeation current). High oxygen permeation rates were obtained with the system $(La, Sr)MO_{3-\delta}$ (M = Fe, Co, Cr) [11.77], but some deterioration over time was noticed. Research continues into this class of materials with regard to long-term ordering of defects, surface exchange kinetics, optimization of oxygen conduction and phase stability under steep oxygen activity gradients. One of the best materials developed to date is the BICUVOX compound, with composition $Bi_2V_{0.9}Cu_{0.1}O_{5.35}$, which shows a particularly large mixed conductivity that enables high oxygen permeation rates at moderate temperature, such as 700 K, at high and intermediate oxygen partial pressures [11.38].

Mixed oxide ion and electronic conductivity is also observed in composites of a solid oxide ion electrolyte and a noble metal, if percolating pathways exist for each component. These mixed conducting oxide ceramic–metal composites (cermets), including Y-stabilized ZrO_2 with Pd [11.84], Sm-doped CeO_2 with Pd [11.85], and rare-earth doped Bi_2O_3 with Ag [11.86], have an appreciable oxygen permeation rate at elevated temperature without degradation and are considered attractive for industrial applications, although they are relatively expensive. Recent work by *Takamura* et al. [11.87] shows promising results based on ceramic/ceramic composites.

11.4.4 Batteries

Power storage requires high energy density batteries. The highest possible energy density is achieved using reactants with high free energies of reaction and low mass, such as lithium or sodium. This also requires that the solid electrolyte remains stable under highly reducing or highly oxidizing conditions. Major advantages of solid electrolytes over liquid electrolytes are the absence of leakage and container problems and, foremost, the possibility of miniaturization; for example using thin solid films.

There is an increasing demand for microbatteries compatible with microelectronics technology, related to the development of laptop computers or portable telephones. This led to the development of high energy density and long life-cycle rechargeable lithium batteries, initially based on metallic lithium anodes. However, systems based on metallic lithium suffered from problems due to metal oxidation and poor rechargeability due to the formation of metallic dendrites. The alternative "rocking chair" concept, proposed in 1980 [11.88], based on two lithium insertion compounds Li_xWO_3 and Li_yTiS_2, replaced the unstable lithium electrode, but was unable to provide sufficiently high energy densities. Improved rechargeable lithium ion batteries based instead on nongraphitic "hard" carbons as the lithium insertion anodes have since been developed [11.89]. This was followed by the successful association of hard carbon insertion anodes with the high-voltage $LiCoO_2$ insertion cathode. Due to the relatively high cost of Co, alternative systems based on other cathode materials, such as $LiNiO_2$ or $LiMn_2O_4$, are currently under investigation [11.90]. Polymer (rather than inorganic) electrolytes are used in these applications (see above). An overview of lithium batteries and polymer electrolytes can be found in books by *Julien* and *Nazri* [11.91] and *Gray* [11.92].

11.4.5 Electrochromic Windows

Electrochromic light transmission modulators – so-called "smart windows" that use solid ionic conductors – may play a significant role in energy-saving by regulating thermal insulation. In such a system, the window is maintained transparent in the visible and reflecting in the IR during the winter, allowing penetration of sunshine but blocking loss of interior heat. On the other hand, the window is rendered partially opaque during hot summer days, reducing the amount of radiation entering the building. The electrochromic elements [11.93], which color or bleach upon insertion/deinsertion of lithium or hydrogen ions, are sandwiched between two transparent thin-film electrodes and are separated by a solid electrolyte. The transparent electrodes are generally indium tin oxide (ITO). Glass electrolytes appear to be promising choices. Research and development on tungsten trioxide-based electrochromic materials started in the 1970s. Ions fill empty tetrahedral sites in the WO_3 structure [11.94]:

$$x\text{Li}^+ + \text{WO}_3 + x\,\text{e}^- \rightarrow \text{Li}_x\text{WO}_3 \,. \tag{11.29}$$

Key requirements include: (1) compatible electrochromic and solid electrolyte thin film materials; (2) the ability to operate near ambient temperature, in the range -40 to $+120\,°\text{C}$; (3) cycle reversibly many thousands of times a year for a lifetime of 20 years, and; (4) exhibit significant shifts in reflectivity with the degree of insertion/deinsertion.

11.5 Future Trends

A rapidly converging interest in thin film oxides has been developing in the microelectronics and solid state ionics communities. In the solid state ionics arena, the desire to reduce the operating temperature of solid oxide fuel cells (SOFC) has been stimulating a shift in emphasis from bulk to thin film electrolytes. Likewise, the trend in recent years has shifted away from bulk ceramics towards miniaturized smart sensor systems in which the sensor elements are integrated with electronics and various MEMS-based components, including microheaters, valves and membranes. Considering the continued drive towards ever smaller submicron lateral dimensions in MOSFET technology, it is likely that future efforts will be directed towards the construction of micro- and nanoscale ionic devices.

Specifically, one can envision the embedding of miniaturized thin film or SOFC structures as sensors or power sources together with microelectromechanical (MEM) components and other active electronics in the same silicon wafer. By applying standard Si technology, such as thin film deposition and photolithography, one accesses methods for tailoring electrolyte and electrode geometry (thickness, active electrode area and triple phase boundary length) with exceptionally high dimensional reproducibility, while retaining the ability to scale to larger dimensions. Attention will need to be focused on the special challenges that the marriage between solid state ionics and electronics implies, including semiconductor-compatible processing, rapid temperature excursions, stress-induced property modifications and interfacial stability.

Finally, one will need to consider how the defect and transport properties of thin films may differ from their bulk counterparts, and how the silicon platform provides opportunities to examine such properties in an in situ manner, and thereby identify novel or distinctive properties associated with low-dimensional structures.

References

11.1   H. L. Tuller: J. Phys. Chem. Solids **55**, 1393–1404 (1994)
11.2   P. Knauth, H. L. Tuller: J. Am. Ceram. Soc **85**, 1654–1679 (2002)
11.3   H. L. Tuller: In: *Ceramic Materials for Electronics*, 3rd edn., ed. by R. C. Buchanan (Marcel Dekker, New York 2004) p. 87
11.4   M. K. Paria, H. S. Maiti: J. Mater. Sci. **17**, 3275 (1982)

11.5 J. B. Goodenough: In: *Solid Electrolytes*, ed. by P. Hagenmuller, W. Van Gool (Academic, New York 1978) p. 393

11.6 F. A. Kröger: *The Chemistry of Imperfect Crystals*, 2nd edn. (North-Holland, Amsterdam 1974)

11.7 J. Hladik (ed.): *Physics of Electrolytes*, Vol. 1 (Academic, New York 1972)

11.8 B. C. H. Steele: J. Power Sources **49**, 1–14 (1994)

11.9 S. Kramer, M. Spears, H. L. Tuller: Solid State Ionics **72**, 59–66 (1994)

11.10 T. Ishihara, H. Matsuda, Y. Takita: J. Am. Ceram. Soc. **116**, 3801–3803 (1994)

11.11 S. Wang: *Fundamentals of Semiconductor Theory and Device Physics* (Prentice Hall, Englewood Cliffs 1989)

11.12 L. W. Strock: Z. Phys. Chem. **B25**, 441 (1934)

11.13 L. W. Strock: Z. Phys. Chem. **B25**, 132 (1936)

11.14 R. Cava, B. J. Wuensch: Solid State Commun. **24**, 411–416 (1977)

11.15 T. Kudo: In: *The CRC Handbook of Solid State Electrochemistry*, ed. by P. J. Gellings, H. J. M. Bouwmeester (CRC, Boca Raton 1997) p. 195

11.16 P. K. Moon, H. L. Tuller: Solid State Ionics **2 8–3 0**, 470–474 (1988)

11.17 J. Wang, M. Kaffari, D. Choi: J. Chem. Phys. **63**, 772 (1975)

11.18 S. A. Kramer, H. L. Tuller: Solid State Ionics **8**, 2, 15 (1995)

11.19 B. J. Wuensch, K. W. Eberman, C. Heremans, E. M. Ku, P. Onnerud, S. M. Haile, J. K. Stalick, J. D. Jorgensen: Solid State Ionics **12**(9), 111–133 (2000)

11.20 R. M. Dell, A. Hooper: In: *Solid Electrolytes*, ed. by P. Hagenmuller, W. Van Gool (Academic, New York 1978) pp. 291–312

11.21 H. L. Tuller, A. S. Nowick: J. Electrochem. Soc. **122**, 255–259 (1975)

11.22 B. C. H. Steele: J. Mater. Sci. **36**, 1053–1068 (2001)

11.23 T. Ishihara, T. Shibayama, M. Honda, H. Nishiguchi, Y. Takita: J. Electrochem. Soc. **147**, 1332–1337 (2000)

11.24 S. Kramer, H. L. Tuller: Solid State Ionics **82**, 15–23 (1995)

11.25 A. A. Yaremchenko, V. V. Kharton, E. N. Naumovich, F. M. B. Marques: J. Electroceram. **4**, 233–242 (2000)

11.26 R. A. De Souza, J. A. Kilner, J. F. Walker: Mater. Lett. **43**, 43–52 (2000)

11.27 J. Wu, L. P. Li, W. T. P. Espinosa, S. M. Haile: J. Mater. Res. **19**, 2366 (2004)

11.28 F. Damay, L. C. Klein: Solid State Ionics **162–163**, 261–267 (2003)

11.29 J. S. Kasper: In: *Solid Electrolytes*, ed. by P. Hagenmuller, W. Van Gool (Academic, New York 1978) pp. 217–235

11.30 T. Matsui, J. B. Wagner Jr.: In: *Solid Electrolytes*, ed. by P. Hagenmuller, W. Van Gool (Academic, New York 1978) pp. 237–252

11.31 T. Takahashi: In: *Superionic Solids and Solid Electrolytes: Recent Trends*, ed. by A. L. Laskar, S. Chandra (Academic, San Diego 1989) pp. 1–41

11.32 F. A. Fusco, H. L. Tuller: In: *Superionic Solids and Solid Electrolytes: Recent Trends*, ed. by A. L. Laskar, S. Chandra (Academic, San Diego 1989) pp. 43–110

11.33 Z. Jiang, B. Carroll, K. M. Abraham: Electrochim. Acta **42**, 2667 (1997)

11.34 J.-M. Reau, J. Portier: In: *Solid Electrolytes*, ed. by P. Hagenmuller, W. Van Gool (Academic, New York 1978) pp. 313–333

11.35 H. A. Harwig, A. G. Gerards: J. Solid State Chem. **26**, 265–274 (1978)

11.36 T. Takahashi, H. Iwahara: Mater. Res. Bull. **1**(3), 1447–1453 (1978)

11.37 K. R. Kendall, C. Navas, J. K. Thomas, H.-C. zur Loye: Chem. Mater. **8**, 642–649 (1996)

11.38 J. C. Boivin, G. Mairesse: Chem. Mater. **1**(0), 2870–2888 (1998)

11.39 F. Abraham, J. C. Boivin, G. Mairesse, G. Nowogrocki: Solid State Ionics **4**(0–1), 934–937 (1990)

11.40 H. L. Tuller, D. P. Button, D. R. Uhlmann: J. Non-Cryst. Solids **42**, 297–306 (1980)

11.41 H. L. Tuller, M. W. Barsoum: J. Non-Cryst. Solids **73**, 331–50 (1985)

11.42 F. A. Fusco, H. L. Tuller: In: *Superionic Solids and Solid Electrolytes: Recent Trends*, ed. by A. L. Laskar, S. Chandra (Academic, New York 1989) pp. 43–110

11.43 D. Ravaine, J. L. Souquet: Phys. Chem. Glasses **18**, 27–31 (1977)

11.44 D. P. Button, R. P. Tandon, H. L. Tuller, D. R. Uhlmann: J. Non-Cryst. Solids **42**, 297–306 (1980)

11.45 D. P. Button, R. P. Tandon, H. L. Tuller, D. R. Uhlmann: Solid State Ionics **5**, 655–658 (1981)

11.46 F. A. Fusco, H. L. Tuller, D. P. Button: In: *Proc. Symp. Electro-Ceramics and Solid State Ionics*, ed. by H. L. Tuller, D. M. Smyth (Electrochemical Society, Pennington 1988) pp. 167–178

11.47 D. P. Button, P. K. Moon, H. L. Tuller, D. R. Uhlmann: Glastech. Ber. **56K**, 856–861 (1983)

11.48 M. A. Ratner, P. Johansson, D. F. Shriver: MRS Bull. **25**, 31–36 (2000)

11.49 B. Scrosati, C. A. Vincent: MRS Bull. **25**, 28–30 (2000)

11.50 D. Y. Wang, D. S. Park, J. Griffiths, A. S. Nowick: Solid State Ionics **2**, 95–105 (1981)

11.51 T. Ishihara, H. Matsuda, Y. Takita: J. Am. Chem. Soc. **11**(6), 3801–3803 (1994)

11.52 H. Iwahara, T. Esaka, H. Uchida, H. Maeda: Solid State Ionics **3**(4), 359–363 (1981)

11.53 A. S. Nowick, Y. Du: Solid State Ionics **7**(7), 137–146 (1995)

11.54 M. Cherry, M. S. Islam, J. D. Gale, C. R. A. Catlow: J. Phys. Chem. **9**(9), 14614–14618 (1995)

11.55 D. Y. Wang, D. S. Park, J. Griffiths, A. S. Nowick: Solid State Ionics **2**, 95–105 (1981)

11.56 R. Gerhart-Anderson, A. S. Nowick: Solid State Ionics **5**, 547–550 (1981)

11.57 H. L. Tuller, A. S. Nowick: J. Electrochem Soc. **126**, 209–217 (1979)

11.58 C. C. Liang: J. Electrochem Soc. **120**, 1289 (1973)

11.59 H. Sata, K. Eberman, K. Eberl, J. Maier: Nature **408**, 946–48 (2000)

11.60 J. Maier: Prog. Solid State Chem. **23**, 171–263 (1995)

11.61 Y.-M. Chiang, E. B. Lavik, I. Kosacki, H. L. Tuller, J. Y. Ying: J. Electroceram. **1**, 7–14 (1997)

11.62 H. Seh, *Langasite Bulk Acoustic Wave Resonant Sensor for High Temperature Applications*, PhD thesis, Dept. Materials Sc. & Eng. MIT, February, 2005.

11.63 L. Heyne: In: *Solid Electrolytes*, ed. by S. Geller (Springer, Berlin, Heidelberg 1977) p. 169

11.64 B. M. Kulwicki, S. J. Lukasiewicz, S. Subramanyam, A. Amin, H. L. Tuller: In: *Ceramic Materials for Electronics*, 3rd edn., ed. by R. C. Buchanan (Marcel Dekker, New York 2004) pp. 377–430

11.65 N. Q. Minh, T. Takahashi: *Science and Technology of Ceramic Fuel Cells* (Elsevier, Amsterdam 1995)

11.66 H. L. Tuller: In: *Oxygen Ion and Mixed Conductors and their Technological Applications*, ed. by H. L. Tuller, J. Schoonman, I. Riess (Kluwer, Dordrecht 2000) pp. 245–270

11.67 C. Julien, G.-A. Nazri: *Solid State Batteries: Materials Design and Optimization* (Kluwer, Boston 1994)

11.68 C. G. Granqvist: In: *The CRC Handbook of Solid State Electrochemistry*, ed. by P. J. Gellings, H. J. M. Bouwmeester (CRC, Boca Raton 1997) pp. 587–615

11.69 P. T. Moseley, B. C. Tofield (eds.): *Solid State Gas Sensors* (Adam Hilger, Bristol 1987)

11.70 W. Göpel, T. A. Jones, M. Kleitz, I. Lundström, T. Seiyama (eds.): *Sensors: A Comprehensive Survey, Chemical and Biochemical Sensors*, Vol. 2nd and 3rd (VCH, New York 1991)

11.71 A. D. Brailsford, M. Yussouff, E. M. Logothetis: Sensor. Actuat. B **44**, 321–326 (1997)

11.72 J. Maier, M. Holzinger, W. Sitte: Solid State Ionics **74**, 5–9 (1994)

11.73 N. Yamazoe, N. Miura: J. Electroceram. **2**, 243–255 (1998)

11.74 T. Takeuchi: Sensor. Actuat. **14**, 109–124 (1988)

11.75 W. Gopel, G. Reinhardt, M. Rosch: Solid State Ionics **136-137**, 519–531 (2000)

11.76 S. C. Singhal: MRS Bull. **25**, 16–21 (2000)

11.77 H. J. M. Bouwmeester, A. J. Burggraaf: In: *The CRC Handbook of Solid State Electrochemistry*, ed. by P. J. Gellings, H. J. M. Bouwmeester (CRC, Boca Raton 1997) p. 481

11.78 I. Riess, M. Godickemeier, L. J. Gauckler: Solid State Ionics **90**, 91–104 (1996)

11.79 S. B. Adler, J. A. Lane, B. C. H. Steele: J. Electrochem. Soc. **143**, 3554–3564 (1996)

11.80 S. A. Kramer, M. A. Spears, H. L. Tuller: Novel Compatible Solid Electrolyte-Electrode System Suitable for Solid State Electrochemical Cells, U. S. Patent No. 5,5403,461 (1995)

11.81 N. J. Long, F. Lecarpentier, H. L. Tuller: J. Electroceram. **3:4**, 399–407 (1999)

11.82 F. Lecarpentier, H. L. Tuller, N. Long: J. Electroceram. **5**, 225–230 (2000)

11.83 B. Ma, U. Balachandran: J. Electroceram. **2**, 135–142 (1998)

11.84 T. J. Mazanec: Solid State Ionics **70/71**, 11–19 (1994)

11.85 K. Huang, M. Schroeder, J. B. Goodenough: J. Electrochem. Solid State Lett. **2**, 375–378 (1999)

11.86 J. E. ten Elshof, N. Q. Nguyen, M. W. den Otter, H. J. M. Bouwmeester: J. Electrochem. Soc. **144**, 4361–4366 (1997)

11.87 H. Takamura, K. Okumura, Y. Koshino, A. Kamegawa, M. Okada: J. Electroceram. **13**, 613 (2004)

11.88 M. Lazzari, B. Scrosati: J. Electrochem. Soc. **127**, 773–774 (1980)

11.89 J. R. Dahn, A. K. Sleigh, H. Shi, B. M. Way, W. J. Weydanz, J. N. Reimers, Q. Zhong, U. von Sacken: In: *Lithium Batteries*, ed. by G. Pistoia (Elsevier, Amsterdam 1994) pp. 1–47

11.90 J.-M. Tarascon, W. R. McKinnon, F. Coowar, T. N. Bowmer, G. Amatucci, D. Guyomard: J. Electrochem. Soc. **141**, 1421–1431 (1994)

11.91 C. Julien, G.-A. Nazri: *Solid State Batteries: Materials Design and Optimization* (Kluwer, Boston 1994)

11.92 F. M. Gray: *Solid Polymer Electrolytes: Fundamentals and Technological Applications* (VCH, New York 1991)

11.93 C. M. Lampert: Solar Energy Mat. **11**, 1–27 (1984)

11.94 M. Green: Ionics **5**, 161–170 (2000)